Military Neuroscience and the Coming Age of Neurowarfare

Krishnan describes military applications of neuroscience research and emerging neurotechnology with relevance to the conduct of armed conflict and law enforcement. This work builds upon literature by scholars such as Moreno and Giordano and fills an existing gap, not only in terms of reviewing available and future neurotechnologies and relevant applications, but by discussing how the military pursuit of these technologies fits into the overall strategic context. The first to sketch future neurowarfare by looking at its potentials as well as its inherent limitations, this book's main theme is how military neuroscience will enhance and possibly transform both classical psychological operations and cyber warfare. Its core argument is that nonlethal strategies and tactics could become central to warfare in the first half of the twenty-first century. This creates both humanitarian opportunities in making war less bloody and burdensome as well as some unprecedented threats and dangers in terms of preserving freedom of thought and will usher in a coming age where minds can be manipulated with great precision.

Armin Krishnan, East Carolina University, USA

Emerging Technologies, Ethics and International Affairs
Series Editors: Steven Barela, Jai C. Galliott,
Avery Plaw, Katina Michael

This series examines the crucial ethical, legal and public policy questions arising from or exacerbated by the design, development and eventual adoption of new technologies across all related fields, from education and engineering to medicine and military affairs. The books revolve around two key themes:

* Moral issues in research, engineering and design
* Ethical, legal and political/policy issues in the use and regulation of Technology

This series encourages submission of cutting-edge research monographs and edited collections with a particular focus on forward-looking ideas concerning innovative or as yet undeveloped technologies. Whilst there is an expectation that authors will be well grounded in philosophy, law or political science, consideration will be given to future-orientated works that cross these disciplinary boundaries. The interdisciplinary nature of the series editorial team offers the best possible examination of works that address the 'ethical, legal and social' implications of emerging technologies.

Most recent titles

1. Social Robots: Boundaries, Potential, Challenges
Marco Nørskov

2. Legitimacy and Drones: Investigating the Legality, Morality and Efficacy of UCAVs
Steven J. Barela

3. Super Soldiers: The Ethical, Legal and Social Implications
Jai Galliott and Mianna Lotz

4. Commercial Space Exploration: Ethics, Policy and Governance
Jai Galliott

5. Healthcare Robots: Ethics, Design and Implementation
Aimee van Wynsberghe

Military Neuroscience and the Coming Age of Neurowarfare

Armin Krishnan

Routledge
Taylor & Francis Group

LONDON AND NEW YORK

First published 2017
by Routledge

2 Park Square, Milton Park, Abingdon, Oxfordshire OX14 4RN
711 Third Avenue, New York, NY 10017

Routledge is an imprint of the Taylor & Francis Group, an informa business

First issued in paperback 2018

British Library Cataloguing in Publication Data
A catalogue record for this book is available from the British Library

Library of Congress Cataloguing in Publication Data
Names: Krishnan, Armin, 1975– author.
Title: Military neuroscience and the coming age of
neurowarfare / Armin Krishnan.
Description: New York, NY : Routledge |
Includes bibliographical references and index.
Identifiers: LCCN 2016015423 | ISBN 9781472473912 (hardback) |
ISBN 9781315595429 (ebook)
Subjects: LCSH: Military art and science – Technological innovations. |
Neurosciences – Government policy – United States. | Neurosciences –
Moral and ethical aspects. | Military research – United States. | Military
weapons – Technological innovations. | Psychological warfare – United States. |
Electronics in military engineering – United States. | Human engineering –
United States. | Neurosciences – Social aspects. | Military art and science –
Technological innovations – Moral and ethical aspects.
Classification: LCC U42.5.K75 2016 | DDC 623–dc23
LC record available at https://lccn.loc.gov/2016015423

ISBN: 978-1-4724-7391-2 (hbk)
ISBN: 978-1-138-36144-7 (pbk)

Typeset in Times New Roman
by Out of House Publishing

Contents

Tables

Acknowledgements

More than a decade ago, I discovered John Marks' fascinating book *In Search of the Manchurian Candidate*, which described CIA MK ULTRA research during the Cold War. This was my first introduction to the topic of 'mind control' and since then it has become almost an obsession for me. Often dismissed as 'conspiracy theory', the reality of mind control is simply undeniable if one dares to look at the evidence. At the same time, a lot of ink has been spilled over the question of whether humans can be brainwashed and their behaviour controlled to the point that they act against their basic drives and moral convictions or whether minds can be 'read'. Luckily, there are good chances that neuroscience will be able to definitively answer these questions within a few decades at the most. I am thankful to Jonathan Moreno for making the issue of mind control and brain warfare a subject of serious academic inquiry and for raising public awareness of the topic. His 2006 book *Mind Wars* eventually encouraged me to pursue research into military neuroscience and its significance to national and international security. I hope that the end product of my efforts can stand on its own and make some contribution to the question of what future neurowarfare could look like and why it should be considered one of the biggest dangers facing humanity. When writing a book, many debts are incurred. Several people have contacted me and sent me relevant information for which I am grateful. The book grew out of conversations and communications I had with many informed individuals. In particular, I would like to thank Robert Bunker for taking the time to explain his ideas of future warfare to me at a conference. I want to thank Jürgen Altmann with whom I had an interesting discussion on directed energy weapons at a conference dinner, which stimulated my interest and determination in writing this book. I am very grateful to Larry Valero and Mark Gorman, my former colleagues at the University of Texas at El Paso, for their generous and continuing support. I want to thank Robert Thompson for his support of my research activities and in particular for keeping my teaching load low, without which I simply would not have been able to manage this project. I am also indebted to the rest of my department at East Carolina University, which has been extraordinarily supportive of their new junior colleague. I am grateful to Jai Galliott, who has encouraged me to work on the topic of military neuroscience and who has been a very helpful collaborator in other publishing projects. I want to thank my editor at Ashgate, Brenda Sharp, for so speedily accepting my book proposal and for working with me on the manuscript. My special thanks go to my wife Svetlana for frequently discussing the rather scary topic of mind control and for constantly motivating me to complete the book at last.

Abbreviations

ADS	Active Denial System
AI	Artificial Intelligence
BBB	Blood-Brain Barrier
BBI	Brain-to-Brain Interface
BCI	Brain-Computer Interface
BW	Biological Weapons
BWC	Biological (and Toxin Weapons) Convention
BWE	Brainwave Entrainment
CNS	Central Nervous System
CW	Chemical Weapons
CWC	Chemical Weapons Convention
DARPA	Defense Advanced Research Projects Agency
DBS	Deep Brain Stimulation
DE	Directed Energy
DEW	Directed Energy Weapons
DHS	US Department of Homeland Security
DIA	US Defense Intelligence Agency
DoD	US Department of Defense
DoJ	US Department of Justice
ECT	Electroconvulsive Therapy
EEG	Electroencephalography
ELF	Extremely Low Frequency
EMF	Electromagnetic Fields
EMP	Electromagnetic Pulse
ENMOD	Environmental Modification
ESP	Extrasensory Perception
FAS	Federation of American Scientists
fMRI	Functional Magnetic Resonance Imaging
HERF	High Energy Radio Frequency
HUMINT	Human Intelligence
IARPA	US Intelligence Advanced Research Projects Activity
IC	Intelligence Community
IRB	Institutional Review Board
IW	Information Warfare
JNLWP	US Joint Non-Lethal Weapons Program
LIC	Low-Intensity Conflict

LRAD	Long Range Acoustic Device
MAE	Microwave Auditory Effect
MEG	Magnetoencephalography
MRI	Magnetic Resonance Imaging
NCW	Network-Centric Warfare
NLW	Nonlethal Weapons
NRC	National Research Council
PSYOPS	Psychological Operations
PTSD	Posttraumatic Stress Disorder
RF	Radio-frequency
RMA	Revolution in Military Affairs
SOCOM	US Special Operations Command
SOD	US Army Special Operations Division
SOF	Special Operations Forces
SRI	Stanford Research Institute
TBI	Traumatic Brain Injury
tDCS	Transcranial Direct Current Stimulation
TMS	Transcranial Magnetic Stimulation
TSS	CIA Technical Services Staff
UAV	Unmanned Aerial System
UN	United Nations
UW	Unconventional Warfare
VEO	Violent Extremist Organizations
VLF	Very Low Frequency
WMD	Weapons of Mass Destruction

1 Introduction

Jonathan Moreno was onto something big when he published the first academic book on military neuroscience in 2006. This was at the very beginning of the trend of the US military exploring military applications for civilian research in neuroscience (Moreno, 2006a). Since then, the American National Research Council and the Royal Society have issued several studies that relate to neuroscience and security, which form the basis for this monograph. The goal is to go beyond this established body of literature and to theorize about the relevance of military neuroscience to contemporary warfare. The main argument presented here is that possible future breakthroughs in neuroscience have the *potential* to fundamentally alter human society, human consciousness, warfare and security. This could make the human mind a distinctive new domain of war. Systematic efforts of dominating this new domain could be termed 'neurowarfare', which will be sketched in this book. The new 'mind control' weapons could turn populations into new WMD, or result in new forms political repression. In anticipation of these threats, the book advocates radical transparency in the field of military neuroscience and international arms control regulation of future neuroweapons.

1.1 Advances in neuroscience

Why is there reason to believe that neuroscience and neurotechnology (neuro S/T) could bring anything new or substantive to the practice of modern warfare? At least some neuroscientists (not all) believe that the secrets of the human brain and the human mind can be eventually unlocked. For example, the recently deceased neuroscientist Richard F. Thompson stated a few years ago in the preface to a revised edition of his book on the brain:

> Will we one day develop instruments with which we can 'read' minds? Will it be possible one day to insert thoughts into minds or transmit them from one mind to another? Will we be able to greatly enhance our intellectual abilities through 'symbiosis' with artificial intelligence? ... Previously I believed that it was unnecessary to discuss these possibilities today. I was wrong. Developments in neuro- and computer sciences are

advancing so rapidly that many of these possibilities that seem akin to science fiction could become reality in our life times.

(Thompson, 2012: X)

It is not difficult to see how such technologies, if they turn out to be feasible, could completely alter human reality quite literally. Humans could control machines and communicate alone by thought; they could be plugged into a *Matrix*-like virtual reality to gain experiences that they could not otherwise have, at no risk; the quality of human life could be vastly improved in numerous ways, including better mental health and life extension; and society may be uplifted by making its members smarter and better able to reach their full potential, resulting in a new golden age of discovery and invention.

Neuroscience is the scientific effort of understanding the workings of the human brain and the human mind, relying on brain imaging and other forms of measurement and modeling. Neuroscience is thus a fairly complicated and diverse scientific field, comprising of diverse disciplines such as 'calculus, general biology, genetics, physiology, molecular biology, general chemistry, organic chemistry, biochemistry, physics, behavioral psychology, cognitive psychology, perceptual psychology, philosophy, computer theory, and research design' (Moreno, 2012: 32). Research that identifies itself as 'neuroscience' receives quite substantial government and corporate funding. The neuroscientist James Giordano estimated in 2013 that the global market for neuro S/T is $150 billion annually and that it is rapidly growing with large investments in Asia and South America, which will supersede American spending by 2020 (Canna, 2013).

Breakthroughs in the field have come from the development of advanced brain imaging technologies like MRI, fMRI, fNIRS, PET, CAT, CT and MEG that have given invaluable insights in the functioning of a living brain and from the development of Brain-Machine Interfaces (BMI) (Kaku, 2014: 9). Particularly important with respect to brain monitoring and BMIs is functional neuroimaging, which either measures the blood flow in the brain or the electromagnetic fields generated by neural activity, as neurons use both chemical and electrical processes for communicating with each other (R.H. Blank, 2014: 49–51). Unfortunately, the 'mapping' of the brain through functional neuroimaging is complicated by 'brain plasticity', or the tendency of the brain to constantly reorganize itself. Science journalist John Horgan claimed that brain plasticity may very well stand forever in the way of fully understanding the human mind since there may not even be any 'code' that can be 'decoded' (Horgan, 2004).

However, there are strong pressures for science to try. Over two billion people suffer from neurological diseases and psychiatric illnesses, who could benefit from neuroscience research (Lynch, 2009: 4). Alzheimer's disease alone will present itself as a major problem to all aging Western societies: almost half of all people over 85 have Alzheimer's and 13 per cent of all people over 65 have it. There are now more people, as young as 30, with an early onset of the disease (Alzheimer's Association, 2015). RAND estimated that current health costs related to treating Alzheimer's alone in the US are in the range of \$159 billion to \$215 billion and could double by 2040 (Hurd et al., 2013).

Apart from this big push factor, there is also a powerful pull factor for expanding neuroscience research: the promise of life extension and, possibly, the immortality of the mind. Transhumanists like Ray Kurzweil, who heads Google's R&D in artificial intelligence (AI), consider the human brain to be little more than a biocomputer that can be emulated on a computer. They claim that not only will it be possible to simulate the mind on a computer, but even to copy or upload a human mind into a computer, thereby becoming immortal (Kurzweil, 2005: 198–205). If we just replicate all the complex connections of a living brain, so the argument goes, consciousness would supposedly emerge. It then mostly becomes a challenge of creating sufficient computing power and memory for 'running' or simulating a mind on a computer.

Roboticist Hans Moravec claimed that the entire brain of a human being could be stored on a computer in less than hundred million megabytes or 10^{15} bits, which means that storing the brains of the entire world population would only require 10^{28} bits (Moravec 1999: 166). Ray Kurzweil similarly argues 'that to functionally simulate the brain would require between 10^{14} and 10^{16} calculations per second (cps) and used 10^{16} to be conservative' (Kurzweil, 2012: 196). Kurzweil thinks that brain plasticity poses no obstacle since it can all be emulated on software. Based on his calculations of brain power and projections of future computing power, it would be possible to model the brain by 2020; to run a nuanced simulation of a mind by 2029; and to have human and machine intelligence increased by a billion-fold by 2045 (Barrat, 2013: 131; Kurzweil, 2005: 199–200).

Neuroscientist Kenneth Hayward has agreed that the basic assumptions made by transhumanists are valid, stating: 'I am virtually certain that mind uploading is possible', while suggesting this is 'probably centuries away'. However, Hayward argued that in principle 'all these perceptual and sensorimotor memories are stored as static changes in the synapses between neurons', which could be preserved in a connectome model of the brain (Shermer, 2016). But it is not likely that the wealthy of the world would want to wait for so long: various Silicon Valley entrepreneurs like Peter Thiel, Sergey Brin and Larry Ellison are now investing significant sums of money in immortality, as this is the ultimate prize (Isaacson, 2015). In Russia, billionaire Dimitry Itskov has launched the 2045 Initiative, which aims to realize mind uploading technology by 2045 (2045.com). Whether these are pipedreams or not, what matters is that globally large amounts of resources are directed towards understanding the brain for different reasons and that eventually the knowledge gained could be used (or abused) for military or political purposes.

1.2 Large-scale government brain initiatives

In 2007, US scientists declared at George Mason University the 'decade of the mind' with the explicit goal of 'mapping' the mind similar to the 'mapping' of the human genome, calling for $4 billion in funding (Kavanagh, 2007: 1321). The proposal stressed the urgency caused by the great economic burden of mental disorders in the US.

1.2.1 The BRAIN initiative

President Obama subsequently announced the $3 billion BRAIN Initiative in April 2013 to revolutionize our understanding of the brain. The President explained that it will be a long-term scientific effort comparable to the human genome project and that it could impact 'the lives of not millions, but billions of people on this planet' (White House, 2013). The original plan was to spend $100 million in federal money and $200 million in private sector money on neuroscience research for ten years, but it has been extended to twelve years and a total of $4.5 billions (Requarth, 2015). The project is led by the NIH, the NSF, the FDA, DARPA and IARPA in conjunction with private sector partners such as the Allen Institute for Brain Science, the Howard Hughes Medical Institute, the Kavli Foundation and the Salk Institute for Biological Studies (Insel et al., 2013: 687). According to the White House:

> The BRAIN Initiative will accelerate the development and application of new technologies that will enable researcher to produce dynamic pictures of the brain that show how individual brain cells and complex neural circuits interact at the speed of thought. *These technologies will open new doors to explore how the brain records, processes, uses, stores, and retrieves vast quantities of information, and shed light on the complex links between brain function and behavior.* (my emphasis)
>
> (White House, 2013)

The BRAIN Initiative is pretty clear on the goal of 'deciphering' the brain and about the extensive involvement of the private sector, which is by no means limited to medical research. It thereby acknowledges that the current neuroscience revolution is driven by civilian applications in such diverse areas as health, security, marketing, finance and politics (Lynch, 2009). In 2010 alone, 800 neurotechnology patents have been filed – a doubling of patents per year from the previous decade. Most patents were filed by the marketing research company Nielsen (100) and by software giant Microsoft (89), which shows that neurotechnology has already gone beyond medical applications and is poised to proliferate across society (Griffin, 2015).

1.2.2 The Human Brain Project and similar projects

The European Union inaugurated a similar neuroscience research effort called the Human Brain Project (HBP) in October 2013. The EU pledged to spend €1 billion over ten years to 'gain fundamental insights into what it means to be human, develop new treatments for brain diseases, and build revolutionary new Information and Communications Technologies (ICT)' (Markram 2012, 8). The project is coordinated by the Ecole Polytechnique Federale de Lausanne (EPFL) and has 13 subprojects, which include: Strategic Mouse Brain Data (SP 1), Strategic Human Brain Data (SP 2), Cognitive Architectures (SP 3), Theoretical Neuroscience (SP 4), Neuroinformatics (SP 5), Brain Simulation (SP 6), High Performance Computing (SP 7), Medical Informatics (SP 8), Neuromorphic Computing (SP 9), Neurorobotics (SP 10), Applications (SP 11), Ethics and Society (SP 12) and Management (SP 13). The HBP project website states:

> The Human Brain Project (HBP) is a European Commission Future and Emerging Technologies Flagship that aims to accelerate our understanding of the human brain, make advances in defining and diagnosing brain disorders, and develop new brain-like technologies...A major goal of the HBP is to deliver a collaboratively built first draft 'scaffold' model and simulation of the human brain by 2023. This will not be a complete simulation of every detail, but rather provide a framework for integrating data and knowledge related to the structure and function of the human brain from research and clinical studies around the world. The model and simulations will provide a community test bed for hypotheses and theories of brain function in health and disease.
>
> (HBP, 2015)

The goal of the HBP to build a working computer model of the brain is apparently inspired by the EPFL and IBM collaboration *Blue Brain Project* that was launched in 2005, which receives most of the funding (Frégnac and Laurent, 2014: 28). Similar large-scale brain research projects have been initiated in several other countries, including Australia, Canada, Japan, Israel and China. While the focus of these research initiatives is clearly civilian in nature with curing Alzheimer's high on the priority list, it is also obvious that some of the research can be repurposed for military usage (Marchant and Gaudet, 2014). Robert McCreight suggested that 'this "competitive environment" could feed into a sort of neurological space race, a contest to control and commoditize neurons' (quoted from Requarth, 2015).

1.2.3 The military interest in neuroscience

Since Moreno's 2006 book there has been a steady growth in literature on military and security-related neuroscience research and technological

Table 1.1 Neuroscience initiatives

Country	Initiative name	Initiated	Funding	Research funding programs
United States	BRAIN Initiative	2014	$3 billion	• Multi-scale integration of the dynamic activity and structure of the brain • Neurotechnology and research infrastructure • Quantitative theory and modelling of brain function • Brain-inspired concepts and designs • BRAIN workforce development
European Union	Human Brain Project	2013	€1 billion	• Strategic Mouse Brain Data (SP 1) • Strategic Human Brain Data (SP 2) • Cognitive Architectures (SP 3) • Theoretical Neuroscience (SP 4) • Neuroinformatics (SP 5) • Brain Simulation (SP 6) • High Performance Computing (SP 7) • Medical Informatics (SP 8) • Neuromorphic Computing (SP 9) • Neurorobotics (SP 10) • Applications (SP 11) • Ethics and Society (SP 12)
Canada	Brain Canada	2010	$200 million	• Brain Repair Program • Alzheimer's Prevention Initiative Program • RBC-Brain Canada Research Partnership in Mental Health for Children and Youth
Australia	Cooperation with the US BRAIN Initiative	2014	$250 million	• Digital brain atlasing • Multiscale modelling • Ontologies of neural structures • Standards for data sharing

Japan	Brain/ MINDS	2014	Not stated	• Structure and functional mapping of the non-human primate brain • Development of novel, cutting-edge technologies that support brain mapping • Human brain mapping and clinical research
Israel	Israel Brain Technologies	2010	Not stated	• Brainnovations: start-up program for BrainTech • Braingels: program to connect investors, start-ups and consumers • Prize programs: reward for excellence in neuroscience • Focus: BMI and neuro-stimulation devices
China	Brainnetome	2003	Not stated	• Identification of Brain Networks • Dynamics and Characteristics of Brain Networks • Network Manifestation of Functions and Malfunctions of the Brain • Genetic Basis of Brain Networks • Simulating and Modelling for Brainnetome

Source: Author's own data.

opportunities. A study by the Pentagon's scientific advisory group, *The Jasons*, titled 'Human Performance' pointed at advances in neuroscience that could make soldiers more effective, especially with respect to exploiting brain plasticity and brain-computer interfaces (BCIs) (Jason, 2008: 12). The National Research Council (NRC) was contracted by the Defense Intelligence Agency (DIA) of the Pentagon to produce a study on *Emerging Cognitive Neuroscience*, which was published in 2008 (NRC, 2008). The report states: 'Important research is taking place in detection of deception, neuropsychopharmacology, functional neuroimaging, computational biology, and distributed human-machine systems, among other areas' (NRC, 2008: 2). The report recommends that the Intelligence Community (IC) consistently monitor advances in these areas in order to avoid technological surprise by some of America's strategic competitors, such as Russia, China and Iran. In 2009, the NRC produced another report titled *Opportunities in Neuroscience for Future Army Applications*, which primarily focused on aspects of warfighter enhancement, such as: training and learning, enhanced cognition, decision-making and sustaining soldier performance before, during, and after battle (NRC, 2009). In 2012, the Royal Society released a report on *Neuroscience, Conflict and Security*, which outlined various research related to neuroscientific enhancement and degradation in the context of existing legal frameworks and ethical concerns that may arise because of progress in neuroscience. The Royal Society was particularly concerned that technological progress in incapacitating agents could undermine or weaken existing international treaty regimes such as the CWC and BWC (Royal Society, 2012: 21–24).

1.2.4 DARPA and other acknowledged military neuroscience projects

The American BRAIN Initiative clearly includes military research by DARPA and other agencies. DARPA is a research organisation within the Pentagon, which was established in 1958 after the 'Sputnik-shock' to prevent technological surprise. The law that created DARPA defined its mission to 'engage in such advanced projects essential to the Defense Department's responsibilities in the field of basic and applied research and development which pertain to weapons systems and military requirements' (quoted from Belfiore, 2009: 52). DARPA operates like a venture capitalist enterprise: it invests in 'blue sky' technology projects, some of which may succeed and many of which may fail to produce tangible results. This is the main reason for DARPA's mixed track record that includes the dismal failure of the Autonomous Land Vehicle (ALV) of the 1980s, but also the invention of the Internet.

Since 2002, DARPA has shown interest in biotechnologies and is currently funding a range of neuroscience-related projects, dealing with diverse approaches and problems, such as 'metabolic dominance' or efforts to keep soldiers at peak performance levels, robotic prostheses that can replace lost limbs with full functionality ('Revolutionizing Prosthetics'), treating PTSD with Deep Brain Stimulation (SUBNETS), cognitive augmentation (C2TWS) that flags subconsciously registered threats to operators, genetic sequencing and editing for genetically engineering new species ('Biological Robustness in Complex Settings' or BRICS), nanosensors for 'continuous physiological monitoring of the warfighter' (In Vivo Nanoplatforms or IVN) and studying narratives that can help perfecting the art of propaganda ('Narrative Networks').

The intelligence community version of DARPA – the Intelligence Advanced Research Projects Activity or IARPA – is pursuing at least four neuroscience research programs: the 'Integrated Cognitive Neuroscience Architectures for Understanding and Sensemaking' (ICArUS), 'Knowledge Representations in Neural Systems' (KRNS), 'Machine Intelligence from Cortical Networks' (MICrONS) and 'Strengthening Human Adaptive Reasons and Problem-solving' (SHARP) (IARPA website).

The different US military service branches have also started to sponsor their own smaller military neuroscience projects. The Army is experimenting with 'Mindfulness Meditation' that trains soldiers 'to "tune" their autonomic nervous system' for improving the soldiers' resilience to stress (Brewer, 2014: 803). The Navy is sponsoring a neuroscience-based 'Mental Toughness' program for its SEALs (Durnell, 2014). The Air Force is experimenting with Non-Invasive Brain Stimulation to enable their airmen to remain concentrated for longer (Shachtman, 2010b). US Special Operations Command (SOCOM) is pursuing 'operational neuroscience' with a view of improving HUMINT. In 2013, SOCOM wanted to create a Center of Excellence for Operational Neuroscience at Yale University in 2013, but Yale withdrew from the initiative because of human rights concerns (Eidelson, 2013). Finally, the Department of Homeland Security has been funding a major neuroscience-related project called Future Attribute Screening Technology (FAST), which aims to detect hostile intent in screened people to stop terrorists before they can act (US DHS, 2008). These are just some of the military and security applications of neuroscience that illustrate the interest in 'weaponizing' neuroscience.

1.3 The scope of the book

Appropriately limiting the scope of the book to only 'military neuroscience' is difficult for a variety of reasons: (1) neuroscience is a large interdisciplinary field, which creates a lot of overlap with relevant subfields, such as behavioural science and computer science; (2) the lines of civilian vs. military applications are not well-defined and it is only up to the imagination of practitioners in the military/ security field how they could take advantage of neuroscience; and (3) since neuroscience is still an emerging technological field, it remains still unclear what kinds of concepts, technologies and applications will prove practical in the future.

The book will cover three main areas that relate to military neuroscience and that are very likely to impact warfare: (1) human enhancement technologies; (2) intelligence and security applications of neuro S/T; and (3) performance degradation technologies or 'neuroweapons'. While human enhancement technologies and many intelligence and security applications are less controversial and have received therefore more academic attention, the aspect of degradation technologies is often absent in the debate, which gives military neuroscience an enhancement spin that greatly underestimates its true offensive potential. The NRC report merely noted that '[t]he neurotechnology degradation market segment is completely underground with only speculative information available' without making much effort to describe related technologies (NRC 2008, 129). The Royal Society report at least discusses two approaches to performance degradation: chemical agents and biological agents (2012). It is thus necessary to think more systematically about what a 'neuroweapon' could be, or do.

1.3.1 The concept of the 'neuroweapon'

Neuroscientists James Giordano and Rachel Wurzman have suggested that '[a] weapon is defined as "a means of contending against another" and "...something used to injure, defeat, or destroy"'. They include into their definition any uses of neurotechnologies in 'intelligence and/ or defense scenarios', enhancement and degradation technologies, as well as intelligence applications of neurotechnologies. They claim:

> The objectives for neuroweapons in a traditional defense context (e.g. combat) may be achieved by altering (i.e. either augmenting or degrading) functions of the nervous system, so as to affect cognitive, emotional and/or motor activity and capability (e.g. perception, judgment, morale, pain tolerance, or physical abilities and stamina). Many technologies (e.g. neurotropic drugs; interventional neurostimulatory devices) can be employed to produce these effects.
>
> (Giordano and Wurzman, 2011: 56)

In a few publications in the field of international law the term 'neuroweapon' refers to a weapon that is neurally triggered by the technical interpretation of subconscious brain responses to a perceived threat without conscious control on part of the individual whose brain is used for this purpose (White 2008; Noll 2014). This is a far too narrow definition that excludes too many other potential offensive or combat uses of military neuroscience.

Former diplomat and consultant Robert McCreight argues that it is difficult to define neuroweapons precisely because there is a wide range of possibilities how one can attack a person's mind, listing the various possibilities identified in the work of military analyst Timothy Thomas (1998). McCreight stated:

> Neuroweapons defy easy explanation and definition. An agreed global definition of this term does not exist and differences about core components, structure, design, and intent will no doubt continue as populations wrestle with the notion that such a category of weapons can be created at all…Neuroweapons are intended to influence, direct, weaken, suppress, or neutralize human thought, brainwave functions, perception, interpretation, and behaviors to the extent that the target of such weaponry is either temporarily or permanently disabled, mentally compromised, or unable to function normally.
>
> (McCreight, 2014: 117–118)

In view of these difficulties, it is argued here for the sake of simplicity that *neuroweapons are weapons that specifically target the brain or the central nervous system in order to affect the targeted person's mental state, mental capacity and ultimately the person's behaviour in a specific and predictable way.* Of course, not all neuroweapons are created equal and there are huge differences in their possible capability and the main mechanisms they could use. Finnish National Defense University Professor Torsti Sirén suggests dividing electromagnetic or *psychotronic* weapons into three subcategories: (1) weapons that produce a 'mild' effect 'similar to inducing sleep or losing one's inhibitions'; (2) weapons that produce 'in-between behavioural effects', e.g. causing aggressiveness or passivity; and weapons that produce 'extreme effects' such as direct mental coercion (Sirén, 2013, 86; Binhi, 2010: VIII-IX). Another way of classifying these weapons proposed by this author is to distinguish between weapons that merely attack mental capacity at the level of mental states, perception and cognitive ability (category 1 neuroweapons) and those that manipulate consciousness at the level of emotion, belief and thoughts (category 2 neuroweapons). Category 2 weapons would have a more long-lasting and more profound effect on the victim than a category 1 weapon, which may have effects that are only transient, but could have much faster onset.

The term neurowarfare shall subsequently apply to the systematic effort of an international actor to utilize neuro S/T for the purpose of gaining an advantage in a political or military conflict by influencing enemy minds. It follows that deemphasizing kinetic effects and emphasizing psychological effects would be a defining characteristic of neurowarfare. Deception and mind manipulation would become the main means of securing victory.

1.3.2 Related areas and other definitions

Since the term neuroweapon is relatively new and since there is a long history of efforts of developing weapons that fit the general description, these shall be mentioned and defined here briefly to more clearly delineate them from what is included in neurowarfare:

- *Brainwashing*: This is a term that was introduced by Edward Hunter in 1950 in his book *Brainwashing in Red China* (Hunter, 1951) and also explained by Joost Merloo's *Rape of the Mind* in reference to the workings of fascist dictatorships (Merloo, 1956). More recently, the term has been used in connection to cult or terrorist brainwashing. According to psychologist Kathleen Taylor, brainwashing can be defined as 'a systematic processing of non-compliant human beings which, if successful, refashions their very identities' (Taylor, 2004: 9). Brainwashing can be accomplished through a combination of indoctrination and systematic abuse. In the future, the practice may be refined through relevant neuroscientific research.

- *Psychotronic weapons*: This is a term that appeared in the late 1960s and can still be found in some of the military literature, especially the Russian, in a sense similar to the concept of the neuroweapon (T. Thomas, 1998). Psychotronics is the Czech word for parapsychology and relates to research into 'psi phenomena' such as telepathy, telekinesis and extrasensory perception (Kumar et al., 2012: 42). As used in the literature of the 1970s (e.g. US DoD, 1972), psychotronic weapons rely on some type of directed energy or light and sound in order to achieve some 'psi' effect, usually manipulating or spying on the mind of a victim from a distance. This includes electromagnetic emanations, infrasound, ultrasound and other 'silent sounds' or subliminals (T. Thomas, 1998; Rothstein, 1998). Although the 'psi' concept has been disproven by science, the general idea that electromagnetic waves, infrasound, ultrasound, or 'silent sounds' containing subliminals can affect mental capacity and behaviour is sound. So one can acknowledge the possibility of psychotronic weapons as a form of directed energy weapon without buying into the existence of paranormal phenomena.

- *Directed Energy Weapons (DEW)*: These are weapons that use energy for producing a weapons effect, which can be anti-personnel or anti-material, or lethal, as well as non-lethal. The most well-known types of DEW are lasers, high-powered microwaves and microwave/ millimetre wave systems. The Pentagon has described DEW as a potential 'game-changer' since these weapons can produce almost instant effects over long distances (US DoD, 2007). DEW are relevant here in so far some of them are anti-personnel/ nonlethal and could be used for targeting the brains and CNS of enemy personnel, which would make them similar to the Soviet psychotronic weapon concept. The book will mostly use the term 'psychotronic weapon' to refer to DEW that are designed to attack mental capacity and to influence behaviour.

- *Nonlethal Weapons (NLW)*: According to the DoD NLW are '[w]eapons that are explicitly designed and primarily employed so as to incapacitate personnel or materiel, while minimizing fatalities, permanent injury to personnel, and undesired damage to property and the environment' (US DoD, 1996). Naturally, neuroweapons would fall into the broad NLW category since they target personnel with the intention of producing less lethal effects. The mission of these weapons is usually described as 'incapacitation'. However, NLW researcher Neil Davison noted that US military interest interest in NLW 'will likely expand beyond solely "non-lethal" effects' into the area of behavioral effects (Davison, 2009: 181).

- *Psychological operations*: A very established form of nonlethal warfare is psychological warfare that aims at influencing the perceptions, emotions and behaviour of an adversary by using communication (radio/TV broadcasts, leaflets, the Internet) as a main tool (US DoD, 2003, 1.1). PSYOPS falls largely, but not completely, into the field of strategic communications, which is meant to influence foreign audiences through dissemination of information (Paul, 2011). While traditional methods of PSYOPS and strategic communication are outside of the scope of the book, neuroscience research can make propaganda much more effective.
- *Military deception*: This is a systematic effort to mislead or misdirect an adversary by manipulating the adversary's perception. Military deceptions have always been part of warfare and examples range from the Trojan horse to the D-Day deception, which was probably the most sophisticated military deception in history (Sirén, 2013). Military deception clearly intersects, like PSYOPS, with the main theme of the book, as it also targets an adversary's mind, or more specifically, the adversary's perception. PSYOPS and military deception can be substantially enhanced, if a better understanding of how the human brain processes information and perceptions can be gained, as specifically stated in DARPA's 'battlefield illusion' project (Shachtman, 2012).
- *Cyber warfare*: Military and intelligence organizations are using computer network attacks for purposes of cyber espionage and high-tech sabotage. In the past militaries have combined conventional military operations with distributed denial of service attacks and it is expected that cyber attacks could enable militaries in the future to hijack or disable networked capabilities such as drones, satellites, or critical infrastructure such as the power grid, mass transportation systems and stock markets (Clarke, 2010). As cyber-neuro systems such as BCI or BMI spread across society, it becomes theoretically possible that cyber attacks could specifically target users opposed to computer software and hardware (Diggins andArizmendi, 2012). Many issues that are discussed in the cyber warfare field, such as the attributability of attacks and the difficulty of internationally regulating cyber warfare have also great relevance to neurowarfare.

The purpose of this book is to bring these different concepts, approaches and techniques together into a more unified theory centred on mind manipulation for achieving various political, military and security objectives in order to provide some more theoretical perspective of future neurowarfare and how it would differ from the Western paradigm of war.

1.4 Chapter overview

The first chapter provides some necessary historical background for military brain research and nonlethal warfare by discussing CIA and military research related to individual 'mind control', psychic spying, germ warfare and psychochemical warfare. The chapter will discuss the origins of CIA mind control research, namely the concern that the communists had perfected 'brainwashing' by the late 1940s and the discovery of the hallucinogen LSD. The CIA closely cooperated with the US Army's Special Operations Division (SOD) and played some part in the development of biowarfare plans. Furthermore, the CIA and DARPA experimented with brain implants in the 1960s and early 1970s, which was based on the published research by neuroscientists Robert Heath and Jose Delgado. By the 1970s the CIA shifted its interests to the investigation of the paranormal. This resulted from fears that the Soviets had developed 'psychotronic' weapons that could interfere with human consciousness using 'psychotronic generators'. It is concluded that MK ULTRA and similar research was probably more successful than has been claimed, which makes it unlikely that the research was ever fully abandoned.

The second chapter will outline some of the current human enhancement initiatives in view of warfighter enhancement. There is some focus on DARPA projects for the simple reason that DARPA publicly announces programs with a transformative potential so that researchers, called 'performers' in DARPA terminology, can apply for funding. Although the programs are announced, research results get subsequently classified. It thus becomes impossible to determine whether a classified program is leading to some real military capability or is merely one of DARPA's 'crazy ideas' that never makes it to the real world. Human enhancement methods discussed include (1) the development of performance enhancing nutrition/ supplements and psychopharmaca, (2) new methods of brain stimulation, (3) brain-computer interfaces (BCIs) and (4) genetic selection and enhancement. Human enhancement is difficult, but breakthroughs may originate from two particular areas: BCIs that closely integrate man and machine into a system and secondly, human genetic engineering that promises the development of humans with superhuman abilities, including desirable behavioural and character traits, such as a greater resilience to stress, sleep deprivation and trauma.

The third chapter will explore applications of neuroscience research for the fields of intelligence and homeland security. The chapter will discuss the areas of (1) strategic intelligence, (2) intelligence analysis and decision-making, (3) threat detection and (4) neuroscience and interrogation. It is argued that strategic intelligence can be improved through cultural neuroscience that explores the differences in the way people from foreign cultures process information and how their values shape their actions. Intelligence analysis could be improved by enhancing intelligence analysts or by using neuro-cyber systems that utilize and replicate human pattern recognition ability. The chapter will also describe efforts of developing advanced threat detection systems aimed at identifying dangerous individuals in a crowd. Finally, the chapter will discuss the uses of neuro S/T for interrogation from deception detection to 'mind reading'.

The discussion of degradation technologies or neuroweapons is divided into two chapters. The fourth chapter will look at chemical/biochemical and biological neuroweapons. Chemical/biochemical agents are divided into calmatives, which are designed to tranquilize individuals, hallucinogens, which aim to distort a person's perception and cause irrational behaviour, hypnosis-inducing drugs, which make it theoretically possible to put a person into a highly suggestible state of mind (scopolamine) and biochemicals that produce other effects, such as to induce trust in people (oxytocin). Biological neuroweapons would be living microorganisms that infect humans or alter their brainchemistry in order to degrade their mental capacity, or to cause behavioural effects. The chapter argues that, due to the great advances in the field of biotechnology, it has become possible to design new disease agents that can target the brain and CNS in order to manipulate mental states and mental capacity, which can result in induced mental disease and changes to personality.

The fifth chapter will discuss the development of nonlethal DEW, in particular acoustic and microwave type weapons, as well as what the Russians refer to as 'information weapons': subliminals, silent sounds and 'mind viruses'. Since the amount of concrete and reliable information is very limited in these areas, the chapter will focus on the known physical principles and behavioural research to support the notion that these types of neuroweapons are viable and could be developed or greatly improved in the future. There is no doubt that microwaves of a sufficient energy density can cook internal organs, such as the brain, and thereby produce behavioural effects. Similarly, it is well-understood that subliminals can influence behaviour to some extent. There is only the question whether such approaches have any military utility.

The sixth chapter will try to establish the larger strategic context in which military neuroscience research occurs by examining major trends in warfare. It is noted that interstate conventional warfare is on the verge of disappearing, as the most important and likely military challenges are going to be unconventional. The chapter thus focuses on intrastate conflict, 'non-obvious warfare' and the challenge of controlling large populations in the face of mass impoverishment, civil unrest and environmental disaster. It will be shown that while neuroweapons would only offer limited advantages in conventional war scenarios, especially since much military decision-making is getting automated, neuroweapons could provide a potentially decisive advantage in unconventional and nontraditional warfare scenarios that are likely to be most relevant in the twenty-first century.

The seventh chapter has the ambition of conceptionalising neurowarfare as a 'new' approach to conducting hostilities between different political groups. It is argued that Western strategic writing is still largely trapped in a Clausewitzian mindset that considers violence to be an integral aspect of war. The concept of neurowarfare proposes instead that other societies could be subverted and assimilated through sophisticated attacks on a collective consciousness. The chapter will draw largely upon Soviet/ Russian strategic literature and the commentary of Soviet defectors to sketch methods of political subversion and information warfare (IW) that could one day be refined into neurowarfare proper – a direct attack on the brains and consciousness of entire populations in order to either cause internal political chaos, or to change their political identity, way of life and allegiance in a stealth conquest without violence.

The final chapter discusses dangers and solutions related to the development of neuroweapons and the rise of neurowarfare. One of the dangers is the likelihood that governments will conduct unethical human experimentation. A second danger is that military neurotechnologies could proliferate across societies and thus lead to the emergence of new forms of terrorism and crime. But the biggest danger by far is the nightmare of a psychocivilized society or a 'scientific dictatorship', as Aldous Huxley called it, where everybody's thoughts, emotions and behaviours can be externally adjusted to fit the societal norms as set by a government or tiny elite. While all these dangers are grave and real, the situation is not hopeless, since there are things that can be done to stop the worst abuses. The final section will sketch some options for neurosecurity and will advocate a 'human right for mental self-determination', which should severely restrict the use of mind reading and mind manipulation technologies that could soon be available to many governments, corporations and a variety of nonstate actors.

2 Cold War brain research and germ warfare

This first chapter will provide some necessary historical context to current military neuroscience research by looking at secret brain research and germ warfare programs of the Cold War era. It will be shown that military neuroscience grew out of the desire of intelligence agencies to find technological methods for mental coercion to be used in interrogations and for the brainwashing and 'mind control' of human agents in human intelligence (HUMINT) operations. This secret research conducted by the CIA in the 1950s and 1960s (largely under the codename MK ULTRA) also overlapped with biological and chemical warfare research of the US Army. The chapter will therefore discuss some aspects of the US Army's biological and chemical warfare programs that aimed to covertly spray enemy populations with nonlethal agents to disrupt their societies. In addition, it will be shown that the CIA and DARPA did not stop at researching hallucinogens, but also developed an intense interest in other methods of behavioural control, such as the influence of electromagnetic fields on the mind, as well as the development of brain implants as pioneered by neuroscientists Robert Heath and Jose Delgado. By the 1970s, the interest of the US defence establishment had shifted towards a very unusual direction: the investigation of the paranormal in view of psychic spying. This was directly inspired by similar research that was pursued in the Soviet Union at the time. It will be argued that the paranormal spying was most likely a cover story to hide Western and Soviet pursuit of new types of nonlethal electromagnetic weapons and biotelemetry. The chapter will conclude that although much of the mind control research of the era was probably unsuccessful, it must have been successful enough to justify continuing research into brain manipulation and new NLW (discussed in chapters five and six) that seem to be based on ideas that were already under investigation half a century ago, e.g. brain chips, hallucinogens, microwave, ultrasound, infrasound and light and sound-based weapons that can potentially affect mental states and mental capacity.

2.1 CIA 'mind control' research

Many details about MK ULTRA remain unknown. Most of the documents relating to CIA mind control were destroyed on orders from Director of Central Intelligence Richard Helms given in late 1972, which means that there is only a very incomplete account of related programs undertaken between 1947 and 1972. In a 1978 interview with David Frost, DCI Helms called it a 'conscious decision' to keep the relationship the agency had with its researchers secret (Frost, 1978: 21). Helms argued that the CIA would have been derelict of duty if they had failed to investigate the area of mind control, since so much was going on in Russia and China (Frost, 1978: 20–21). It seems clear enough that CIA mind control has its internal roots in 'truth serum' and hypnosis research conducted by its predecessor organization OSS during the Second World War (Marks, 1979: 3–22). It also grew out of several other research programs within the US military related to chemical and biological warfare. By the early 1950s, it became a massive and coordinated secret research effort. According to researcher Walter Bowart,

> [t]he testing of drugs by the CIA was just a part of the United States government's top-secret mind-control project, a project which had spanned thirty-five years and had involved tens of thousands of individuals. It involved techniques of hypnosis, narco-hypnosis, electronic brain stimulation, behavioral effects of ultrasonic, microwave, and low-frequency sound, aversive and other behavior modification therapies.
>
> (Bowart, 1978: 19)

There have been several Congressional investigations into CIA drug testing on Americans and related activities that have shed some light on many aspects of CIA mind control research. Congress discovered MK ULTRA during the Church Committee investigations after the Watergate scandal. The Rockefeller Commission report from June 1975 stated: 'As part of a program to test the influence of drugs on humans, research included the administration of LSD to persons who were unaware that they were tested. This was clearly illegal' (US Senate, 1975: 38). In 1977 Congress conducted a special hearing on CIA behavioural research, in which DCI Stansfield Turner testified on the wide range of behavioural research activities by the CIA's technical division, also in collaboration with the US Army (US Congress, 1977). Many more details were added by research John Marks, who managed to obtain 20,000 documents, mostly financial records found through FOIA request in 1977, which the CIA somehow forgot to destroy. Marks published his excellent book *In Search for the Manchurian Candidate* in 1979 on which much of this chapter is based.

2.1.1 Origins of modern mind control

Brainwashing and modern 'mind control' have their origins in the emergence of totalitarian governments of the Soviet Union, Nazi Germany and

Communist China in the twentieth century. The Soviets had an early breakthrough with Pavlov's behavioural theory (Hunter, 1956: 17–42). They attempted to use it for socially reengineering humans and for creating a Soviet 'New Man'. According to Edward Hunter, who introduced the term 'brainwashing' in 1950 (a loose translation of the Chinese term 'thought reform'), Pavlov was kept inside the Kremlin to work on a secret manuscript that outlined behavioural control. He once met Lenin, who told him that 'he had "saved the Revolution," and that his findings guaranteed the future for world communism' (Hunter, 1956: 40). Hunter was an OSS veteran paid by the CIA to popularize brainwashing and to alert the public to the new threat (Albarelli, 2009: 188).

The Dutch medical doctor Joost Merloo published in 1956 a groundbreaking study of Soviet mind control methods titled *The Rape of the Mind*. According to Merloo, at the core of Pavlovian theory of conditioning is that a particular stimulus can be replaced by any other unrelated stimulus through simultaneous occurrences (Merloo, 2009: 41). Mass conditioning can be achieved through speech or propaganda with particular words triggering associated emotions (this is nowadays called 'neuro-linguistic programming'). Conditioning allows bypassing the mind or consciousness by focusing on conditioned physical or bodily response like Pavlov's dogs salivating when they hear the bell. Similarly, humans could be, according to this theory, politically conditioned by a combination of relentless indoctrination and traumatisation, effectively turning humans into robots or programmable machines. Over time the terrorising force could be replaced by propaganda and indoctrination: 'repeat mechanically your assumptions and suggestions, [and] diminish the opportunity of communicating dissent and opposition. This is the simple formula for political conditioning of the masses' (Merloo, 2009: 47).

The interest of the US defence establishment in behavioural research and 'mind control' began during the Second World War. Research greatly expanded in the years after the war when US intelligence learned about mind control experiments that had been conducted in Nazi Germany and in Communist countries. The Nazis had no scruples in using concentration camp inmates for cruel and often 'terminal' medical experiments in relation to a variety of battlefield medical problems, including: 'typhus, phosphorus burns, bone and muscle grafts, the use of new drugs to treat infections, sterilization techniques' and hypothermia experiments performed at the concentration camp Dachau (Moreno, 2000: 59–60).

According to Marks, the Nazis were also intensively investigating hallucinogenic drugs and other psychoactive substances. Mescaline was administered to concentration camp inmates in view of using it as a truth drug for interrogations (Marks, 1979: 5). Around the same time, the OSS was conducting a few experiments with Cannabis as a truth drug and carried out a field test on an underworld figure from the 'Lucky' Luciano family in New York, which they deemed a success (Ranelagh, 1986: 203–204).

After the end of the war, the Americans subsequently seized documents about Nazi drug experiments that were related to the breeding program *Ahnenerbe*. The documents were brought to the US, where they immediately aroused the interest of American chemical warfare specialists (Albarelli, 2009: 371–372). The US government arranged the transfer of selected Nazi scientists to the US under Project Paperclip, amongst them notorious war criminals such as Hubertus Strughold, who had run the Nazi flight medicine research (Moreno, 2000: 62). An estimated number of 1,800 German scientists were brought to the US, where they could continue their wartime research (Gimbel, 1986: 443). At least six of them were medical doctors, who had conducted unethical human experiments at KZ Dachau (Jacobson, 2014: 64). Annie Jacobson argues in her book *Operation Paperclip* that 'physicians and chemists from Operation Paperclip would work on jointly operated classified programs code-named Chatter, Bluebird, Artichoke, MK ULTRA, and others' (Jacobson, 2014: 302). It is hard to avoid the conclusion that Nazi science inspired American mind control research years before the interest shifted towards Soviet and Chinese mind control techniques.

2.1.2 Bluebird and Artichoke

The first post-war program was Project Chatter started by the US Navy in 1947. It was aimed at developing drugs for interrogation and for the recruitment of agents. According to the 1977 Senate hearing on MK ULTRA, '[t]he research included laboratory experiments on animals and human subjects involving Anabasis aphylla, scopolamine, and mescaline in order to determine their speech-inducing qualities' (US Congress, 1977). It is also often stated that the CIA became interested in mind control research after the show trial of Cardinal Mindszenty in Hungary in 1949 during which the cardinal confessed out of character to all charges, seemingly under Communist mind control (Ranelagh, 1986: 203; Merloo, 2009: 28–31).

The CIA's Office of Security started Project Bluebird during the same year. It included experiments with drugs and hypnosis, to test whether better results can be achieved with these methods in interrogation. In October 1950 Bluebird teams were sent to North Korea to experiment on captured North Koreans (Ranelagh, 1986: 204). The CIA's Office of Scientific Intelligence started Project Artichoke in 1951, which had the goal to 'exploit along operational lines, scientific methods and knowledge that can be utilized in altering the attitudes, beliefs, thought processes, and behavior patterns of agent personnel' (quoted from Ranelagh, 1986: 204).

The CIA's concern about Communist brainwashing techniques had been fuelled by Edward Hunter's book Brain-Washing in Red China, as well as by the fact of mass treason of American PoWs in North Korea (Frost, 1978: 20). It is estimated that an astonishing 70 per cent of American PoWs in North Korea were willing to go on radio or TV to denounce alleged American war crimes (Marks, 1979: 134). To uncover the mystery CIA Director Allen Dulles decided to contract two psychologists, Harold Wolff and Lawrence Hinkle to conduct a study on communist brainwashing techniques in 1953 (Project QK HILLTOP). Although the study suggested that the communists used conventional methods for brainwashing such as isolation, indoctrination, sleep deprivation and peer pressure, it was still assumed that there could be technological approaches that could produce results much faster (Marks, 1979: 133–135).

Project Artichoke thus investigated some novel and more scientific methods than were apparently used by the Chinese communists, such as 'psychological harassment' and 'total isolation' (CIA 1975). Later, in 1956, the agency's methods for total isolation would include sensory deprivation tanks or floatation tanks, which had been invented by brain researcher John C. Lilly. In a condition of zero external stimuli, severe hallucination results after only a few hours – the condition becomes intolerable after three hours (Victorian, 1996).

Several Bluebird/ Artichoke documents make it clear that the CIA was interested in creating a 'human robot' or a so-called 'Manchurian candidate' (Ranelagh, 1986: 204). John Marks argued that the CIA's desire was to find a way 'to remove the human element', which results in randomness and uncertainty in HUMINT operations (Marks, 1979: 49). The agency wanted to know whether people can be hypnotically programmed to carry out a task, even against their will and without conscious knowledge, and then made to forget everything about it afterwards. Artichoke experiments included the use of hypnosis and drugs like morphine. A 1952 CIA memorandum stated the program's purpose succinctly as a question to be answered: 'Can we get control of an individual to the point where he will do our bidding against his will and even against fundamental laws of nature, such as self-preservation?' (CIA, 1952a) The CIA also wanted to know:

> [c]an we in a matter of an hour; two hours; one day, etc., induce an H[ypnotic] condition in an unwilling subject that he will perform an act to our benefit?…Can we create by post H[ypnotic] control an action opposed to an individual's basic moral principles?…Could we seize a subject and in the space of an hour or two by post H[ypnotic] control have him crash an airplane, wreck a train, etc.?
>
> (CIA, 1951)

A few documents found by researchers indicate that many of these questions could be answered in the affirmative, although there is indeed no proof that hypnotic mind control has been operationally used. Marks described a case, where CIA hypnotist Morse Allen put a CIA secretary into trance and hypnotized a second secretary:

> Miss…was then instructed (having previously expressed fear of firearms in any fashion) that she would use every method at her disposal to awaken Miss…(now in a deep hypnotic sleep) and failing this, she would pick up a

pistol nearby and fire it at Miss…She was instructed that her rage would be so great that she would not hesitate to "kill" for failing to awaken. Miss… carried out these suggestions to the letter including firing the (unloaded pneumatic pistol) at…and then proceeding to fall into deep sleep.

(CIA, 1954; Marks, 1979: 194)

Only the psychiatrist and hypnotist George Estabrooks, who was a consultant to the OSS, later wrote several books in which he claimed to have successfully created Manchurian candidates and that his techniques were used in OSS operations during WW II, including the use of 'hypnotic couriers' (C. Ross, 2007: 28). Hypnotic couriers can be hypnotically programmed to remember a message that can only be recovered through a post-hypnotic cue, such as a code word (C. Ross, 2006: 164).

2.1.3 MK ULTRA

The main problem with Bluebird/ Artichoke research was that it was too amateurish and lacked scientific validation. In an effort to have a more scientific basis for mind control, DCI Allen Dulles authorized Project MK ULTRA in 1953 (US Congress, 1977: 86). It was thus decided that the CIA's Technical Services Staff (TSS later renamed TSD), under the direction of Sidney Gottlieb, should outsource the research to universities. In the end a large number of scientists, about 88 institutions, amongst them 44 colleges and universities, 15 research foundations and chemical/ pharmaceutical companies, 12 hospitals, and 3 prisons, in the US, Canada and the UK contributed to MK ULTRA research. Some of the institutions participating significantly in the research were Harvard University, the Allen Memorial Hospital, Tulane University, the University of Maryland, University of Denver, the Vaccaville State Prison and others. Altogether there were overall 149 subprojects for MK ULTRA, which covered a wide spectrum of brain and behavioural research (US Congress, 1977: 5).

Monies were dispersed using front organisations such as the Human Ecology Society, the Geschickter Foundation and the National Institutes of Health (Marks, 1979: 209–230). Many scientists working on these projects were, in fact, 'unwitting': they did not know that their funding came from the CIA and that their results would be used for national security purposes (US Congress, 1977: 3). Others, such as psychiatrists Ewen Cameron, Martin Orne or Jolyon West are believed to have been willing collaborators, although the agency avoided a paper trail proving these connections (as argued by C. Ross, 2006). In the case of Ewen Cameron, the CIA collected all his files after he retired from the Allen Memorial Institute in 1964 (G. Thomas, 2008: 273).

John Marks suggested that during the MK ULTRA era 'nearly every scientist on the frontiers of brain research found men from the secret agencies looking over his shoulders [and] impinging on the research' and may have thus unwittingly contributed to CIA research (Marks, 1979: 151). Some neuroscientists did not like the idea that the CIA could use their research for nefarious purposes. John Lilly was so dismayed about the possibility of the CIA abusing his work on sensory deprivation and other human brain research that he quit his job at the National Institutes of Health in 1958 to work with dolphins instead (Marks, 1979: 151–154).

MK ULTRA was mostly focused on experiments with drugs, but also continued Project Artichoke-type research into the operational uses of hypnosis. MK ULTRA subprojects 5, 25, 29 and 49 carried out by Alden Sears at the University of Denver involved work on hypnotic couriers and hypnotic mind control (C. Ross, 2006: 63). Later MK ULTRA research involved also the investigation of electromagnetic fields for mind control. In late 1963, MK ULTRA was shut down after it was discovered by the CIA's Inspector General (Richelson, 2001: 11). It was succeeded in 1964 by the smaller in-house MK SEARCH program, which ended in 1972.

2.1.4 The LSD experiments

A great number of MK ULTRA subprojects dealt with the use of LSD, which is a strong hallucinogenic synthetic drug that was first produced by the Swiss scientist Albert Hofmann from ergot alkaloids at the Swiss pharmaceutical company Sandoz in 1938 in the hope of obtaining 'a novel, improved circulatory stimulant', similar to Coramin (Hofmann, 1996). Animal tests were disappointing, but in 1943, Hofmann ingested LSD by accident and discovered its strong hallucinogenic qualities (Hofmann, 1996). Psychiatrists became increasingly interested in the drug because it could induce psychotic states in subjects, which suggested that there could be also a drug that could switch a psychotic person to a normal state (Marks: 1979: 57). After the first scientific papers on LSD were published in the late 1940s, it aroused the interest of the CIA and the US Army. The CIA's TSS was in particular interested in LSD because '[t]he most fascinating thing about it … was that such minute quantities had such a terrific effect' (Marks, 1979: 58). A 'normal' dose of LSD is between 100 and 500 microgram.

What must have also peaked the CIA's interest in the drug was the poisoning of bread in the French town of Pont-Saint-Esprit with a fungus similar to LSD in 1951. Hundreds of poisoned people ran crazy on the streets of the town for hours, suggesting that a society could be successfully disrupted by introducing LSD into the food system (Albarelli, 2009: 349–351). The CIA eventually ordered 10 kg of LSD from Sandoz in 1953, which was at the time the only producer of the drug, to have complete control over the world's entire LSD supply. However, the order was a mistake, as Sandoz had only 100g on hand. Since there are 10,000 doses in a gram LSD, it meant that the CIA purchased altogether over a hundred million doses of LSD – enough to drug almost the entire US population in the 1950s (Bowart, 1978: 77). Not surprisingly, several MK ULTRA projects involved research related to LSD, including experiments on animals and often unwitting humans. For example, the New Yorker psychiatrist Jolyon West conducted MK ULTRA subproject 43 in which he tested a high dosage of LSD on the elephant Tusko of the Oklahoma Zoo in 1962, which, in combination with other drugs, killed the animal (West et al, 1962: 1102).

Most bizarrely, the CIA operated safe houses in New York and San Francisco posing as brothels, where unwitting 'Johns' were given drugs like LSD to observe the effects during Operation Midnight Climax (Marks, 1979: 94–112). The safe houses were run by a former OSS and at that time narcotics officer George White. Apart from studying bedroom behaviour under drugs, Operation Midnight Climax also provided an opportunity to perfect the art of sexual blackmail (Segel, 2002). But by the early 1960s, the CIA had already lost much of its interest in LSD and the secret drug testing. A CIA memorandum from 1963 stated: 'The experience of TSD to date indicates that both the research and the employment of the materials [LSD] are expensive and unpredictable in results...some of the testing of substances under simulated operational conditions was judged to involve excessive risks to the Agency' (CIA, 1963).

2.1.5 *Psychic driving*

The Canadian psychiatrist Ewen Cameron was running MK ULTRA subproject 68, which aimed at behavioural modification through 'the repetition of verbal signals'. The experiments took place at the Allen Memorial Institute in Montreal between 1957 and 1964 and were initially funded by CIA front organisations. Cameron called his method 'depatterning' that he defined 'as breaking up existing patterns of behavior, both the normal and the schizophrenic, by means of particularly intensive electroshocks, usually combined with prolonged, drug-induced sleep' (Marks, 1979: 133). The goal was to wipe the brains of test subjects clean of memories, so that they could be reprogrammed by Cameron using tapes with verbal suggestions.

Cameron kept some patients in a drug-induced sleep for up to 65 days in a 'sleep room' during which they would receive 20–40 electroshocks three times a day. When he achieved a state of amnesia in his patients through this torture, he would then switch to the method of 'psychic driving', during which the patients were subjected to continuous verbal messages to create a new personality with new behavioural traits. 'From tape recordings based on interviews with the patient, he selected emotionally loaded "cue statements" – first negative ones to get rid of unwanted behavior and then positive to condition in desired personality traits' (Marks, 1979: 136). Cameron was evidently very successful in causing amnesia and destroying the personality of some of his patients to the point that they had to be permanently institutionalized. In the end, his research proved worthless to the CIA and the funding was discontinued in August 1960, although Cameron kept sending research papers to the Society for the Investigation of Human Ecology (a CIA front) (G. Thomas, 1989; 232).

2.1.6 Other MK ULTRA-related projects

There were several projects that were related to mind control and behavioural research, but were not part of MK ULTRA. For example, MK DELTA was the operational arm of MK ULTRA. TSS tested various drugs on foreign agents of questionable loyalty overseas for the purpose of interrogation and also collected biological materials for 'harassment, discrediting, and disabling purposes' (C. Ross, 2006: 68). MK NAOMI was a collaboration with the Army's Special Operations Division aimed at developing chemical and biological agents that could be used in intelligence operations, e.g. for assassination.

Table 2.1 Mind control programs during the MK ULTRA era

Program name	Period active	Objectives
Project Chatter	1947–53	US Navy program for the development of truth drugs
Project Bluebird	1949–51	CIA program for the development of truth drugs, which was later renamed 'Artichoke'
MK DELTA	1950–72	CIA program for the development of truth drugs, including the procurement and testing of these drugs overseas
MK NAOMI	1950–70	CIA/ US Army program for the research of biological and chemical agents and delivery systems also in view of using them in assassinations and other covert operations
Projekt Artichoke	1951–57	CIA program for the development of truth drugs, most closely associated with the creation of 'Manchurian candidates'
MK ULTRA	1953–64	CIA umbrella program for behavioral research, parapsychology research and 'mind control'. Ended officially in 1964
QKHILLTOP	1954	CIA program for researching Chinese brainwashing techniques carried out at the Cornell Medical Human Ecology Program
Third Chance/ DERBY HAT	1960–61	CIA program for testing truth drugs on captured agents in Europe and the Far East
OFTEN/ CHICKWIT	1967–73	CIA/ US Army research programs related to the procurement and testing of drugs relevant for 'psychochemical warfare'

Source: Author's own data.

Not much is known about MK SEARCH, since it was run in-house by the CIA and not many related files survived. Gordon Thomas claims that MK SEARCH reactivated many formerly abandoned MK ULTRA projects, including those related to hypnosis, RF experiments and biowarfare research. In excess of $1 million was spent on MK SEARCH between 1964 and 1972 (G. Thomas, 2008: 275–278). The connected projects MK OFTEN and MK CHICKWIT seem to have been also drug-related research for the purpose of behavioural control. MK CHICKWIT's description gives as its objectives to 'identify new drug developments in Europe and Asia' and to 'obtain samples'.

MK OFTEN involved the study of dopamine and ibogaine in order to create 'new pharmacologically active drugs affecting the central nervous system [to] modify men's behavior' (Drummond, 2010). However, there seems to have been a much darker aspect to MK OFTEN that is striking: a direct connection to Satanism and paranormal research. According to Thomas, Sidney Gottlieb wanted to 'explore the world of black magic' and 'harness the forces of darkness and challenge the concept that the inner reaches of the mind are beyond reach' (G. Thomas, 2008: 295). By 1971 the CIA employed three astrologers full-time with a salary of $350 dollars plus expenses per week (G. Thomas, 2008: 296). The CIA thus investigated the strange world of the occult in view of leveraging it for mind control and 'intelligence'.

2.2 Chemical and biological warfare research connections

Early on, the CIA mind control research became closely connected to chemical and biological warfare research conducted by the US Army. A US Army chemist and technical director at the Edgewood Arsenal by the name L. Wilson Greene wrote a classified report titled 'Psychochemical Warfare: A New Concept of War', in which he advocated the use of chemical incapacitants as a new method of war. He argued:

> Throughout recorded history, wars have been characterized by death, human misery, and the destruction of property; each major conflict being more catastrophic than the one preceding it...I am convinced that it is possible, by means of the techniques of psychochemical warfare, to conquer an enemy without the wholesale killing of his people or the mass destruction of his property.
>
> (quoted from Khatchadourian, 2012)

In effect, Greene suggested that incapacitating drugs that could induce mass hysteria and he even provided a list of 61 compounds that could be suitable. The US military awarded him a $50,000 grant to conduct further research into psychochemical warfare (Jacobson, 2014: 289). Around the same time, the CIA began to cooperate with the US Army Special Operations Division (SOD) at Fort Detrick to conduct extensive human experimentation on thousands of US soldiers, as well as numerous biological warfare simulations in several American cities such as San Francisco, St. Louis and New York City, affecting millions of unwitting Americans.

2.2.1 Project MK NAOMI

According to a 1977 declassified DoD memo from 1977, MK NAOMI 'began in the 1950s and was terminated, at least with respect to biological projects in 1969. This may have been a successor-project to MK DELTA. Its purpose was to stockpile severely incapacitating and lethal materials, and to develop gadgetry for the dissemination of these materials' (US DoD, 1977: 2). The CIA was interested in developing biological and chemical toxins that could be used for assassination and that would be extremely difficult to detect during an autopsy. During the Church Committee Hearing on 16 September 1975 DCI Colby presented to the committee a 'nondiscernible microbioinculator' (a dart gun) that could fire 'a toxin-tipped dart' with the width of a human hair accurately up to 250 ft. for 'executive actions' (a euphemism for assassination), which was probably produced by SOD and was linked to MK NAOMI (Time, 1975). Another example is the shellfish-toxin pills that were carried by Gary Francis Powers when he was shot down over the Soviet Union in May 1960, which were also developed by SOD (Marks, 1979: 79).

The US Army chemist Frank Olson, who was working with the CIA on Project MK NAOMI, became the most famous victim of illegal CIA mind control experimentation. After a trip to Europe involving a 'terminal inter-rogation' in summer 1953 Olson came back greatly disturbed (G. Thomas, 2008: 169). The CIA spiked a glass of Cointreau with LSD on 19 November 1953, which put him in a state of deep depression. The CIA eventually moved him on 28 November to a room (1018A) in the New York Hotel Statler, which he exited the same night through a closed window on the tenth floor (G. Thomas, 2008: 12). The incident was covered up under the guise of national security. Details regarding Olson's mysterious death first emerged in a 1976 newspaper story, which led to a lawsuit by Olson's family against the CIA. The Agency admitted to negligence. They had to apologize to the family and offered $750,000 in an out-of-court settlement, which the family accepted (Bowart, 1978: 89). An autopsy of the exhumed body of Frank Olson conducted in 1994 by James E. Starrs from George Washington University seems to have provided some ambiguous forensic evidence that the scientist was indeed murdered: he 'had an unexplained lump on his forehead'. Starrs commented, 'I am exceedingly skeptical of the view that Dr. Olson went through the window on his own' (Los Angeles Times, 1994: 17). Journalist Gordon Thomas suggested that George White murdered Olson on Sidney Gottlieb's orders because he knew too much to retire (G. Thomas, 2008: 170–173).

2.2.2 Edgewood Arsenal experiments

According to a General Accounting Office (GAO) report on human experimentation, the US military has tested hundreds of different radiological, chemical and biological agents on hundreds of thousands of people between 1940 and 1974 in support of weapons development programs and for identifying vulnerabilities and health threats to US personnel (US GAO, 1994). Most of the chemical and biological warfare human experiments post-World War Two took place at the Edgewood Arsenal in Maryland. Classified research for the development of incapacitating agents was conducted there between 1952 and 1975. The GAO report on the experiments stated: 'The chemicals were given to volunteer service members at the Edgewood Arsenal, Maryland, and four other locations. Army documents identify a total of 7,120 Army and Air Force personnel who participated in these tests, about half of whom were exposed to chemicals' (US GAO, 1994: 5).

The soldiers were administered LSD, THC derivatives, benzodiazepines and BZ. The GAO report acknowledges that many of the test subjects complained that they had not been fully informed about risks involved and that many developed health problems years after the exposure to the substances, which they attributed to the experiments (US GAO, 1994: 6–7). Walter Bowart wrote that LSD was administered to 585 soldiers and the even more powerful hallucinogen BZ was given to 2,490 'volunteers' at Edgewood Arsenal (Bowart, 1978: 91, 83). Some soldiers were ordered to participate in the tests without any knowledge of what substances they were exposed to (Bowart, 1978: 94–98).

2.2.3 Biological warfare simulations in the US

An aspect of MK NAOMI was the development of effective methods of biological and chemical warfare. The US military therefore not only tested biological and chemical agents on individuals, but also on populations across the US in order to develop effective ways for dispersing agents and for vulnerability testing. According to GAO, between 1949 and 1969, the Army conducted several hundred biological warfare simulations (according a US Army testimony before Congress in 1977, they conducted 239 such open-air tests), in which unaware populations were sprayed with bacterial tracers or simulants that the Army *thought* at that time were harmless (US GAO, 1994: 6). For example, hallucinogenic drugs like LSD were sprayed from aircraft in open-air tests in Maryland and Utah in an attempt to test the feasibility of 'psychochemical warfare' (Khatchadourian, 2012). The CIA may have released, according to Albarelli, LSD in a subway car in New York on two occasions in November 1950 and in 1952 (Messing, 2010). Some of the tests involved spraying large areas, such as the cities of St. Louis and San Francisco, and others involved spraying in closed environments, such as the New York City subway system and the Washington National Airport (US

GAO, 1994: 6). Biological agents and fluorescent compounds (zinc cadmium sulphide particles – now known to be carcinogenic) were sprayed over Panama City, Florida and Key West, Florida (Carlton, 2001: 1).

The US Army tried to develop nonlethal biological agents that could be used as safe simulants for a real bioweapons attack on the US population. The most commonly used simulants were *serratia marcescens* and *bacillus subtilis* (*bacillus globii*) (Cole, 1988, 167). These two microorganisms were sprayed as aerosols in combination fluorescent particles from ships on six separate tests, lasting 30 minutes each, in the San Francisco Bay in a simulated biowarfare attack during 20th and 27th September 1950 (Cole, 1988: 79). The biological agents travelled over 23 miles from the source, covering much of the San Francisco bay area. As a result, 'San Francisco residents were inhaling millions of the bacteria and particles every day during the week of testing' (Cole, 1988: 81). The US military did not inform the public about the test and made no effort of monitoring the health of the affected population. The people of San Francisco only learned about the test 26 years later on 22 December 1976 through an article in the *San Francisco Chronicle* (Cole, 1988: 75).

Another test took place in the New York City subway on 6–10 June 1966. The Army concluded: 'Test results show that a large portion of the working population in downtown New York City would be exposed to disease if one or more pathogenic agents were disseminated covertly in several subway lines at a period of peak traffic' (Cole, 1988: 68). A PBS special feature claimed that the US Army contemplated the spraying of whole countries with chemical and biological agents in the 1950s:

> The next stage was to increase dispersal patterns, dispensing particles from airplanes to find out how wide of an area they would affect. The first Large Area Concept experiment, in 1957, involved dispersing microorganisms over a swath from South Dakota to Minnesota; monitoring revealed that some of the particles eventually traveled some 1200 miles away. Further tests covered areas from Ohio to Texas and Michigan to Kansas. In the Army's words, these experiments 'proved the feasibility of covering large areas of the country with [biological weapons] agents'.
> (PBS, 2013)

What became clear from the chemical and biowarfare simulations of this era is that populations are highly vulnerable to such attacks and that attacks could be conducted covertly without anybody noticing until it was too late.

2.3 Electromagnetic 'mind control' research

In the 1950s, the primary emphasis of CIA mind control research was on the use of psychoactive drugs, especially LSD. However, several MK ULTRA subprojects also investigated the influence of electromagnetic fields and waves on brains and behaviour. For example, MK ULTRA subproject 62, managed by Maitland Baldwin, researched the effects of electromagnetic waves on monkeys. In one published experiment he exposed monkeys to microwaves of a frequency of 388 MHz and the power of 100 V. He noted several effects such as changes in the EEG of the exposed monkeys, as well as arousal and drowsiness. Interestingly, he even observed lethal effects as a result of microwave exposure: 'It was possible to kill monkeys within a few minutes' exposure...when the head was elevated and the chin was fixed' (Baldwin, 1960: 185). Some of Ewen Cameron's experiments also included the use of microwaves on patients in a specially created Radio Telemetry Laboratory, where patients were exposed to a wide range of RF radiation to monitor behavioural changes (S. Taylor, 1992). This kind of research seems to have been inspired by the mysterious 'Moscow signal' that was discovered by the CIA in the mid-1950s.

2.3.1 The 'Moscow signal'

In 1952, the US established, for the first time, an embassy in the Soviet Union and moved their diplomatic representation from a small office near the Kremlin to a much bigger building in Chekovsky Street (Pollack, 1979:1182). The CIA's TSS discovered in 1953 that the US embassy in Moscow was periodically subjected to microwave radiation in a frequency range of 2 GHz and 7 GHz coming from an opposite building (Pollack, 1979: 1183). The energy level for the microwaves remained fairly low at a maximum of only up to 18 μW per sq cm, never exceeding 4mW per sq cm (Brodeur, 1977: 117).

Initially, it was believed that the microwave bombardment was related to a Soviet attempt to eavesdrop on conversations in the embassy. Indeed, a sophisticated microwave-powered bug hidden in a carved seal of the United States that had been given by the Soviets as a present to the American ambassador in 1946 was discovered in 1952 in a routine 'bug sweep' (Wallace et al., 2009: 162). It seemed logical that there could be more passive bugs of that kind. What made this theory more questionable, however, was the fact that the microwave bombardment had gradually increased in intensity and frequency by the late 1950s. An early study on the incident claims that: 'From 1953 until May 28, 1975 there was a single source beam illuminating the west facade of the chancery building starting at about the sixth floor, peaking at around the tenth floor and the roof' (Pollack, 1979: 1183). By 1975, the embassy was illuminated by a second microwave source from the other side of the building with maximum power levels of 18 μW/cm² (Pollack, 1979: 1184).

The CIA thus consulted an outside expert, Milton Zaret – an ophthalmologist – and placed devices for the measurement and the recording of microwave radiation exposure at all four corners of the embassy building in 1962. The CIA pursued several theories about the 'Moscow signal'. It could be: 1) a new method of eavesdropping; 2) an attempt to interfere with US electronic eavesdropping from the embassy; 3) an attempt to damage the health of US embassy personnel in retaliation of US espionage from the embassy; or 4) an attempt to influence the minds and behaviour of US diplomats – this idea was based on Soviet medical literature that suggested that 'microwaves can cause nervous tension, irritability, even disorders' (J. Anderson, 1972: B15).

However, when looking into these possibilities, the CIA did not inform embassy personnel or warn them about any potential health hazards. Columnist Jack Anderson 'broke' the story of potential Soviet 'brainwashing' with microwaves in May 1972, which resulted in Soviet accusations that Spassky had lost the world chess championship due to the use of 'non-chess means of influence' such as 'electronic devices and chemical substance' on behalf of competitor Bobby Fisher (Edmonds and Eidinow, 2004: 237–238). A cover story in *Time* magazine from March 1976 claimed that many members of the embassy staff returned home sick and that two US ambassadors had died from cancer and a third one, Walter Stoessel, suffered from leukaemia (Time, 1976: 19). The State Department initially tried to downplay the incident, but eventually agreed to a 20 per cent salary increase for Moscow embassy staff for hardship (Brodeur, 1977: 105).

2.3.2 *Project Pandora*

In 1965, DARPA decided to contract out research to the Walter Reed Army Medical Research Institute and Johns Hopkins University under codename Project Pandora to determine possible bioeffects on humans as a result of microwave exposure. Until then the scientific consensus in the West has been (and still remains) that only ionising radiation is dangerous, but not non-ionising radiation like microwaves. Furthermore, the energy levels used by the Soviets on the US embassy were far below the levels considered dangerous in the US. Interestingly, the Soviets had set their safety standards a thousand times lower (a 1 μW/cm^2 standard compared to a 1,000 μW/cm^2 standard in the US), which was initially believed to be a mere Soviet deception (Becker and Seldon, 1985: 314).

At the same time, the Moscow signal raised at least the possibility that Western science may have misunderstood the true dangers of electromagnetic radiation. Pandora researchers thus exposed monkeys to all kinds of RF and microwave radiation and also analyzed tissue samples from Moscow embassy personnel. An analysis of Soviet literature by Milton Zaret in the context of Project Pandora revealed the following:

> The primary emphasis of Soviet-bloc research on the biological effects of non-ionizing electromagnetic radiation is concerned with induced neurophysiological or behavioral aberrations. These may be either inhibitory or excitory; the primary locus of action may be in either the peripheral or central nervous system, and the effects may be reversible or irreversible. Super-imposed upon these three sets of variables are the concepts of

thermal versus non-thermal effects, the role of continuous versus pulsed waveform, and the degree of wavelength specificity. Although the various reports appear, at first glance, to represent a confused and conflicting mass of data, a logical, orderly and meaningful pattern can be evolved in evaluating their data. The basic factors which must constantly be borne in mind are (1) that microwave radiation is electromagnetic in nature, (2) that the nervous system functions as an electronic network normally shielded or protected from spurious fields and (3) that when extraordinary electromagnetic fields are created around neural elements this can produce functional or organic neurological anomalies.

(Brodeur, 1977: 297–298)

Two Pandora scientists, Joseph Sharp and Mark Grove, specifically examined the possibility of the Soviets trying to influence the moods and behaviour of the embassy staff. In particular, they investigated the possibility whether the Soviets were testing psychotronic weapons on the embassy staff. This idea was supported by Soviet scientific literature on the bioeffects of microwaves, as well by intelligence reports that indicated a massive Soviet paranormal and psychotronic weapons research effort (US DoD, 1972). According to columnist Jack Anderson, '[o]fficial reports concluded that the Soviets may have been trying "mind control" or "electronic induction of illness." As far as anyone can determine, the attempt failed' (J. Anderson, 1985: C9).

However, some Pandora researchers have made public statements about the possibility that electromagnetic psychotronic weapons are, in principle, possible. Pandora researcher Robert O. Becker, who was twice nominated for a Nobel Prize in medicine for his work on electric healing, stated in a BBC broadcast in 1984:

I believe that there is very little question in that you do produce central nervous system disturbances by microwave exposure. I don't believe that you could at the present level of technology put someone to sleep instantaneously like that. But one could interfere with decision making capacity. One could do, produce, say, a situation of chronic stress in which your embassy personnel do not operate quite as efficiently as they should. This would obviously lead to the Soviets' advantage.

(Jones, 1984)

Becker published two well-received books that touched upon possibilities for weaponizing electromagnetic fields, *The Body Electric* (with Gary Selden, 1985) and *Cross Currents* (1990).

To a limited extent, the Moscow signal incident started a wider scientific and public debate on the potential dangers of microwaves. The journalist Paul Brodeur wrote in 1977 *The Zapping of America*, in which he assembled a wealth of evidence of the various health hazards of microwaves to the human body, including genetic damage, damage to the eyes and harms to general human well-being (Brodeur, 1977). Johns Hopkins University conducted an epidemiological study on the health effects on US Moscow embassy staff in 1976, using health records, questionnaires and death certificates as data, which was completed in 1978. The study 'concluded that personnel working at the American Embassy in Moscow from 1953 to 1976 suffered no ill effects from the microwaves beamed at the Chancery' (Lilienfeld, 1978). A second study was conducted by a team from George Washington University and indicated 'a high rate of mutation in white blood cells in a sample of exposed workers' (Goldsmith, 1996: 83). John Goldsmith reviewed the two studies and published an article, in which he claimed that the dangers of microwave exposure at the Moscow embassy had been deliberately downplayed and that there had been a cover-up (Goldsmith, 1996).

2.4 Brain implants

Psychiatrists have used electrical stimulation of the brain since the late nineteenth century to cure mental illness (Keiper, 2006: 6). The Hungarian neuroscientist Ladislav Meduna developed the electroconvulsive therapy (ECT) in 1934, which stimulates the brain with electroshocks applied to the frontal lobes, to cure schizophrenia, depression and catatonic states (R.H. Blank, 2013: 27). In the 1950s, researchers began to implant electrodes deep inside the brain to electrically stimulate particular brain regions. The first neuroscientist to do so was Robert Galbraith Heath from Tulane University. Heath discovered pleasure centres in the brain that he could stimulate in order to change moods and personality (Keiper, 2006: 14). According to MK ULTRA researcher Gordon Thomas,

> Dr Heath had actually implanted 125 electrodes in the brain and body of a single patient – for which he claimed a world record – and had spent hours stimulating the man's pleasure centers…the neurosurgeon concluded that ESB could control memory, impulses, feelings, and could evoke hallucinations as well as fear and pleasure. It could literally manipulate the human will – at will.
>
> (Thomas, 1989: 265)

In one case, Heath inserted a brain implant into a 19 year old male to cure him from his illness – homosexuality – by stimulating him electronically via brain implant to orgasm while watching heterosexual pornography (C. Ross, 2007: 27). Heath's work was partially funded by the US military and the CIA. A history of Tulane University suggests that Heath was likely a witting MK ULTRA collaborator (Mohr and Gordon, 2001: 120–123). Another important brain researcher who developed brain implants, but with far less obvious CIA connections, is Jose Delgado. Their work and relevant CIA research will be discussed briefly below.

2.4.1 Animal experiments

Most of the experiments involving the implantation of electrodes and radio receivers were obviously carried out on animals, also in view of effectively roboticising them. The CIA researchers had some mixed success with this approach. MK ULTRA subproject 142 was 'a small biological program of electrical brain stimulation [of animals] involving some new approaches to the subject...Some of the uses proposed for these particular animals would involve possible delivery systems for CW/BW or for direct executive action type operations as distinguished from eavesdropping application' (CIA, 1962). The document further states: 'it appears that certain practical guidance systems involving more detailed behavioral control of both positive and negative sorts may be possible than are attainable in the warm-blooded animals being investigated' (CIA, 1962).

Subproject 94 aimed for the remote control of dogs using implants and a related document claims that such a capability has been demonstrated: 'At the present time we feel that we are close to having a debugged a prototype system whereby dogs can be guided along specified courses through land areas out of sight and at distance from an operator' (CIA, 1961b). More famous is the 'acoustic kitty' experiment during which the CIA attempted to turn a cat into a living bug that could be sent to listen inconspicuously to conversations. According to former CIA officer Victor Marchetti,

> [o]ne of the problems with an audio device in the wall or under the mat is that, like cameras, they take a picture of what they see and not what you see in your mind's eye. Human beings have a cochlea in our ears masking out noise, so we can have conversations at a cocktail party...Then they got this idea: let's stop trying to make a cochlea – let's use a real cochlea. Cats have cochleas. So if you wired up a cat, he could mask everything out. That's what they did. They trained him to listen to conversations and not to listen to all the background noises.
>
> (Ranelagh, 1986: 208)

The cat was implanted with a microphone inside the ear, an antenna along the spine and carried a transmitter with a power supply in its belly. In the end, the experiment turned out a complete failure. The cat was released in a New York park, but instead of listening to conversations, it ran on the street and was hit by a car. In a final assessment of the project, TSD determined that transmitters could be implanted in animals 'without damage or discomfort', but that the acoustic kitty had little operational value without the handler having the ability to control the cat itself (Wallace et al., 2009: 201).

The Spanish-born neuroscientist Jose Delgado demonstrated in a famous experiment in 1968 that direct control of large animals was technologically in reach at the time. He implanted self-contained electrodes with a radio receiver that enabled him to activate brain regions by remote control to influence moods and trigger actions. Delgado implanted a bull with a device that he called *stimoceiver* that could receive a radio signal and then stimulate a particular part of the brain. Delgado let the bull charge at him in order to disrupt the bull's movement in the last moment by pressing a button that activated the stimoceiver (Pines, 1973: 39). A *New York Times* article commented about this experiment: 'though their [the neuroscientists'] methods are still crude and not always predictable, there can remain little doubt that the next few years will bring a frightening array of refined techniques for making human beings act according to the will of the psychotechnologists' (Schrag, 1978: 171).

Delgado also stimulated the brains of monkeys, which enabled him to control them like robots. By electrically stimulating a particular brain area he could make a monkey perform the exact same sequence of actions 20,000 times over (Pines, 1973: 40). He could activate pleasure centres to reward them or make them submissive or aggressive, thereby altering the social structure of the monkey group at will (Pines, 1973:40). Delgado wrote: 'In animals... offensive activity could be evoked by ESB [electric stimulation of the brain]... today it is clear that both sham and true rage can be elicited by ESB depending on the location of stimulation' (Delgado, 1969: 124–125). Delgado's research, however, did not stop with animals.

2.4.2 Remote brain control

Delgado also conducted several experiments where he implanted electrodes into 25 human subjects suffering from schizophrenia and epileptics as a method of treatment (Horgan, 2005: 68). His results were as much remarkable from a scientific point of view as they were disconcerting from a moral and philosophical point of view. In one experiment he stimulated with implanted electrodes a particular part of a man's brain that caused him to make a fist with every electric stimulus. Delgado asked the man to keep his fingers extended, but he was forced to make a fist once the stimulus was received regardless. The man replied: 'I guess, Doctor, that your electricity is stronger than my will!' (quoted from Pines, 1973: 41) In another experiment he electrically forced a man to turn his head. Every time after Delgado sent the stimulus and asked the man why he did it he gave different answers that made it appear as if the man believed he turned his head voluntarily (Pines, 1973: 42).

Delgado could not only control behaviour, but also emotions and thinking. Delgado describes the case of female patient, who, when an area in her thalamus was stimulated, would show a fearful facial expression and express that she felt a threat and believed something horrible would happen. This phenomenon could be reproduced with great reliability regardless of environmental and other conditions (Delgado, 1969: 135). In another case he could get a female patient to express gratitude to her doctor every time her brain was stimulated. Although Delgado could not know in advance what kind of behaviour the stimulation of a particular part of the brain would produce, he could consistently produce the same response with the same stimulus. Like Heath, Delgado also used the stimulation of pleasure centres as an approach to behavioural control. He wrote in his book: '...studies in human subjects with implanted electrodes have demonstrated that electrical stimulation of the brain can induce pleasurable manifestations, as evidenced by the spontaneous verbal reports of patients, their facial expression and general behavior, and their desire to repeat the experience' (Delgado, 1969: 142–143). He claims that he could induce sexual thoughts and shift 'mood from dysphoria to contentment and euphoria, usually with concomitant sexual motivation and some "orgastic sensations"' by the push of a button (Delgado, 1969: 143). Furthermore, Delgado could induce hallucinations by stimulating the frontotemporal region of the brain. He claims: '[t]he fact that stimulation of the temporal lobe can induce complex hallucinations may be considered well established' (Delgado, 1969: 151).

In the early 1970s, Delgado wanted to develop an implantable 'cerebral pacemaker' that 'would maintain people's mental stability at all times and thus preserve peace' (Pines, 1973: 42). So it is not surprising that Delgado eventually teamed up with MK ULTRA scientist Jolyon West in 1972 to establish the *Center for the Study and Reduction of Violence* in California (Schrag, 1978: 6–8). The centre planned to conduct experiments on prisoners at the Vacaville prison in California, including the implantation of neural devices in their brains for monitoring their violent tendencies. Among the planned research programs were 'genetic, biochemical, and neurophysiological studies of violent individuals, including prisoners and "hyperkinetic" children; experiments in the "pharmacology of violence-producing and violence-inhibiting drugs"; studies on "hormonal aspects of passivity and aggressiveness in boys"...and, most significantly, the development of a test "that might permit detection of violence-predisposing brain disorders prior to the occurrence of violent episode' in view of '"implanting tiny electrodes deep within the brain," connecting them to radio transmitters, and monitoring (perhaps even controlling) the behaviour of violence-prone individuals or probationers – or indeed, anyone else – by remote control' (Schrag, 1978: 2–3). The plan was endorsed by Governor Ronald Reagan, but failed to convince the Californian legislature (G. Thomas, 1989: 285).

2.4.3 *Field testing Cameron's and Heath's research in Vietnam*

The CIA conducted human experiments on Vietnamese patients and prisoners at the Bien Hoa Hospital in Saigon in 1966–68. In 1966, the CIA sent psychiatrist Lloyd Cotter to Vietnam to field test some of the methods developed by Cameron (McCoy, 2006: 65). So he subjected patients there to extensive electroconvulsive therapy (ECT), where electrodes are attached to a patient's head that give stimuli of 70 to 150 V at 500 to 900 milliamperes in an effort of 'pacifying' the patients (Schrag, 1978: 151).

Gordon Thomas alleges that experiments on prisoners went even further than involuntary ECT and included the implantation of electrodes into their brains:

> Each man was anesthetized and the neurosurgeon, after he had hinged back a flap in their skulls, implanted tiny electrodes in each brain. When the prisoners regained consciousness, the behaviorists set to work. The prisoners were placed in a room and given knives. Pressing the control buttons on their handsets, the behaviorists tried to arouse their subjects to violence.
>
> (G. Thomas, 1989: 264–265)

The purpose of these experiments was obviously to develop a technology that could be useful in CIA operations and to provide solutions to some of the questions raised in relation to Project Artichoke.

The CIA and DARPA were clearly interested in the development of human brain implants in the context of MK ULTRA/ MK SEARCH research. An apparent goal was the development of a remote control for humans that enabled external control over their thoughts and actions. In 1967, a former FBI agent wrote under the pseudonym Lincoln Lawrence the book *Were We Controlled?* about the Kennedy assassination. With the help of a brain implant it would be possible to remotely change the emotions of a subject and even induce partial amnesia. The technique was called 'radio-hypnotic inter-cerebral control' and 'electronic dissolution of memory' (RHIC-EDOM). Lincoln described it in the following way:

> Under RHIC, a 'sleeper' can be used years later with no realization that the 'sleeper' is even being controlled! He can be made to perform acts that he will have no memory of ever having carried out. In a manipulated kind of kamikaze operation where the life of the 'sleeper' is dispensable, RHIC processing makes him particularly valuable because if he is detected and caught before he performs the act specified ... nothing he says will implicate the group or government which processed and controlled him... By electronically jamming the brain, acetylcholine creates static which blocks out sights and sounds. You would then have no memory of what you saw or heard; your mind would be a blank.
>
> (quoted from G. Thomas, 1989: 261–264)

Further confirmation about this research project came from the Tennessee journalist James L. Moore, who claimed that he had come into the possession of a 350 page CIA manual from 1963 that explained RHIC-EDOM (Bowart, 1978: 262–264).

The only official mention of RHIC-EDOM is the questioning of Sidney Gottlieb by Senator Richard Schweiker during a 1977 Congressional hearing, where he partially admitted to its existence. Gottlieb testified: 'As I remember it, there was a current interest, running interest, all the time in what effects people's standing in the field of radio energy have, and it could easily have been that somewhere in many projects, someone was trying to see if you could hypnotize someone easier if he was standing in a radio beam. That would seem like a reasonable piece of research to do' (US Congress, 1977b). In the light of the published experiments with brain implants by Heath and Delgado, the RHIC-EDOM story seems at least plausible.

It appears from this short review that the CIA has indeed sought complete electronic control of animals and human agents, using implants and other methods. It is unclear whether anything of practical value came out of these efforts.

2.5 Extrasensory perception (ESP) research

Some MK ULTRA subprojects dealt with the esoteric topics of hypnosis and ESP. Especially in the Soviet Union there were very close connections between mind control research, parapsychology and nonlethal weapons research to the point that they were indistinguishable. The main Soviet concept was expressed in the buzz word of the era: 'psychotronics', which means 'parapsychology' in Czech and which also elegantly combines the words psychology and electronics. Psychotronics is thus 'the field dealing with the construction of devices capable of enhancing and/or reproducing certain psi phenomena (such as psycho kinesis in the case of "Psychotronic Generators"' (Kumar et al., 2012: 42).

The Soviets believed that they could build machines that could be based on paranormal phenomena, such as ESP, and that could be turned into mind control weapons (US DoD, 1972: 26). The Soviets were apparently influenced by German Ahnenerbe documents and conducted extensive psychotronics experiments in the 1960s and 1970s based on an assumed EM influence on biological objects (Kernbach, 2013: 6).

Parapsychology is now usually considered a 'pseudo-science', as it proclaims itself to use scientific methods while investigating phenomena that defy any conventional scientific explanation. However, there was a time, not so long ago, when the discipline seemed respectable. The first academic parapsychology research program was set up at Duke University by John Banks Rhine, who is the father of modern parapsychology. He coined the term 'extrasensory perception' and claimed to have proven its existence as a reproducible phenomenon (Rhine, 1966: 218). His research attracted some small funding from the US military (e.g. a grant from the Navy to research how homing pigeons find their way), but overall there was scant interest in ESP and other phenomena by the US military.

This changed during the 1960s because of a story reported by a French newspaper *Constellation* in 1958 (US DoD, 1972: 24). The newspaper claimed that the US Navy conducted a 16-day telepathy experiment involving the nuclear submarine *Nautilus* at the Arctic Circle and a researcher at Duke University. According to the newspaper, the experiment was supposedly a great success and led to the recommendation by the RAND Corporation to President Eisenhower to fund further telepathy research in view of military applications. Although the story was immediately refuted by the US Navy, it apparently created enough Soviet curiosity to stimulate Soviet parapsychology research (Ostrander and Schroeder, 1997: 31).

2.5.1 Soviet parapsychology research

There is a long history of unorthodox scientific research related to hypnosis, telepathy and ESP in Russia that goes back to the late nineteenth century. When the communists took control in 1917 some of the research became sponsored by the secret police OGPU/ NKVD that set up several laboratories for 'neuro-energetic' research. The main focus of this research was to identify the type of energy that is responsible for paranormal phenomena such as telepathy. An initial theory that was later disproven was that it could be linked to electromagnetic waves in general and to brain waves in particular (Kernbach, 2013: 4). Later, the unknown energy source was called 'psi'. In 1937, all paranormal research was forbidden by Stalin, as it did not conform to the communist ideology of materialism (US DoD, 1972: 23).

After Stalin's death, the political climate changed and there was a renewed interest by the Soviet authorities to invest in parapsychology research, also in view of possible military applications. Roboticist Serge Kernbach, who reviewed Russian science literature, found that the Soviets poured over $1 billion into unorthodox research during the Cold War period (Kernbach, 2013). The Soviets thought that an understanding of biophysics and biolectronics could enable them to manipulate the brain and CNS using electromagnetic fields, to either enhance psychic abilities or to mentally coerce or harm enemies. Some Soviet psychics allegedly demonstrated their ability to communicate telepathically, to move small objects without touching them and to stop the hearts of animals (Ostrander and Schroeder, 1997). Some of the feats of Soviet psychics were reported in the West and led to growing concerns of a widening 'psychic gap' as columnist Jack Anderson jokingly called it (J. Anderson, 1981: 11).

A classified DIA intelligence report titled 'Controlled Offensive Behavior – USSR' from 1972 alleged that the Soviets may have had a breakthrough in parapsychology research and that the West needed to carefully monitor Soviet research in this area. It summarises all kinds of unorthodox research, including paranormal research and the development of NLW based on psychopharmacology, sound, light and electromagnetics. The report claims that the Soviets had over 20 parapsychology research centres and would spend over 21 million annually on parapsychology research in the early 1970s (US DoD, 1972: XI). The conclusion argues:

> The Soviets are attempting to apply ESP to both police and military use...some years ago a project was begun in the USSR to apply telepathy to indoctrinate and re-educate anti-social elements. It was hoped that suggestion at a distance could induce individuals, without them being aware of it, to adopt the officially desired political and social attitudes... Obviously, telepathy and clairvoyance would make ideal additions to a spy arsenal and such undercover groups are constantly said to be supported ESP research in the USSR...Soviet efforts in the field of psi research, sooner or later, might enable them to do some of the following: a) Know the contents of top secret US documents, the movements of our troops and ships and the location and nature of our military installations. b) Mold the thoughts of key US military and civilian leaders at a distance. c) Cause the instant death of any US official, at a distance. d) Disable, at a distance, US military equipment of all types including space craft.
>
> (US DoD, 1972: 39–40)

This DIA report, which was declassified soon after it was written, may have provided the CIA with the political justification to also move into the contentious area of paranormal research. In 1980 Colonel John B. Alexander published his article 'The New Mental Battlefield' in the journal *Military Review*, in which he openly argued for the reality of ESP/ remote influencing and for the necessity to get into the psychic arms race (Alexander, 1980). Alexander was apparently not the only one in the military with an interest for the paranormal: in 1981, Lt. Col. Jim Channon would write down the First Earth Battalion handbook, after having explored all kinds of alternative knowledge for creating the Army's Jedi knights (Squires, 1988: 47). However, many circumstances surrounding this unusual military interest in the paranormal point in the direction that it was merely a smoke screen or cover story for the continuation of CIA brain and behavioural research for the creation of mind controlled 'super soldiers', also involving research along the lines of biotelemetry/ biocommunication and NLW.

2.5.2 *CIA parapsychology research*

CIA interest in the paranormal began shortly after the Second World War because of anecdotal reports about Nazi ESP research. Andrija Puharich, a captain in the US Army Medical Corps, who served at Fort Detrick and the Edgewood Arsenal at the time, gave a lecture at the Pentagon titled 'On the Possible Usefulness of Extrasensory Perception in Psychological Warfare' in 1952 (Kress, 1977: 7). Puharich would go on to become a prominent investigator of the paranormal with an interest in extremely low frequency

(ELF) waves for mind control (Begich, 2006: 96–97). The CIA funded several ESP projects during the MK ULTRA/ MK SEARCH period.

For example, in 1961 the agency awarded an unnamed researcher $8,579 to conduct parapsychology research in MK ULTRA subproject 136. The only surviving document related to subproject 136 states that the purpose of it was an 'experimental analysis of Extrasensory Perception' and that the 'research effort is moving beyond the question whether ESP exists. He [the researcher] is attempting to approach the twin questions of what are the functional relationships between other personality factors and ESP ability and what are the factors that must be considered in using ESP as a method of communication' (CIA, 1961). It remains unknown what the results of the project were. If it was unsuccessful, it would beg the question – why continue on such a path?

In 1972, the CIA officially began a scientific investigation of paranormal phenomena that lasted several years. The research was mainly conducted in collaboration with the Stanford Research Institute (SRI) in Menlo Park, California. CIA officers met with the physicists Russell Targ and Harold Puthoff from SRI in April 1972 in an initial consultation (Kress, 1977). The CIA decided to contract Targ and Puthoff to conduct parapsychology research on several promising psychics to validate or disprove their paranormal abilities. They found two very talented psychics Ingo Swann and Pat Price, who would go on to develop 'technical remote viewing' for the CIA (Richelson, 2001: 177).

The basic technique consisted in giving the psychics geographic coordinates on which they concentrate. The psychics would then draw what they saw in their minds. SRI also tested an Israeli psychic with the name Uri Geller, who was invited to SRI towards the end of 1972. Puthoff and Targ conducted several months of experiments, where Geller was asked to move objects with his mind (telekinesis), to influence a magnetic compass needle, bend metal objects and guess zener cards (Margolis, 2013: 36–41). At the end of the test series the head of SRI announced on TV: 'We have observed certain phenomena for which we have no scientific explanation', which created the impression to the world that Geller's paranormal abilities were scientifically proven to be genuine (Margolis, 2013: 35).

At SRI psychics like Ingo Swann and Pat Price demonstrated in numerous experiments that they could describe faraway places they never visited using only their mind. In one experiment, Swann and Price were given the coordinates of the secret Sugar Grove NSA facility in West Virginia with no further information as to what the object was. Quoting a memo evaluating the psychics' performance, US intelligence expert Jeffrey Richelson states: '[a] map drawn by Swann was "correct," while the terrain was "exactly as drawn" by Price, and was "not otherwise accessible to non-base personnel"…"an astonishing similarity between [Price's] description of the facility, some dissimilarities, but most of the important ones do match"' (Richelson, 2001: 180).

2.5.3 *Project Stargate*

In 1976, the CIA remote viewing research program was turned over to the Defense Intelligence Agency (DIA) and moved to Fort Bragg, North Carolina. Christened *Project Grill Flame*, the DIA not only researched ESP, but actually tried to operationally use it. Psychics were employed to spy on Soviet and Chinese nuclear test sites, to find the American hostages during the 1980/81 Iranian hostage crisis, find kidnapped Brigadier General James Dozier in Italy in 1981, confirm the status of kidnapping victim William Buckley in Beirut in 1984 and to locate Muammar Gaddafi during the 1986 bombing of Libya (Rifat, 2002: 39–55; Squires, 1988). *Project Aquarius* was tasked with locating Soviet nuclear submarines in the oceans.

Grill Flame was later designated *Center Lane* in 1983 and then *Sun Streak* in 1985. These and similar programs remained active until 1995, when they were turned over to the CIA and renamed *Project Stargate*. The Agency conducted an external review of *Project Stargate* and then decided to close it down with great fanfare (Richelson, 2001: 261). The 1995 review by the American Institutes for Research stated that: 'A statistically significant laboratory effort has been demonstrated in the sense that hits occur more often than chance', but at the same time raised doubts that this could have been due to actual paranormal abilities of the remote viewers (Mumford et al., 1995: E-3). The report claimed that the accuracy of the remote viewers was only 15 per cent and that most of the time they were not able to produce operationally useful intelligence (Mumford et al., 1995).

2.5.4 *How real is ESP?*

Many people, who had been directly involved in the psychic spying, including former head of Army Intelligence General Albert Stubblebine, NLW concept developer for the US Army, John B. Alexander, psychic spy David Morehouse, psychic spy Joseph McMoneagle and psychic spy Ed Dames, all publicly claimed that the respective programs were far more effective than presented in the official review and that the reviewers had been biased against them from the beginning. The parapsychologist Edwin May from SRI wrote a long critique of the reviewers' flawed methodology that, according to him, did not enable them to accurately evaluate the psychic spying program. The reviewers had used a too limited dataset, failed to speak to program participants, did apply inconsistent criteria for success or failure and failed to consider previous reviews and research (May, 1996). May also emphasized that members of the psychic spying unit had received commendations for their service, quoting from Joseph McMoneagle's citation:

> ...He [McMoneagle] served most recently as a Special Project Intelligence Officer for SSPD, SSD, and 902d MI Group, as one of the original planners and movers of a unique intelligence project that is revolutionizing

the intelligence community. While with SSPD, he used his talents and expertise in the execution of more than 200 missions, addressing over 150 essential elements of information [EEI]. These EEI contained critical intelligence reported at the highest echelons of our military and government, including such national level agencies as the Joint Chiefs of Staff, DIA, NSA, CIA, DEA, and the Secret Service, producing crucial and vital intelligence unavailable from any other source....

(May, 1996)

An earlier CIA assessment of parapsychology research undertaken at SRI shows that at least *some* in the CIA were believers in ESP. It concludes that 'sufficient understanding and assessment of parapsychology has not been achieved. There are observations such as the original magnetic experiments at Stanford University, the OSI remote viewing, the OTS code-room experiments, and others done for the Department of Defense that defy explanation. Coincidence is not likely and fraud has not been discovered' (Kress, 1977: 17).

There are clearly very contradictory views on the effectiveness of psychic spying and other military applications of parapsychology. The official view is that it does not work and that earlier interest in these phenomena by Western and Eastern defence establishment was merely the result of Cold War paranoia and the unfounded suspicion that the other side may have had, or could have, a breakthrough in using ESP in the future. At the very least it is notable that remote viewing has survived to the present day – it was merely privatized (Ronson, 2004: 89–116).

There are several remote viewing companies, such as PSI TECH or Mankind Research Unlimited (MRU), that claim to still provide services to the government and major corporations. They may well be 'fronts' for unorthodox government research. PSI TECH has a higher profile because notable individuals such as Major General Stubblebine, Colonel John B. Alexander and Major Dames were at one point associated with it (Rifat, 2002: 65). MRU has received comparatively less attention. A 1980 article on MRU reveals that the company is a wholly owned subsidiary of Systems Consultants, Inc. – a company specializing in military electronics – and that its main areas of interest are (1) 'biophysics' or 'biological effects of electromagnetic fields', (2) 'to produce a device that emits waves which causes mental confusion', (3) '[b]ehavioral science' that includes 'Metapsychiatry and the Ultraconscious Mind', and (4) '[p]sychophysics' that includes 'bioluminescence' (aura photography) (Weberman, 1980: 16). The article suggests that MRU has numerous government connections and might be a CIA front, which the organization has vehemently denied. At the same time, the company seemed to have shared some of the same research interests with the CIA, namely electromagnetic mind control (Begich, 2006: 81–82).

Several mind control researchers have alleged that the open interest of the superpowers in ESP during the 1960s and 1970s was little more than a cover story for protecting very advanced biotelemetry technology that was being developed during this time period. Targ and Puthoff had previously worked in laser technology before they switched to parapsychology (Richelson, 2001: 176–177). Journalist Alex Constantine quoted civil rights activist Ken Lawrence, who said: 'it takes considerable mind-bending to suppose the CIA hired men with these skills to try to read the Kremlin's classified zener cards' (Constantine, 1995: XII). By claiming that they were using ESP a somewhat plausible explanation was provided as to how certain sensitive information was obtained.

The use of elaborate and highly public cover stories is not unusual in the intelligence world. For example, Vaughan Purvis has claimed that the *Nautilus* story was a carefully concocted CIA disinformation operation to misdirect the Soviets with respect to the newly developed US technology of using VLF for underwater communication with submarines (Purvis, 1997). Other examples include Project Azorian (the secret salvaging of a Soviet submarine under the cover story of deep sea mining), or the CIA using UFOs as a cover story for its secret aircraft operations in the 1950s. If, on the other hand, the CIA/Pentagon paranormal research was indeed not a cover story, it would follow that tricksters such as Uri Geller managed to fool *teams* of scientists at SRI and other elite institutions for years, which seems unlikely.

2.6 Conclusion

The chapter has argued the following: (1) the CIA became interested in mind control because of research in this direction in totalitarian regimes at the time and wanted to create methods for mentally coercing people to do their bidding. It seems probable that CIA mind control research was inspired both by Nazi research and by evidence of communist mind control that surfaced in the late 1940s and early 1950s. A particular concern was the apparently successful brainwashing of US PoWs by North Korea, which caused fears about the communists having achieved some sort of technological breakthrough. (2) CIA mind control research was intrinsically linked to biological and chemical warfare research conducted by the US Army SOD, which involved simulated attacks on populated areas, also for 'psychochemical warfare'. MK ULTRA, especially in the earlier years, concentrated primarily on drug experimentation with LSD and other hallucinogens. (3) An area on which the CIA and DARPA focused in the 1960s was electromagnetic mind control or more generally the influence of electromagnetic fields on behaviour. This was apparently inspired by the 'Moscow signal' and Soviet medical literature that suggested that electromagnetic waves could influence mental states and cause disorders. (4) During this time there was also substantial interest on part of the CIA with respect to brain implants as pioneered by Robert Heath and Jose Delgado. These two neuroscientists conclusively proved the concept that mental states and human behaviour can be controlled through electrical stimulation of specific brain areas through implanted electrodes. (5) When MK

ULTRA type research was phased out in the early 1970s because of the fear of exposure following the Watergate scandal, the US defence establishment's interest shifted into the unusual area of paranormal research. Most likely, this was a cover story for other kinds of research relating to biotelemetry and NLW, not unlike research into psychotronics as carried out in the Soviet Union. This is indicated by the fact that the psychic spying was continued for over 20 years, which makes no sense if it was all nonsense. Furthermore, it was privatized with the companies (PSI TECH and MRU) having notable government connections.

3   Neuroscientific enhancement

This chapter will review human enhancement technologies as they relate to neuroscience, as well as to brain and behavioural research that can be used for sustaining and enhancing the performance of warfighters. The chapter will look at four enhancement approaches that for many reasons seem to be most promising in terms of improving human performance: (1) neuropharmaceutical enhancement, (2) brain monitoring and stimulation, (3) brain-computer interfaces (BCIs) and (4) genetic selection and enhancement.

DARPA has been interested in warfighter enhancement since at least 2002 when DARPA program director Joseph Bielitzki announced an ambitious human enhancement program in relation to the Army's desire of creating a 'Future Force Warrior' (now discontinued) with the motto '[b]e all you can be and a lot more' (Burnam-Fink, 2011). Eventually, DARPA shifted its efforts to a more modest 'metabolic dominance' program, which relies on a 'nutraceutical' approach to improve soldier endurance (Moreno, 2004: 7). By 2007, the term favoured by DARPA was 'Peak Soldier Performance' and aimed at keeping soldiers at their peak performance for longer.

However, since the Iraq War of 2003, DARPA saw itself confronted with a somewhat different problem: how to make soldiers cope with traumatic brain injury (TBI)/ PTSD and how to restore limbs lost in battle? TBI and PTSD have reached epidemic proportions with over 600,000 veterans who could be affected (Tanielian and Jaycox, 2008: XXI). DARPA has several research programs for finding better treatments for TBI and PTSD, including Systems-Based Neurotechnology for Emerging Therapies (SUBNETS), Restoring Active Memory (RAM) and Electrical Prescriptions (ElectRx). DARPA started the program *Revolutionizing Prosthetics* in 2005 to develop high-tech robotic prostheses that would enable some wounded soldiers to return into battle after losing a limb (Belfiore, 2009: 14). This research has already led to several significant breakthroughs in the fields of prostheses, BCIs and brain implants, which will be reviewed in the chapter. The chapter will conclude that human enhancement will be most certainly an aspect of future warfare, although it will remain difficult to enhance human capability much beyond the natural limits, unless some more radical approaches are used. Breakthroughs in human performance could result from implanted

BCIs that seamlessly connect human minds to computers (commonly associated with 'cyborgs'), or from a complete genetic redesign of humans, possibly by combining human and animal DNA (transgenics).

3.1 Neuropharmaceuticals

It is now a well-established fact that drugs can affect 'all aspects of human psychology, including cognition, emotion, motivation, and performance' (NRC, 2008: 41). Psychopharmacology has made considerable advances in the last few decades and there is now a plethora of drugs that can manage mental states and emotions, influence behaviour and potentially improve performance. The use of drugs seems thus a good starting point for building better soldiers. In fact, militaries around the world are currently evaluating various known psychotropic drugs for performance enhancement. They could be administered to troops before, during, or after battle for a variety of purposes: to improve learning, to calm anxiety, to increase wakefulness, perceptiveness, concentration and to reduce the impact of traumatic memories.

3.1.1 Nutrition and supplements

It is logical to start thinking about enhancement by looking at improving nutrition and supplements before more drastic, more expensive and riskier methods of cognitive enhancement are used. This was the foundation of DARPA's 'metabolic dominance' program, which aimed at changing the metabolism of soldiers for high performance (Shachtman, 2007). It makes sense. Soldiers should be considered high-performance athletes, who have to cope with extraordinary stress. This means that like professional athletes, they should have a special nutrition regime that can make them healthier and more resilient. Unfortunately, Western soldiers have poor diets that are common in Western societies and that lead to a variety of long-term health hazards and a short-term reduction in performance, such as obesity, poor cardiovascular health, heart disease and cognitive decline (Ford and Glymour, 2014: 49).

The US military is trying to address the problem by developing healthier diets and by providing soldiers with natural supplements instead of leaving it up to the individual and the expanding alternative health industry that often lacks scientific foundations. For this purpose, the US Army Institute for Environmental Medicine has created a Military Nutrition Division, which '[c]onducts nutritional research that provides a biomedical science basis for developing new rations, menus, policies and programs to enable Warfighter health-readiness and optimal performance'. Among the research carried out by the Military Nutrition Division is an investigation of an 'optimal omega-3 diet', which promises to improve cognition, mental health and mental resilience. Deficiency in the omega-3 fatty acid docosahexaenoic acid (DHA), which is together with EPA, a major component of neuronal membranes, is linked to 'cognitive decline, increased rates of telomere shortening (implicated in premature aging), diminished cardiovascular health, and increased susceptibility to and poorer recovery prospects from brain injury' (Ford and Glymour, 2014: 49). Furthermore, ketogenic diets are investigated that change a metabolism from burning sugar to burning fat-based ketones (Ford and Glymour, 2014: 50; Shachtman, 2007). In practice, it is a diet that systematically minimises the intake of carbohydrates and instead relies on healthy saturated fats as a main source of calories, which has been shown to promote brain health and cognitive function (Perlmutter and Loberg, 2013).

Another supplement that is of interest in terms of cognitive enhancement, is piracetam, which is a synthetic supplement that may improve longevity and is considered to be also a nootropic drug that can improve cognition by boosting overall brain function. Piracetam is currently not licensed for medical use or as a dietary supplement in the US, but it is available in Europe and other parts of the world. Piracetam has been shown to improve cognitive deficits associated with brain injuries and may have a neuroprotective effect (Malykh and Reza, 2010: 287–312).

Kenneth Ford and Clark Glymour suggest in an article of the *Bulletin of the Atomic Scientists*:

> At a minimum, improvements to the diet available to warfighters could have an immediate impact on physical and cognitive fitness and a long-term impact on health with respect to common but avoidable maladies such as coronary disease, diabetes, and cancer. Science-based improvements in the efficiency of cellular metabolism, managed through dietary changes and supplementation, could have beneficial impacts on physical, cognitive, and psychological health and resilience, and they certainly warrant further research.
>
> (Ford and Glymour, 2014: 50)

Although good nutrition and supplements can certainly be beneficial to anybody and are likely to promote mental health and cognition, it will only marginally improve the performance of healthy individuals. Nevertheless, nutrition and supplements are integral to sustaining the health and performance of soldiers and should therefore not be neglected.

3.1.2 Stimulants

Several studies on stimulant drugs have indicated that 'there is no doubt that cognitive performance can be maintained at a higher level, and for a longer duration with stimulating drugs than without'; the question is merely whether they are safe and reliable so that they could be used in an operational setting (Baranski and Pigeau, 1997: 90). There is already a long history of militaries drugging their fighters to stay awake longer, to better cope with stress and fear and to improve stamina. Used substances included coca leaves, cocaine, alcohol, caffeine and nicotine amongst others (Tracey and Flower, 2014: 825–834).

Table 3.1 Stimulants and supplements for combating fatigue and enhancing cognitive functions

Class of drug	Example	Effect	Military usage
Amphetamines	Dexedrine	Increases stamina, physical strength and alertness; euphoria; improves working memory	'Go pill' – maintaining alertness and vigilance; reduced need for rest
Nervous system stimulants	Ritalin	Combats fatigue; improves working memory; improves performance on boring and difficult tasks	Improved concentration in boring tasks
Eugeroics	Modafinil	Promotes wakefulness; treats sleep disorders; enhances cognitive abilities	Reduced need for sleep
Nootropics	Piracetam	Combats cognitive problems and decline; improves memory	Improved cognitive performance in tasks that require concentration
Omega-3 fatty acids	Docosahexaenoic acid (DHA)	Combats cognitive decline; in high dosages improves cognitive functions; combats depression	Sustaining soldier performance long-term

Source: Author's own data.

During the Second World War, many armed forces experimented with synthetic drugs like amphetamines/ methamphetamines (CNS stimulants) and barbiturates (anti-anxiety drugs). The German military developed the 'miracle pill' Pervitin for pharmacologically enhancing their soldiers. The amphetamine-based drug was put into chocolate called 'pilot's chocolate' when handed out to pilots and 'tanker's chocolate' when handed out to tank crews. It was reported that the drug made soldiers to better withstand extreme cold and to march for longer periods of time with spirits improved and greater alertness. The Nazi doctors were working on even more refined drug cocktails, using concentration camp inmates as test subjects. A compound called D-IX was given to prisoners at the concentration camp Sachsenhausen. With the help of the drug, inmates were able to march an astounding 90 km per day with 20 kg backpacks, although some test subjects died in the process (Farren, 2010: 27–28).

The US military has also used amphetamines during the Second World War and continues to use them (under the brand name Dexedrine) as a stimulant to the present day (Tracey and Flower, 2014: 826). According to Farren, 225 million doses of amphetamines (mostly Dexedrine) were administered to

US soldiers during the Vietnam War alone (Farren, 2010: 39). To a far lesser extent, Dexedrine was also handed out to troops during the 1991 Gulf War and the 2003 Iraq War.

Since then, concerns over amphetamines have grown substantially. There have already been cases where soldiers under the influence of performance enhancing or psychiatric drugs have behaved recklessly, resulting in the loss of life. In a well-known incident, an F-16 pilot under the influence of Dexedrine accidentally bombed Canadian troops in Afghanistan in 2003 (Bonné, 2003). Another concern is that amphetamines may be a contributory factor to PTSD (Farren, 2010: 39).

The US military has been looking for a replacement of Dexedrine that is safer. One of the drugs that is being investigated is the stimulant drug *modafinil*, which is typically prescribed for treating narcolepsy. It can also be potentially used for 'promoting wakefulness, attention and executive function and memory' in healthy persons so that they can work at optimal mental performance for longer periods of time (Tracey and Flower, 2014: 826). Apparently, the British government has already issued modafnil under the brand name Provigil to British troops as indicated by a $100,000 dollar order for the drug in 2002 (Farren, 2010: 41). While modafinil seems a safer drug, it is clearly not without risk. There is concern that modafinil may also cause overconfidence, resulting in errors of judgment (Barranski and Pigeau, 1997: 84–91).

Another drug of interest is *methylphenidate* (MPH), which is a CNS stimulant. It is typically prescribed under the brand name Ritalin for treating attention deficit disorder (ADD) and attention deficit and hyperactivity disorder (ADHD) in children. Some studies have claimed that methylphenidate could improve mental stamina, concentration and memory; although a metastudy conducted by Repantis et al. suggests that there is 'no consistent evidence for neuroenhancement effects of MPH' (Repantis et al., 2010: 203).

In the past, militaries have even administered cocaine to their soldiers. Cocaine was actually widely used by many militaries, officially and unofficially, during the First World War and by the Wehrmacht during the Second World War (in the early nineteenth century Germany was the world's leading manufacturer of cocaine, where it had been first isolated by Albert Niemann in 1860 (Streatfield, 2001: 58; 151–166)). Cocaine is still a popular illegal drug that is also taken for performance enhancement by warfighters. However, systematic drug-testing since the Vietnam War era has massively reduced illegal drug use in Western militaries.

More recently there have been reports that LSD in very small doses may have a performance-enhancing effect. Many IT professionals in Silicon Valley apparently use LSD in doses of 10 to 15 micrograms (a tenth of a 'normal' dose) to enhance their performance and enable them to remain happy and relaxed despite high workloads or boring activities. There seems to be no scientific studies on LSD micro-dosing effects, but anecdotal evidence suggests that it works; although the risks are unclear (Boult, 2015).

3.1.3 Psychiatric drugs

A psychiatric drug is a medication that affects brain chemistry and CNS to treat mental illness and other mental disorders. They can be grouped into antidepressants, anti-anxiety drugs, mood stabilisers and anti-psychotics, which all regulate brain chemistry to some extent to allow patients to manage emotions and to better control their behaviour. While it makes sense to use these drugs for treating major mental illness, there are concerns that these dangerous drugs are often overprescribed and misused for performance enhancement. Currently one in five American adults is taking psychiatric drugs to cope with their daily lives (Friedman, 2013).

Psychiatrist Peter Breggin claims that the practice of over-drugging troops with psychotropic prescription drugs would be 'disturbingly rampant'. The Pentagon spent $6.8 billion on drugs in 2011, of which $2.7 billion were spent on anti-depressants (Reno, 2014). Psychiatric drugs are also prescribed to active soldiers as treatment of PTSD or other mental disorders. Soldiers may also take psychiatric drugs to manage anxiety, stress and fatigue. The problem is that these drugs are dangerous, as they can produce unpredictable changes in personality and behaviour. Psychiatric drug use by military service members has been linked to higher rates of PTSD in a major study (Friedman, 2016). Psychiatric drugs are also dangerous because they are de facto addictive: they have to be constantly taken to keep the brain chemistry and psyche stable. According to psychiatrist Peter Breggin, '[w]hen these antidepressants or antianxiety drugs are stopped, the brain can be slow to recover from its own biochemical adjustments or compensatory effects. In effect, the brain cannot immediately keep up with the withdrawal from the drug. This can produce distressing and dangerous withdrawal effects' (Breggin, 2013: XXIII).

Furthermore, all pharmaceuticals can damage organs (kidney and liver are particularly vulnerable) and may interact negatively with other drugs and foods. Psychiatric drugs often cause permanent brain damage and lower longevity. They may provide temporary relief, but at a high cost. In other words, 'Minor short term gains can result in major long term losses' (Coker, 2013: 230). Military theorist Christopher Coker therefore cautions against drugging troops: 'Prolonged drug use can induce paranoia, agitation, and hallucinations. It can lead…to episodes of even greater unmediated (or psychotic) violence' (Coker, 2013: 230). Although pharmaceutical treatment and enhancement seems desirable since it is a relatively cheap and easy way to modify the brain, it carries tremendous inherent risks, especially from a long-term perspective. However, neuroscience research may lead to improvements in nutrition to enhance brain function, may lead to the development of new psychotropic drugs that are less dangerous, may result in better delivery methods for existing drugs to produce more precise effects and may enable better monitoring of the brain chemistry of an individual in order to mitigate some of the risks.

3.1.4 Modelling the brain for drug development

Many of the psychotropic drugs currently in use have been developed already in the 1970s and 1980s. Progress in the development of new neuropharmaceuticals has been very slow despite an improved understanding of the brain (Markou, 2009). As a matter of fact, drug development is very hard. It is a very slow process that has to go through several stages from preclinical modelling to clinical trials and eventually FDA approval, which takes many years and costs millions of dollars. Drugs have to be safe before they can be introduced as treatments or performance enhancers. Before the stage of clinical trials it is very hard to predict what exactly a drug will do to the brain. Current drug models do not work very well and often it is difficult to measure whether drugs work at all (Jogalekar, 2014).

Progress may come through more use of and improvements in brain imaging technologies, such as the MRI, fMRI, PET and MEG. Brain scans can be used for gaining a better understanding of the changes they cause in the brain. Ultimately, the goal is the development of a model that can simulate how body and brain interact and what specific effects certain drugs will have on the brain. The NRC suggests: 'Changes in models of brain function may create new and surprising ideas about how, when, where, or why drugs produce their effects; about what those effects are; about the kinds of chemicals that function as drugs to alter human functioning; and about ways to enhance, minimize, or counteract drug effects' (NRC, 2008: 41–42). Scientists can now grow 'mini brains' in Petri dishes, where they can observe how drugs affect brain cells (Yuhas, 2016). The effects of new drugs could be much better predicted, which could speed up drug development and may lead to more effective drugs than those that are currently known.

Modelling the brain may also result in knowledge how exactly brain chemistry manifests itself in behaviour, which could make it possible to control behaviour by precisely managing brain chemistry. Brain imaging and other methods of medical monitoring could make it also possible to tailor drugs to the individual such as personalized anti-depressants (Holsboer, 2008: 638–646). Nanotechnology may enable the development of nanosensors that could be injected into a body and monitor neural processes in view of controlling them precisely. Michio Kaku suggests 'nanoprobes might be able to receive and transmit signals from the handful of neurons that are involved in specific behaviors' (Kaku, 2014: 97).

Researchers from the University of California, Berkeley are working on nanoprobes they call 'neural dust', which are thousands of microsensors the size of a tenth of millimetre that would be sprinkled in brain tissue and that could monitor brain processes using ultrasound, which also powers the tiny sensors (Marcus and Koch, 2014). This is at the moment only a theoretical study, but may become one day a viable approach for an implanted BCI.

3.1.5 *New methods of drug delivery*

The blood-brain barrier (BBB) protects the brain from dangerous toxins in the blood and it prevents the uptake of most pharmaceuticals into the brain. By finding a way to overcome the BBB, existing drugs could be made much more effective. A prospective approach is nanotechnology: nanoparticles are small enough to cross the BBB and drugs could be reengineered accordingly. The NRC report also mentions nanotubes to which drugs could be attached or the use of biodegradable nanoshells or nanospheres that are vehicles for transporting drugs across the BBB (NRC, 2008: 49).

Microchips implanted in the brain containing a drug reservoir could control the release of a drug to a local area in the brain. Multiple psychotropic drugs could be on that chip and regulate brain chemistry so that physicians and patients can better control the treatment (Maloney et al., 2005: 244–255). Very small dosages of a drug would be effective if administered to a local brain area. As a result of precise administration of drugs in the brain, there would be less risk of drug interactions. Patients would no longer need to remember to take their medications. Overall, less side effects from the drugs can be expected since they would not end up going through the bloodstream into other organs. The chip could be inserted with a syringe through a tiny opening in the skull and would be fully biodegradable. A company in Massachusetts, *Microchips Biotech* has recently developed such an implantable drug release chip and is now partnering with a pharmaceutical company, which has invested $35 million to get the chip quickly through clinical trials (Matheson, 2015).

The drugs on the chip could be released either in the form of a timed release or through activation by a wireless signal. One application that is being discussed is the use of a chip implanted under the skin for birth control, which could arrive as soon as 2018 (Lee, 2014). Another possibility is its use as a substitute for a pacemaker: the chip 'sends an endocrine or chemical signal, instead of an electrical signal' (Matheson, 2015). Soldiers could be implanted with neural sensors and microchips that release drugs in response to changing mental states. For example, when they experience high levels of stress the implanted chip might release a calming or anti-anxiety drug, or if they suffer fatigue the chip may release a drug that enhances their cognition.

To conclude, drugging soldiers may offer some temporary benefits, but it seems to have more downsides than upsides. Some researchers believe that 'the kind of substances presumed by many arguments to make us significantly smarter without serious adverse effects do not exist, and will not exist in the foreseeable future' (R.H. Blank, 2013: 217). Drugs are likely to negatively affect the long-term physical and mental health of soldiers and they can result in lapses of judgment and even outright psychotic behaviour with no warning. Neuroscience might improve drug safety to some extent, but will probably not solve fundamental problems that result from altering brain chemistry, namely the issue of addiction and the potential for brain damage.

3.2 Brain monitoring and stimulation

Neuroscience can enable the brain monitoring and the external brain stimulation of soldiers. This would allow commanders and the soldiers themselves to monitor and control mental states for better performance. Some of the relevant technologies already exist and are at a testing stage. In terms of portable brain monitoring devices electroencephalography (EEG) and functional near-infrared spectroscopy (fNIRS) make most sense. The US military is considering electromagnetic and ultrasound brain stimulation as a potentially healthier and more effective way of cognitive enhancement than psychotropic drugs. There are indications that brain stimulation can enhance cognition and alter mental states. In fact, electromagnetic brain stimulation might be able to produce all the effects of drugs in the brain, but without the side effects. The Pentagon is considering integrating brain monitoring and stimulation devices into soldiers' helmets that could monitor the mental states of soldiers, keep them awake, reduce stress and relieve pain. Alternatively, soldiers might be implanted with brain chips that regulate their brain functions. Obviously, non-invasive technologies would be preferable to invasive technologies that may require medically risky procedures and that could permanently alter personalities of implanted soldiers. Some of the techniques that are currently researched include brainwave entrainment (BWE), transcranial magnetic stimulation (TMS), transcranial direct current stimulation (tDCS), transcranial pulsed ultrasound (TPU) stimulation and deep brain stimulation (DBS). Only DBS is invasive and it is currently only researched in view of therapeutic applications.

3.2.1 Brain monitoring

Mental states could be monitored through a variety of sensors such galvanic skin response, heartbeat, eye movements, pupilometry, EEG, near infrared spectroscopy (NIRS), blood oxygen saturation (NIRS) and facial expression by optical computer recognition. This could enable commanders to better determine the level of readiness of their soldiers (NRC, 2009: 76). For example, commanders would have a much better idea whether a soldier is mentally capable of completing a task. A neural monitoring device can indicate when the soldier is getting tired or suffers great pain or stress. The technology could be used for monitoring the mental states of pilots and drivers to alert them automatically if they fall asleep or if they are on the verge of losing consciousness. As noted in a medical study, '[d]rowsy driving is a significant contributor to death and injury crashes on our nation's highways' to which the EEG monitoring of drivers could provide a solution (Brown et al., 2013). If high stress loads are detected, 'adaptive automation agents' can assist humans in complicated tasks such as piloting an aircraft (NRC, 2008: 93–94).

The most simple and non-invasive brain monitoring device, which was already developed in the 1920s, is the EEG. Electrodes are placed on the scalp that measure millisecond fluctuations of voltage resulting of neural activity or neuronal or action potentials, which are called 'brain waves'. Current EEG research focuses on the brain's electrophysiological response to a stimulus (NRC, 2008: 55). The EEG sensors are usually built into a scalp cap that the subject wears, which can measure brain activity in different parts of the brain. Since measurements are only taken from the surface of the scalp EEGs can only very approximately determine where brain activity occurs and are thus inferior to brain imaging techniques. On the upside, EEGs are cheap and small, which means they can be easily integrated into a helmet. A primitive EEG costs as little as $100 and it can give some general indication about mental states and brain activity. Of course, it has a low spatial resolution and cannot monitor deeper brain regions (Kaku, 2014: 26).

A different approach to brain monitoring is called fNIRS. It can measure blood flows in the brain, which are indicative of neural activity. Usually fNIRS measures the concentration and ratio of oxygeneated and deoxygeneated haemoglobin in the blood, using near-infrared light waves that are backscattered to sensors at the side of the light source (NRC, 2008: 62). fNIRS can only penetrate 2–3 cm into the skull, but in the future they could reach up to 5 cm deep (NRC, 2008: 65). It is also possible to use fNIRS for creating 3D maps of the brain. FNIRS can be miniaturized and thus made portable and it is cheap compared to fMRIs or MEGs, which could make it suitable as a method for monitoring the brains of soldiers in the field (NRC, 2008: 53).

3.2.2 Brainwave entrainment (BWE)

The simplest way of stimulating the brain for high performance and of altering mental states is called brainwave entrainment, which can be used in conjunction with a biofeedback device. Brain activity and brainwaves can be demonstrably changed through meditation or through mental training, which can be measured with an EEG. The human brain uses a frequency spectrum of 0.1 Hz to up to 80 Hz, which is divided into delta, theta, alpha and beta states. Brainwaves adjust to external frequencies that are in the brainwave range and can thus be altered by such external frequencies, which is called 'frequency following response'. This includes visual or auditory stimuli and electromagnetic ELF waves. Brain entrainment can be achieved by electrical cranial stimulation, light pulse systems and sound systems, also in combination (Begich, 2006: 198).

The technique of binaural sounds works through stereo headphones that have sounds of a slightly different frequency coming from each speaker to adjust the brain to the frequency difference, e.g. the left speaker may use a 200 Hz frequency and the right speaker a 207 Hz frequency, which would adjust the brain to 7 Hz, which is in the theta spectrum and characteristic for a meditative state. Biofeedback machines usually provide audio-visual stimuli and measure brainwave response with an EEG to adjust the brain to a desired

Table 3.2 Brainwaves and corresponding mental states

Brain wave	Frequency range	Corresponding mental state
Delta	1–4 Hz	Sleep
Theta	4–7 Hz	State between asleep and awake; meditation/ trance; lucid dreams
Alpha	8–12 Hz	Relaxed awake state; ideal for learning and creative work
Beta	12–22 Hz	Concentrated awake state; thinking; actively engaged
High Beta	22–35 Hz	Less focused behavior, emotional state, excitement/ panic
Gamma	35 Hz-80 Hz	Simultaneous processing of information from different brain areas

Source: Author's own data.

mental state. With experience, people become able to create brainwave patterns at will. Such devices cost only a few hundred dollars and are often used by therapists and mental health counsellors to treat depression or other mental disorders (Begich, 2006: 200). They are also typically used for relaxation and enable people to manage their mental states. Several studies indicate that brainwave entrainment can potentially enhance cognition by putting a person into a relaxed 10 Hz alpha state, which improves learning, heightens concentration and achieves greater perceptiveness (e.g. Cruceanu and Rotarescu, 2013). Brain entrainment using biofeedback is a fairly low-cost and low-risk method for cognitive enhancement: it can both produce relaxation and boost alertness and concentration.

3.2.3 *Transcranial magnetic stimulation (TMS)*

TMS works with a magnetic coil with several thousands of volts being placed outside of the head to electromagnetically activate or stimulate a particular region of the brain. The stimulation by the magnetic field only lasts 100 to 200 microseconds and could be applied in pulse, paired pulse, or repetitive pulse treatments (R.H. Blank, 2013: 30). The effect is similar to the electrical stimulation of the brain (ECT or DBS), but has the advantage of being neither painful, nor invasive. The mechanism itself was discovered by Silvanus Thompson in London in 1910 and was scientifically validated in a study at the Royal Hallamshire Hospital and the University of Sheffield in 1985 (Chicurel, 2002: 114).

Since the mid-1980s, there has been research into using TMS as a psychiatric treatment method. TMS is since 2008 an FDA-approved treatment method for major depression. Several studies published in recent years indicated that TMS has been successful in the treatment of patients with PTSD.

TMS is also investigated for the treatment of schizophrenia, autism spectrum disorder and traumatic brain injury (R.H. Blank, 2013: 30). Up to now, there are not many adverse effects that have been observed in connection with TMS and it seems a fairly safe treatment. A private clinic in California with the name *Brain Treatment Center* has refined the technology of TMS and now offers a treatment it calls Magnetic Resonance Therapy (MRT) to persons with mental disorders and brain injuries (Leiby, 2015). It uses a proprietary algorithm for applying the TMS for daily treatments of 30 minutes for up to two months. The clinic offers treatments to veterans for free and claims to have already successfully treated over a hundred veterans. Up to now, the treatment is unproven because of its novelty, but there are hopes that it might be effective with treating autism, Alzheimer's disease and other neurodegenerative diseases.

As pointed out by Canli et al., there are many potential applications of TMS: 'Depending on the TMS stimulation parameters, activation in the cortex can be increased or reduced. In practice, TMS can influence (either improve or diminish, depending on the parameters and target region) many brain functions, including directing physical movement, visual perception, memory, reaction time, speech and mood' (Canli et al., 2007: 4). Neuroscientists are already using TMS in conjunction of BCIs and brain-to-brain interfaces. For example, TMS can be used to stimulate the motor cortex and thereby enable external control over the movements of test subjects or it could stimulate the auditory cortex for transmitting messages, which will be discussed further below. However, TMS has a low spatial resolution, which means that it is not very precise (Tufail et al., 2010). The other downside of TMS is that it requires a large coil and power source, which are difficult to miniaturize and to make portable. This means TMS could not be integrated into a soldier's helmet, but might be fitted into larger pieces of military equipments such as tanks or aircraft. More suitable as a mobile device for brain stimulation would be tDCS.

3.2.4 *Transcranial direct current stimulation (tDCS)*

tDCS applies a weak current through electrodes to the scalp, which can achieve an effect comparable to TMS in terms of brain plasticity induction (Nitsche and Paulus, 2011: 463). It stimulates the brain with a low constant direct current that is applied through electrodes on areas of interest on the scalp. The current increases the excitability of neurons in the area where it is applied and thereby alters brain function (R.H. Blank, 2013: 31). It is used as a treatment for patients with brain injuries such as strokes and is evaluated for treatment of PTSD, Parkinson's and Alzheimer's disease. There are several studies that indicate some cognitive enhancement resulting from tDCS treatment, especially with respect to visuo-auditory attention tasks, working memory tasks and improved learning (Nitsche and Paulus, 2011: 481). A study conducted at Oxford University in 2010 found that it was possible to improve math skills in test subjects for up to six months using tDCS through treatments lasting 15 minutes (Bland, 2010). The treatment improves brain plasticity and makes learning easier, which would allow healthy people to learn knowledge or a skill faster (Murphy, 2013).

At the moment it may be still premature to declare tDCS an intelligence booster, but there are many advantages that could make this technology useful in many contexts. Unlike TMS it is fairly cheap and portable, which makes it possible to use it continuously and at home. Brain stimulation methods could have numerous benefits in terms of treatment and enhancement for people across society and the technology could spread very quickly as indicated by the great commercial success of tDCS device called *Foc.us* that is being marketed as a 'gaming device'. tDCS is still awaiting FDA approval as a treatment device. The US Air Force has already tested 'external stimulant technology to enable the airman to maintain focus on aerospace tasks and to receive and process greater amounts of operationally relevant information' and has found that 'it can help pilots better pick out targets from radar images' (Shachtman, 2010b). tDCS could be fit into a soldier's helmet, possibly combined with a brain monitoring device such as an EEG or fNIRS) and might make soldiers and other personnel more effective in tasks that require concentration and learning. However, it remains at the current time unclear how great the actual benefits are and whether there are negative long-term effects of 'zapping' the brain with electroshocks, such as an increased risk of brain tumours and other cancers.

3.2.5 *Transcranial pulsed ultrasound (TPU) stimulation*

TPU is the newest method of non-invasive brain stimulation and there are currently not many studies that show therapeutic benefits or the safety of TPU. Research on TPU has been primarily military in nature with DARPA funding research into TPU at Arizona State University. Ultrasound is, for humans, inaudible acoustic wave of a frequency above 20 kHz to several GHz, which can penetrate soft tissue. It is used for ultrasound imaging (sonography) and for underwater range finding (sonar). It has been shown that ultrasound can stimulate and inhibit neuronal activity in the brain and may thus be suitable as an approach to brain stimulation (Tufail et al, 2010). This means that TPU may be able to do all the things that can be done with TMS or tDCS, but with much greater precision and deeper reach into the brain (Dillow, 2010). Unlike TMS it does not require any external coil and large amounts of electricity to function, which means it could be integrated into a soldier's helmet, which is what ASU researchers led by William Tyler are trying to do. According to a *Popular Science* article:

> Tyler's technology, packaged in a warfighter's helmet, would allow soldiers to flip a switch to stimulate different regions of their brains, helping them relieve battle stress when it's time to get some rest, or to boost

alertness during long periods without sleep. Grunts could even relieve pain from injuries or wounds without resorting to pharmaceutical drugs. More importantly, in the periods after brain trauma ultrasound technology could reduce swelling and metabolic damage that is often the root cause of lasting brain damage.

(Dillow, 2010)

The method is still at the stage of animal research and is not well-tested on humans. It might therefore take longer before there is any practical application of TPU. In animal experiments TPU has shown that mice brains can be stimulated to activate specific brain regions (Tufail et al., 2010).

3.2.6 Deep brain stimulation (DBS)

DBS has been under development since the 1950s and it is essentially a pacemaker for the brain: an implanted stimulation device delivers electrical impulses through small electrodes to specific areas of the brain to change brain activity in a controlled manner (R.H. Blank, 2013: 32). It is an FDA-approved treatment method that has been used for the treatment of clinical depression, epilepsy, Parkinson's, Gilles de la Tourette syndrome, alcoholism and Alzheimer's disease. More than 110,000 people worldwide have received DBS brain implants so far (Regalado, 2014). In 2013 DARPA announced that it would spend $70 million on developing DBS brain implants for treating veterans with PTSD (Gorman, 2013: A16).

The project description states: 'The project builds on expanding knowledge about how the brain works; the development of microlectronic systems that can fit in the body; and substantial evidence that thoughts and actions can be altered with well-placed electrical impulses to the brain'. The main idea behind this DARPA SUBNETS project is that several electrodes and a chip is implanted in a brain that can send electrical impulses to different areas of the brain with information also relayed wirelessly back to an external information centre (Jacobson, 2015: 425). According to a DARPA announcement, the brain chips have been already tested on dozens of volunteers (DARPA, 2015). The implants would be able to regulate the emotions of the veterans and would suppress depression and negative thoughts. The technology could be used for treating a variety of mental illnesses, including addiction, depression and borderline personality syndrome. Furthermore, DBS could also enhance memory and learning and treat memory disorders through stimulation of the temporal lobe.

However, there are numerous ethical concerns that DBS could severely affect a person's personality and thoughts and medical concerns that a brain pacemaker may permanently alter or damage brain circuitry (R.H. Blank, 2013: 35). A neuroscience study also raised the concern that DBS treatment can potentially result in 'mind control' and suggested that DBS treatment should therefore be always voluntary (Koivuniemi and Ott, 2014: 1). Unfortunately, non-invasive options have their limitations and may not always be preferable to invasive

Table 3.3 Brain stimulation methods

Type of stimulation	Invasive/ non-invasive	Advantages	Disadvantages
Brainwave Entrainment (BWE)	Non-invasive	Cheap; suitable for mobile use; can be used for relaxation or for improving concentration; no known side-effects or risks	Effectiveness varies from individual to individual
Transcranial Magnetic Stimulation (TMS)	Non-invasive	Targeted stimulation of specific brain areas; few potential side effects or risks	Requires large and expensive equipment; cannot reach deeper brain areas; mostly therapeutic applications
Transcranial Direct Current Stimulation (tDCS)	Non-invasive	Cheap; can be integrated into a helmet; improves cognitive functions; few potential side-effects or risks	Difficult to target specific brain areas; transient effects; optimal or safe levels may vary from individual to individual and are not known
Transcranial Pulsed Ultrasound Stimulation (TPU)	Non-invasive	Suitable for integration into a helmet; can reach deeper areas of the brain; stimulation can be very targeted	Still experimental; unknown side-effects and risks
Deep Brain Stimulation (DBS)	Invasive	Stimulation can be very targeted and precise; can regulate brain functions; suitable as a BCI	Risky surgical procedure necessary; implant may shift in the brain; can have serious side-effects; currently only suitable for therapeutic applications

Source: Author's own data.

technology that is implanted into the brain. Implants may offer more precise brain stimulation and they do not require the individual to carry or wear some external devices, which may be in some situations inconvenient or impractical.

At the moment, brain implants are only considered for therapeutic purposes in special cases since they require a risky medical procedure that includes drilling holes into the skull, which could damage brain tissue, result in infection, brain bleeding and electrode misplacement due to the brain shifting inside the skull. However, it might be the case that in the long run, brain implants could become a fairly standard and risk-free procedure that could be used on healthy individuals.

3.3 Brain-computer interfaces (BCIs)

Most commonly, a BCI is considered to be 'a system that captures and transforms signals originating from the human brain into commands that can control external applications or instruments' (McKendrick et al., 2015). The first BCI was conceived by British neurophysiologist William Grey Walter in 1964. He inserted electrodes into the motor cortex of a patient that recorded brain activity and enabled the patient to advance a slide projector show using the mind. Brain activity that occurred when the patient pressed the button of the slide projector was used as a command for advancing the slides even before the patient pressed the button (Graimann et al., 2010: 2). In the 1970s the University of California received a NSF grant followed by a DARPA grant to interface the human brain with a computer. The researcher on this grant, Jacques Vidal, coined the term Brain-Computer Interface (BCI). He used an EEG for measuring action potential at different points of the scalp in response to visual stimuli to determine the eye-gaze direction, which indicated where a person wanted to move a cursor (He et al., 2013: 87).

In principle, there are two main possibilities how BCIs could be used for human performance enhancement: 1) to '[d]irect signals from the brain could be used to direct or alert external equipment, as an auxiliary to direct human actions'; or 2) '[b]rain-computer interfaces could be used for enhanced sensory input, information input, or control signals to enhance the performance of a combatant' (Jasons, 2008: 73–74). The first application area would include the use of BCIs for the control of robotic prostheses or weapons and the second application area sensory implants, augmented cognition and implanted ICT for 'synthetic telepathy'. BCIs can be invasive or non-invasive, meaning they either require neurosurgery to implant a device inside the skull or they can use some external device to measure or control brain activity. Obviously, in most cases non-invasive approaches are preferable since they do not require risky medical procedures and since they are easily reversible. Furthermore, BCIs could be one way or two-way in terms of information flow. Two-way BCIs can provide feedback, which is important for robotic prostheses that replace a hand since it allows for greater control.

A US Air War College research paper argues that BCIs will impact greatly on warfare:

> This technology [of BCIs] will advance computing speed, cognitive decision-making, information exchange, and enhanced computational power, resulting in substantially enhanced human performance. A direct connection between the brain and a computer will bypass peripheral nerves and muscles, allowing the brain to have direct control over software and external devices. The military applications for communications, command, control, remote sensors, and weapon deployment with BCI will be significant.
>
> (Moore, 2013: III)

This section will review some of the BCIs that are already in use and will also look at the future development and uses of BCIs, which includes robotic and sensory prostheses, neutrally triggered weapons, brain-to-brain interfaces, the potential for external control of the mind via BCIs and implanted ICT.

3.3.1 Robotic prostheses and sensory implants

Most BCI research is aimed at medical applications to make the lives of disabled people easier. Typically, these BCIs use EEG scalp caps that measure action potentials at many points of the scalp simultaneously to determine intent. Research has focused on creating communication tools for persons with 'locked-in syndrome', who are fully paralyzed but conscious. The mu-rhythm of 8–12 Hz, which can be measured by EEG over the central sulcus has been used as a data input for controlling a cursor on a screen. Test subjects could be taught to control their mu-rhythm (increase or decrease), which would then be translated into a cursor movement on a screen (Rao, 2013: 129–130). For example, in an early experiment by Wolpaw et al. test subjects were asked to move a cursor to a particular position indicated on the screen. High mu-rhythm produced an upward movement of the cursor and low mu-rhythm no movement. After some training the test subjects were able to complete the task in an average of three seconds (Wolpaw et al., 1990: 252).

In 2002, DARPA launched the Brain-Machine Interface program and the Human Assisted Neural Device (HAND) programs, which were aimed at developing 'sensorimotor control of prosthetic devices, facilitation of memory encoding, decoding of visual inputs, development of dynamic neural algorithms, as well as the development of new devices for high-resolution neural imaging' (Miranda et al., 2015). The researched methods included both non-invasive and invasive approaches to BCIs.

Brown University in collaboration with DEKA Research and Development Corp. funded by DARPA developed in 2006 the BrainGate chip, which has to be surgically implanted in the brain. One hundred microelectrodes measure brain activity of the motor cortex (Royal Society, 2012: 40). The data is then sent to an external decoding device that translates the brain activity in commands that can be used for controlling a robotic limb (Kaku, 2014: 80–81). The subject simply imagines a movement, which produces brain activity that can be decoded as intent and translated as a command. Researchers from Brown University demonstrated in 2006 that a tetraplegic patient with an implanted BrainGate chip was able to move a cursor on a screen to open an e-mail, to operate simple hardware and to control two robotic arms to open and close the robotic hands at will (Hochberg et al., 2006).

DARPA launched its program *Revolutionizing Prosthetics* in 2005 to develop advanced robotic prostheses for amputated upper limbs (Belfiore, 2009: 4). Two teams participated in the project: DEKA Research and Development Corp. and the Advanced Physics Lab of Johns Hopkins University (Jacobson, 2015: 425). Johns Hopkins University researchers have recently demonstrated a robotic arm with 26 joints known as Modular Prosthetic Limb (MPL) that can be controlled by thought by connecting it to the CNS. Remaining nerves of the amputated arms are surgically remapped so that the electrical signals from the brain can be translated into movements of the prosthetic limb. Johns Hopkins University project manager Robert Armiger said in an interview: 'The long-term goal for all of this work is to have noninvasive – no extra surgeries, no extra implants – ways to control a dexterous robotic device' using, for example, an EEG cap (Canepari, 2015). Currently the MPL costs $500,000 dollars, but less advanced robotic prosthetics could be much cheaper and would become more widely available in society (New York Times, 2015).

There has been also considerable progress with respect to sensory implants that restore human senses. Over 300,000 people worldwide have received cochlear implants that restore their hearing (Marcus and Koch, 2014). The implants consist of a microphone with a speech processor for filtering out background noises that converts sound into electrical impulses and transmits it to a stimulator inside the head for stimulating auditory nerves for some auditory representation of external sounds. Retinal implants use a similar approach for restoring sight: they use an external sensor, in this case a video camera, which produces visual information that is transformed into electrical signals for stimulating the retina. The ARGUS II retinal implant has been recently approved by the FDA and is intended to restore some visual function in individuals with a condition called advanced retinitis pigmentosa (Marcus and Koch, 2014).

Cochlear and retinal implants are still no substitute for natural sensory ability, but in the long-term they may enhance human sensory perception, enabling them to hear beyond the human auditory spectrum and to see at night. The interest of the military in developing advanced prosthetics should therefore not be seen as merely an effort for providing better healthcare to seriously injured veterans, but as an effort of building better man-machine interfaces that have direct military utility. Of course, no healthy human would volunteer to have healthy limbs amputated or natural senses replaced by sensory implants, but there could be wearable technology that could work better with some sort of BCI as is developed in programs like DARPA's *Revolutionizing Prostheses*.

A strategy paper of the British MoD reviewing future developments suggests: 'Brain-machine interfaces may allow direct control of prostheses, exoskeletons and systems remote from the body. Control of simple devices by thought is already a reality' (UK MoD, 2014: 89). Robotic exoskeletons as they are being developed by DARPA could give human soldiers supernatural strength and endurance.

For example, the 2013 movie *Elysium* featured a robotic exoskeleton that was directly connected to the spine of the hero and enabled him to perform well above human limits. The technology is no longer science fiction, as the military already works on an exoskeleton controlled by a BCI, ideally using an EEG integrated into a helmet (Evans, 2013). The technology could also be of great advantage for telepresence tasks, such as the remote control of robots, UAVs, or other machinery. For example, the US DoD has awarded a $400,000 grant to researchers at the University of Texas at San Antonio in 2014 to develop an EEG-based control system for drones, as it would make the load of soldiers lighter since they would not have to carry additional equipment (Russon, 2014). It thus becomes a likely prospect that the robot armies of the future could be remotely controlled by human soldiers via neural interfaces.

3.3.2 Neurally triggered weapons

The brain is still the most energy-efficient and most powerful pattern-recognition machine in existence. Computers tend to do poorly in pattern recognition tasks that are fairly easy for a human such as recognizing faces. Up to now computers can only do it well under optimal conditions: the person has to look directly into the camera, the lighting needs to be right and the face should not be partially obscured or altered. This means that autonomous weapons systems (AWS) that could rely on pattern recognition and that would operate within civilian-occupied territory would be prone to making mistakes, which makes it potentially illegal to use such AWS (Sharkey, 2012: 787–799). As a result, it might be always preferable to have a human brain in the loop.

The human brain is so good at pattern recognition because of its massive parallel processing, which enables to recognize a face much quicker and with greater accuracy. The problem is that the vast majority of the perception information remains subconscious – the brain may recognize something, but the person may not be consciously aware of it. If the recognition of an object should result in a decision how to respond, many seconds will pass before the person is aware of the object in question and is able to make a conscious decision. It thus makes sense to leverage the human brain for pattern recognition tasks and to wire it directly to a computer to cut down on response times.

A man-machine system could therefore perform better than an unenhanced human or a fully automated system. This is the main idea behind DARPA's *Cognitive Technology Threat Warning System* (CT2WS), which uses an EEG scalp cap measuring a P-300 response of an operator, who looks at a computer screen (the P-300 wave is an event-related potential that occurs in decision-making and is used in lie detection). The screen shows footage from a high-resolution camera, which is placed on the battlefield. An image processing software highlights any objects on the screen that are potential threats and the brain response to them would then indicate whether they are real threats or irrelevant. The CT2WS can display 10 images per second to the operator, which means that the brain can reliably identify an object in an image in just 0.1 seconds.

What DARPA eventually wants to develop out of CT2WS has been known as 'augmented cognition' or 'Luke's binoculars': Special Forces would have binoculars or head-up-displays that automatically highlight threats to them detected by their own subconscious minds. DARPA states that their augmented cognition binoculars are able to successfully detect threats in 91 per cent of cases, while soldiers with regular binoculars have a much reduced rate of 53 per cent of successfully detecting threats (Drummond, 2012). A DARPA presentation on CT2WS explains some of the challenges that the system could help soldiers overcome in 'distributed operations': they would need to recognize people that are partially obscured, that may or may not carry weapons, or that are camouflaged; they need to find indications of movement, man-made structures and man-made effects; they would need to spot disciplined and purposeful action, discover changes in an environment and detect unusual elements in scenes. The CT2WS could be adjusted to look for specific threats and could learn about the importance of new types of objects (Kime and Lyons, 2007).

There are concerns by Stephen White and other international law scholars that augmented cognition could lead to neurally triggered weapons (sometimes referred to as neuroweapons), which simply use cerebral processes as input for targeting without having any conscious mind in the loop to speed up the kill chain (White, 2008). The general idea of wiring a human brain to a fire control computer for targeting even might point towards the possibility of inserting organic brains or organic brain tissue into weapons systems as relatively cheap and potentially superior control systems for the operation of military platforms and weapons systems. For example, it has been demonstrated that brain tissue from a rat could be taught how to fly a F-22 jet flight simulator. In an experiment at the University of Florida 25,000 rat brain cells forming a biological neural network were placed in a Petri dish with sensors measuring the neuronal activity, which were connected to a desktop computer. The brain cell could also receive feedback from electrodes that can stimulate various parts of the neural network. Over time the rat brain was able to learn to control the pitch and roll of the F-22 in various weather conditions (Viegas, 2012). Amazingly, the brain cells had short-term memory and were able to gradually improve performance in flying the simulator. One could thus imagine a robot that is partially controlled by a biological brain (animal or human brain tissue) that has been grown in a Petri dish and trained for a specific task.

3.3.3 Brain-to-brain interfaces (BBIs)

Researchers from the University of Washington led by neuroscientist Rajesh Rao have demonstrated in 2013 the feasibility of a brain-to-brain interface, enabling one person to take control of the hand movements of another person.

The brain-to-brain interface detects motor imagery in EEG signals recorded from one subject (the 'sender') and transmits this information over the internet to the motor cortex region of a second subject (the 'receiver'). This allows the sender to cause a desired motor response in the receiver (a press on a touchpad) via TMS.

(Rao et al., 2014)

In other words, the EEG measurements taken from one person is used as data input for a TMS device that stimulating the brain of a second person. The TMS can 'tickle' the motor cortex of the other person and thereby produce the hand movement in the second person.

Using the same experimentation setup, researchers from the University of Barcelona claim to have demonstrated 'the conscious transmission of information between human brains through the intact scalp and without intervention of motor or peripheral sensory systems' (Grau et al., 2014). The EEG monitored brain activity of an 'emitter' related to motor-imagery tasks and translated them into binary signals that were sent by e-mail from Strasbourg to Thiruvananthapuram in India, who received the information in the form of TMS stimulation. After a training period the receivers were able to correctly decode the sent information. The binary information sent were words such as 'hola' or 'ciao' with a transmission rate of two bits per minute. This approach is still very primitive and hardly of practical use, but it is a significant step towards the goal of what the US Army calls 'synthetic telepathy'. The term appeared first in a presentation of General Sidney Shachnow of SOCOM on technology projected for 2020 (Toffler and Toffler, 1993: 96).

In 2008, the US Army awarded the University of California, Irvine, Carnegie Mellon University and the University of Maryland a $6.3 million grant to develop a method for deciphering and transmitting thoughts from one person to another for the purpose of silent communication. It uses the invasive technology of the electrocorticogram (ECoG) instead of an EEG, which offers 'unprecedented...accuracy and resolution, since signals are directly recorded from the brain and do not pass through the skull' (Kaku, 2014: 68). This capability of silent communication could be useful for Special Forces, who

could creep into the caves of Tora Bora to snatch Al Qaeda operatives, communicating and coordinating without hand signals or whispered words. Or a platoon of infantrymen could telepathically call in a helicopter to whisk away their wounded in the midst of a deafening firefight, where intelligible speech would be impossible above the din of explosions.

(Piore, 2011)

The basic idea is to find a way of translating thoughts into text and transmit it wirelessly to another person, who could have the message displayed on their HUD or might hear it in their heads through the stimulation of their auditory cortex. The scientific problem is similar to speech recognition used in dictation software: every word that is thought produces a unique brainwave response. By systematically cataloguing the brainwave response for every single word one could systematically decipher thoughts through analysis of brainwave patterns.

There is of course the obvious problem that brains differ from individual to individual and change over time because of brain plasticity, but this may not be insurmountable provided enough brainwave data was collected on enough people to find commonalities. One of the researchers from the University of California, Irvine, Mike D'Zmura, suggests that it would take 15 to 20 years to have a complete brainwave dictionary for the English language necessary for synthetic telepathy (Bland, 2008). According to Kaku, 'ECOG experiments have demonstrated that it is possible to communicate mentally on the battlefield' using a 'telepathy helmet' (Kaku, 2014: 71). This may not be far off. A recent breakthrough by a group of Japanese neuroscientists has demonstrated that by analysis of brainwaves it is possible to decipher words before they are spoken. The team found that each syllable produced a specific brainwave response, making it possible to decipher words that are thought with 80 to 90 per cent accuracy (Hoffman, 2016).

3.3.4 External control of human minds via BCIs

Neuroscientist Rajesh Rao has expressed concerns that BCIs could enable direct external control over another person (Rao, 2013: 274). Heath and Delgado have demonstrated already 50 years ago that DBS allows control over mental states and behaviour. TMS can be also used for stimulating the motor-cortex to cause specific body movements or the auditory or visual cortex for inducing hallucinations, all of which has been already demonstrated to work. In the future, even more advanced technology for precise brain stimulation may be available. Michio Kaku suggested that this might be optogenetics. He explains: 'a light-sensitive gene that causes a cell to fire can be inserted, with surgical precision, directly into a neuron. Then, by turning on a light beam, the neuron is activated...this allows scientists to excite these pathways, so that you can turn on or off certain behaviors by flicking a switch. Although this technology is only a decade old, optogenetics has already proven successful in controlling certain animal behaviors' (Kaku, 2014: 29–30). Of course it may not stop at making mice run in circles, but could be used for behavioural control of humans, according to neuroscientist Duy Phan from Johns Hopkins University. He has recently claimed in relation to optogenetics that 'the ability to manipulate specific neurons confers not only powerful clinical applications to treat brain disease but also perhaps the science-fiction-like ability to control somebody else's mind. Although I have to admit that mind control seems far-fetched and fictional, I argue that the rise of new neuroscience technologies do show significant promises for the future of brain manipulation' (Phan, 2015). Phan argues that it would not even be necessary to implant optic fibers into the brain to make it work – red-shifted light could penetrate deep into the brain, activating neurons with precision. As a result, a non-invasive device for exercising control over human behaviour through brain stimulation seems possible.

3.3.5 Implanted ICT

A British MoD long-range forecast predicted that by 2040 '[w]earable and implanted wireless ICT is *likely* to become available to all that can afford it' (UK MoD, 2010: 137). In 2015, it seems that the forecast is far off in terms of the time frame: implanted ICT will arrive a lot sooner than 2040 and might become very affordable and widespread. Major ICT companies such as Intel, Apple and Google are already working on BCIs as computer input devices. For example, Google works on brain implants for disabled people. Google executive Larry Page said in an 2013 interview that the ultimate goal was to insert a chip into a user's brain 'for the most effortless search engine imaginable' (Burrell, 2013). Google CEO Eric Schmidt also envisions that in a near future people might be wirelessly connected to the Internet all the time using a chip in their head. They would be able to access all the world's knowledge at the speed of thought. According to Google's Ray Kurzweil, humans will be cyborgs by 2030. At this time '[o]ur thinking then will be a hybrid of biological and nonbiological thinking' (Matyszczyk, 2015).

Google has certainly made some inroads towards the goal of merging the user with ICT with Google Glass, which has wireless Internet access and allows a user to browse websites with the blink of an eye. DARPA has already announced a project to build a Google *Glass*-type device, but without any external hardware that needs to be worn by a user. The goal is to create a $10 dollar chip or 'cortical modem' that can directly interface with the visual cortex and that is powered by the spine. The Internet could be wirelessly projected into a person's head. An article in the transhumanist *H+ Magazine* suggests:

> Consider a more advanced version of the device capable of high fidelity visual display. First, this technology could be used to restore sensory function to individuals who simply can't be treated with current approaches. Second, the device could replace all virtual reality and augmented reality displays. Bypassing the visual sensory system entirely, a cortical modem can directly display into the visual cortex enabling a sort of virtual overlay on the real world. Moreover, the optogenetics approach allows both reading and writing of information. So we can imagine at least a device in which virtual objects appear well integrated into our perceived world. Beyond this, a working cortical modem would enable electronic telepathy and telekinesis. The cortical modem is a real world version of the science fiction neural interfaces envisioned by writers such as William Gibson and more recently Ramez Naam.
>
> (Rothman, 2015)

It is not clear when the cortical modem would be ready, but there are a variety of ideas of how to turn this vision into reality. Some of these ideas do not require any surgical intervention such as the drilling of holes into the skull. Researchers from the University of California, Berkeley have proposed the development of 'neural dust', which would consist of biologically neutral nanoparticles that are inserted into the brain and that would act as sensors within the brain. Each nano-sensor would measure the electrical activity of neurons nearby and would be able to send and receive information using ultrasound (MIT, 2013). MIT researchers have injected magnetic nanoparticles into a brain that could do deep brain stimulation by alternating electromagnetic fields (D.L. Chandler, 2015). There are numerous applications with respect to improving brain imaging, therapy, memory enhancement and so on. The same technology of magnetoelectric nanoparticles (MENs) could be also used for creating a new nanotechnology-based BCI.

The cortical modem could become real with the help of DARPA. The agency recently announced a new research program called 'Neural Engineering System Design' (NESD), which

> aims to develop an implantable neural interface able to provide unprecedented signal resolution and data-transfer bandwidth between the human brain and the digital world. The interface would serve as a translator, converting between the electrochemical language used by neurons in the brain and the ones and zeros that constitute the language of information technology. The goal is to achieve this communications link in a biocompatible device no larger than one cubic centimeter in size, roughly the volume of two nickels stacked back to back.
>
> (DARPA, 2016)

In other words, DARPA wants to directly connect human brains to computers, making it possible to both use brain activity as computer input and to transmit digital information directly into the brain. In order to make it work, DARPA anticipates that it 'will require integrated breakthroughs across numerous disciplines including neuroscience, synthetic biology, low-power electronics, photonics, medical device packaging and manufacturing, systems engineering, and clinical testing' (DARPA, 2016). DARPA is planning to spend $60 million on NESD over the next four years.

3.4 Genetic selection and enhancement

Biotechnology makes it in principle possible to not only improve some of the natural human abilities, such as cognition and behavioural control, but also to completely redesign human beings according to specific requirements and turn them into 'mutants'. Synthetic biology or transgenics enables the combination of organisms of different species through DNA splicing, creating entirely new organisms such as 'spider-goats' or GloFish. At the moment there are still ethical concerns over human cloning and human genetic engineering, but it is foreseeable that current ethical standards will change once more concrete benefits materialise, such as greatly increased performance and longevity (Stock, 2002).

The transhumanist movement is gaining traction and several scientists have advocated embracing biotechnological enhancement rather than trying to prohibit it, which would have little chances of success and would result in a loss of control over the direction and impacts (Walker, 2009). Up to now, the US military has been reluctant to 'modify' its soldiers biologically, but there is already the concern that other powers or adversaries might do so and might gain a decisive advantage over the US (Axe, 2012).

In 2008, the House Foreign Affairs Committee held a hearing on the dangers of 'human genetic technologies', which discussed issues such 'trait selection', 'human cloning' and 'human genetic modification', also in relation to nonstate threats. The resulting report recommended 'effective regulatory oversight and control will be needed at both national and international levels' to address the dangers from the proliferation of biotechnologies (Hayes, 2008: 1). However, others fear that effective control of biotechnology will hardly be possible at all (Kay, 2003). In any case, it is a reasonable assumption that many modern militaries will try to leverage biotechnologies for improving soldier performance, including via genetic selection and potentially via human genetic engineering.

3.4.1 Trait selection

The Human Genome Project was funded by the US Department of Energy and the National Institutes of Health and was officially launched in 1990. The goal was to map the entire human genome consisting of 22,500 genes that contain all biological information of an individual ranging from sex to complex traits such as intelligence and personality. The project was completed in 2003 and represents a scientific achievement on par with the Apollo missions.

DNA sequencing that determines the precise order of nucleotides (A, G, C, T) within the DNA can be used for medical diagnosis, identification and genetic engineering. For example, genetic sequencing could be used for identifying individuals with particular traits, including behavioural traits. The discipline of behaviour genetics studies the relationship between genetic variation and behaviour, mostly in view of heredity. Although the scientific debate over nature vs. nurture has been ongoing for a very long time, there are more and more indications that genetic make-up may play a bigger role in personality than previously believed. A study with 800 sets of twins conducted by the University of Edinburgh found a strong influence of genetics on key character traits such as self-control, decision-making and sociability (Collins, 2012).

The 2009 NRC report suggests that genetic sequencing and neural monitoring could be used for personnel selection:

> Genetic markers, neurohormones, and brain imaging are emerging as sources for biomarkers that may prove to be reliable indicators of neural state when individuals make choices – that is, they can signify behavior underlying the emotional or subjective elements during decision making... Genetic markers are particularly relevant for identifying stable traits. Research data suggest that some genetic markers can identify individuals at greater risk of reacting to chemical agents or suffering from PTSD. It is also known that hormonal state – specifically, an individual's hypothalamo-pituitary axis responsivity – influences decision making as well as fitness for duty...The cost of genetic testing for such traits will decrease over the next decade, and the selectivity and specificity of the tests will improve. As this happens, the Army should position itself to take advantage of the new tools.
>
> (NRC, 2009: 39)

Individuals wishing to join the military could be screened for desirable characteristics, such as a high tolerance to sleep deprivation or to stress. This could result in better recruitment and overall greater military performance. The viability of neural and genetic screening for behavioural traits is supported by substantial research.

For example, neuroscientist Laurence Tancredi has argued that some behavioural dispositions may be hardwired into the brain. He claimed '[t]his might suggest that under certain conditions "immoral" behavior is not necessarily the product of wilful acts. By controlling behavior, brain biology might be responsible for some of the extreme manifestations of behavior' (Tancredi, 2005: 9). It would follow that a psychopath has a malfunctioning or different brain from a 'normal' person (Tancredi, 2005: 47). This suggests that psychopathy could be detected by brain scans, which has been confirmed by a study led by Nigel Blackwood from King's College London (Gregory et al., 2012: 962–972).

For obvious reasons it would be important to screen out psychopathic personalities in the military and for any leadership positions. Furthermore, militaries may use DNA sequencing and brain imaging for selecting individuals with specific desirable traits for specific roles. The desirable traits would vary, but it seems reasonable that militaries may want to select individuals according to three main character traits that are crucial for the military profession: 1) disposition towards violence, 2) risk-taking behaviour and 3) obedience.

DARPA has specifically sponsored research that looks into how soldiers respond to threats in terms of fight or flight in view of improving training (Beckhusen, 2012). It is imaginable that the research could also be leveraged for soldier selection into Special Forces. However, since genetic selection would raise serious fairness and equality issues, the military might want to improve the genetic make-up of its soldiers to create or strengthen in them desirable traits.

3.4.2 Gene doping

Some of DARPA's human performance enhancement programs look into the genetic modification of soldiers. Its 'Living Foundries' program is meant to reduce the costs of gene therapy and to create 'tools for rapid physical construction of biological systems, editing and manipulation of genetic designs' (DARPA, 2012). In other words, DARPA is researching the technology for changing the genetic make-up of soldiers through gene therapy to make them stronger, give them more endurance and to make them more resilient to biological and environmental threats or conditions.

The main idea of gene therapy is to introduce new DNA into the body that fixes the damaged DNA through a specifically engineered recombinant virus that changes the DNA contained in all body cells. Specific genes could be switched on or switched off as a treatment to disease. Although still expensive, the technology exists and several diseases have been successfully treated with gene therapy, including cancer, monogenic disease, cardiovascular disease, infectious diseases, ocular diseases and inflammatory diseases (Wirth et al., 2013: 165). The basic technology of gene therapy could be potentially used for genetic improvement. For several years there has been concern that athletes may use gene doping that influences gene expression for promoting muscle growth or for increasing red blood cell production for greater endurance (Brzezianska et al., 2014: 251–259).

Currently, gene doping is very risky, as it could cause cancer and do other damage to an athlete's health (Brzezianska et al., 2014: 252). Of course, it would be also a permanent and irreversible modification of a person's genetic make-up. In addition, there are important limitations that cannot be overcome by better science: gene doping can improve certain traits such as strength and endurance, but it is unlikely to work for more complex traits such as intelligence and character, especially since changes in DNA do not affect the brain except in early brain development. A more radical approach would be needed, which is to genetically engineer humans from scratch.

3.4.3 Human genetic engineering

DARPA's *Developing Advanced Tools for Mammalian Genome Engineering* program, launched in 2013, has the objective to '[i]mprove the utility of Human Artificial Chromosomes (HACs) by developing new selectable metabolic markers for use in human cells, new high-fidelity methods for inserting DNA constructs of at least 50,000 base pairs (bp) in length into defined genomic loci, and new methodologies for facile intercellular genome transplantation'

(DARPA, 2013). The proposal suggests that it would be a development tool for 'advanced therapeutics, vaccines, and cellular diagnostics, as well as for basic biological and biomedical research'. However, the most interesting suggestion in the proposal is '[t]he successful development of technologies for rapid introduction of large DNA vectors into human cell lines will enable the ability to engineer much more complex functionalities into human cell lines than are currently possible'.

Apparently, DARPA wants to use the method of human artificial chromosome (HAC) as a delivery vehicle for inserting new genes that could contain new traits. Most basically it means adding an additional 47th chromosome to the human DNA. However, radical genetic modification of living humans for creating super soldiers is far beyond existing science, would not be possible in adults and would thus require embryonic intervention, thereby establishing a 'pre-specified warrior class' to which the US military is still opposed (Ford and Glymour, 2014: 46).

In the long-term there is the ethical concern of inserting animal DNA into human DNA to create animal-human hybrids or chimeras, which is more real than is often admitted. For example, it was revealed a few years ago that British laboratories had created 155 animal-human hybrid embryos between 2008 and 2011 in an effort to develop new treatments for incurable diseases (Martin and Coldwell, 2011). By mixing DNA human organs can be grown in pigs or other mammals to produce custom-made organs for people in need of an organ transplant. Using genetic intervention scientists have already grown human brain tissue in mice, which has significantly increased the intelligence of these mice (Coghlan, 2014). This technology that is currently being developed all around the world could lead to entirely new human-animal hybrid or chimera species, some of which might be granted the moral and legal status of humans and some might not (DeGrazia, 2007: 323).

The Oxford transhumanist Nick Bostrom has pointed out the theoretical possibilities of equipping humans with new features:

> [t]he current human sensory modalities are not the only possible ones... Some animals have sonar, magnetic orientation, or sensors for electricity and vibration; many have a much keener sense of smell, sharper eyesight, etc...There is no fundamental block to adding say a capacity to see infrared radiation or to perceive radio signals and perhaps to add some kind of telepathic sense by augmenting our brains with suitably interfaced radio transmitters.
>
> (Bostrom, 2005: 7)

A report by the European Union *Global Governance 2025* discusses the genetic modification of humans and argues that the 'direct modification of DNA at fertilisation is currently widely researched with the objective of removing defective genes; however, discussions of future capabilities open the possibility for designing humans with unique physical, emotional or cognitive abilities' (2010: 56).

Super soldiers may be genetically engineered to see at night like cats or to have the strength of gorillas by adding pieces of animal DNA to their genome. This creates some interesting possibilities for future chimera animal armies. Animals could be roboticized by inserting chips into their brains and spines that allow for remote control. For example, DARPA started the Hybrid Insects Micro Electro-Mechanical Systems (HI-MEMS) in 2005, which aimed to insert chips and other payloads into insects in view of remotely controlling them, using RF, optical or ultrasound signals (Bogue, 2010: 21).

Using more advanced animals as cyborgs is clearly possible. Apes could be engineered as animal soldiers: they could be made smarter by adding some human DNA that enhances their cognitive abilities to a certain extent. If more than 50 per cent was animal DNA, they would probably not be granted human rights status and could be used as slaves or as expendable fighters. Scientists in Britain have therefore already warned of a 'Planet of the Apes' scenario, where a genetic class of human-ape hybrids could turn like Frankenstein's monster on their masters (Academy of Medical Sciences, 2011: 71–72).

Genetic modification of humans and animals is an interesting technology in view of military applications, but it raises some very serious ethical issues that are unlikely to be solved satisfactorily. In the long run it could even split the human race into two or more separate species in which one species may enslave or eradicate other species. Bioethists George Annas, Lori Andrews and Rosario Isai argue: 'The new species, or "posthuman," will likely view the old "normal" humans as inferior...It is ultimately this predictable potential for genocide that makes species-altering experiments potential weapons of mass destruction, and makes the unaccountable genetic engineer a potential bioterrorist' (Annas et al., 2002: 162). This means that any substantial genetic modification of humans or the creation of new species of human-animal hybrids should be strictly prohibited.

3.5 Conclusion

The general idea of warfighter enhancement is a very old one and goes back to ancient times, where breeding, training and stimulants were used to create superior fighters. Over the last decade there has been a renewed interest in human enhancement as a result of progress in biotechnology and neurotechnology. The chapter has reviewed four main approaches for enhancing the minds of soldiers for improving their performance: (1) Neuropharmaceutical enhancement: Drugs are already used by militaries as performance enhancers, but they are also highly problematic because of their side effects. Neuroscience might result in the development of safer enhancement drugs and more effective delivery methods, including implanted biodegradable chips that release drugs when needed. (2) Brain stimulation: This might be preferable to the use of drugs because it promises fewer side effects and can be done non-invasively with methods such as TMS, tDCS and TPU. This is still at the development stage, but within ten years soldiers may have brain monitoring and brain stimulation integrated into their helmets that will help them to control their mental states and alert them of dangers. (3) Brain-computer interfaces: These currently show the biggest promise for a real breakthrough in human performance enhancement. One day it will be possible to connect human minds seamlessly to computers with a 'cortical modem', enabling people to control machinery by thought and to access digital information that is directly uploaded into their brain. This might take several decades to realise: the British MoD estimated that it would happen by 2045. (4) Finally, there is the possibility of using trait selection and genetic modification for performance enhancement, which raises the most troubling ethical questions. It is hard to say if or when biotechnology will be used to radically alter humans, but many biologists have expressed serious concerns about the potential consequences of the tinkering with the human genome and the creation of human-animal chimeras.

4 Intelligence and prediction

This chapter will review how neuroscience can contribute to intelligence analysis and prediction, discussing relevant research programs within the US Department of Defense, US Department of Homeland Security and the US intelligence community. These initiatives can be divided into the areas of (1) strategic intelligence, (2) intelligence analysis and decision-making, (3) threat detection and (4) neuroscience and interrogation. It will be shown that neuroscience can substantially contribute to these areas and that this can lead to a vastly improved understanding of social dynamics and even individual behaviour. Some of the findings of neuroscience already question our current approaches towards threats and this may lead to techniques for better mitigating threats such as terrorism and for avoiding conflicts and crises in the first place. Some of it, most importantly, 'mind-reading' using brain scans will remain controversial and may require more legal regulation in the future.

4.1 Strategic intelligence

Strategic intelligence is not clearly defined, but most generally it deals with estimating the overall capabilities and intentions of an actual or potential adversary. It is intelligence on a strategic level and thus 'intelligence necessary to create and implement a strategy, typically a grand strategy' (Heidenrich, 2007). While it tends to be easier to estimate capabilities, especially when it concerns a state actor, there are still huge problems with respect to accurately reading intentions of leaders and predicting social dynamics such as uprisings and revolutions. Intelligence analysis naturally involves a high degree of uncertainty, which can be reduced by better analytical techniques designed to overcome bias. Although the US intelligence community has greatly improved its analytical techniques and put analysis on a more scientific basis, many traditionally minded intelligence practitioners consider intelligence analysis to be more of an art than a science. For example, former Director of Central Intelligence for Analysis and Production Mark Lowenthal argues 'Intelligence is not and never will be a science and anyone who tries to promote it will be doing our profession a grave disservice. ...If we start equating intelligence to a science the unrealistic expectations will only go up. Intelligence is an art. It is an intellectual activity. An art' (quoted from Marrin, 2012: 533).

The art of making political forecasts consists of intelligence analysts extrapolating from historical patterns and known political trends to arrive at their conclusions. This generally works, but in the past US intelligence has also made major mistakes in this area with the most famous intelligence failures being the failure to predict the Iranian Revolution in 1979 and the failure to correctly assess Iraqi WMDs pre-war and predict a difficult post-war situation (Jervis, 2010). These failures can be attributed, at least in part, to cognitive bias and the inability of analysts to understand the cultural dimension of political dynamics.

4.1.1 Better understanding of foreign cultures

Military analysts and practitioners have often stressed the importance of cultural knowledge for warfare, especially in the context of military operations that take place within foreign societies of a different culture. Without cultural awareness and understanding, it is often not possible to successfully interact with foreign cultures, as there will be misunderstandings, wrong expectations and surprises in how the other culture perceives and responds to own actions. At a strategic, level lack of cultural understanding can lead to major political miscalculations about the enemy's overall intentions and motivations. A failure to understand the operational, social and cultural environment can lead to wrong estimations with respect to the likelihood, nature and intensity of resistance. Military analyst Montgomery McFate therefore suggests that '[c]ultural knowledge of adversaries should be considered a national security priority' (McFate, 2005: 43).

One of the greatest supporters of cultural awareness and cultural intelligence has been former DIA director General Michael Flynn. Flynn heavily criticized the military for its inability to understand the Afghan culture, which led to many mistakes made it in the war and nation-building in Afghanistan. He wrote in 2010:

> Eight years into the war in Afghanistan, the US intelligence community is only marginally relevant to the overall strategy. Having focused the overwhelming majority of its collection efforts and analytical brainpower on insurgent groups, the vast intelligence apparatus is unable to answer fundamental questions about the environment in which US and allied forces operate and the people they seek to persuade. Ignorant of local economics and landowners, hazy about who the powerbrokers are and how they might be influenced, incurious about the correlations between various development projects and the levels of cooperation among villagers, and disengaged from people in the best position to find answers – whether aid workers or Afghan soldiers – US intelligence officers and analysts can do little but shrug in response to high level decision-makers

seeking the knowledge, analysis, and information they need to wage a successful counterinsurgency.

(Flynn et al., 2010: 7)

What should have addressed the issues stated so clearly by General Flynn was called Human Terrain System (HTS), which was started in 2007 and which would end up costing taxpayers $600 million during the five years of its existence (Jacobson, 2015: 392). HTS were built on the research conducted by Human Terrain Teams (HTTs) comprised of anthropologists, sociologists, linguists and other social scientists, who would 'map' the human terrain by going into local Afghan villages to collect sociocultural, anthropologic and ethnographic data (Gusterson, 2009). The concept of the HTS was initially proposed by anthropologists Montgomery McFate (cited above) and Andrea Jackson in 2005 (McFate and Jackson, 2005).

Unfortunately, the 'militarization of anthropology' was a failure on many levels: much of the data produced seemed to have little relevance to military operations and academics complained that it was unethical to 'weaponize anthropology' and send scientists into war zones. A National Defense University study on HTS concluded: 'Ultimately, the HTTs failed to ameliorate growing cross-cultural tensions between US forces and Afghans and were unable to make a major contribution to the counterinsurgency effort' (Lamb et al., 2013: 22). At least the Afghan HTS was closed down in 2014 because of the US troop drawdown. At the same time, the study affirmed the permanent need for HTS-like capability, even as the US military transitions back into peacetime.

4.1.2 Cultural neuroscience

Neuroscience might help to significantly improve HTS or a similar cultural intelligence capability. According to the NRC report, '[c]oncepts found in cultural research serve as intervening variables in neuroscience research, providing an understanding of how culture impacts human cognition and affect with respect to brain functioning, meaning, and behavior in diverse social and political situations' (NRC, 2008: 9). The subdiscipline of 'cultural neuroscience' claims that there are physiological differences in the organisation and workings of the brains of people belonging to different cultures. Culture is part of the 'software' of the mind and is encoded in the 'wetware' or neural wiring of the brain. The information in the brain is encoded in the connections of neurons with each other through synapses. Every experience and everything an individual learns is reorganising the brain as a result of brain plasticity. Changes in the neural wiring can also have consequences with respect to how the brain processes perceptions. While it may seem that human perception is universal and that everybody, regardless of cultural background, would perceive the same things the same way, cultural neuroscience shows that this is not the case and that there are astounding differences in 'object processing, colour discrimination and taste' (Ames and Fiske, 2010: 72). Various studies indicate that East Asians and Westerners have different perceptual styles or mental habits how they process perceptions. Brain imaging technology makes it possible to show that presenting the same picture to a Chinese person and a Westerner will activate different areas of the brain, indicating culturally distinctive ways of processing the same perception (Ames and Fiske, 2010: 73). 'Further research may lead to better understanding of the neural basis of cultural differences in fundamental processes of cognition, emotion, and motivation' (NRC, 2008: 111). As pointed out by Giordano and Wurzman, a better understanding of other cultures, symbolic representations and cultural perceptions can also be used for improving PSYOPS and more easily winning the 'hearts and minds' of a foreign population (Giordano and Wurzman, 2011: 58).

4.2 Intelligence analysis and decision-making

Many aspects of intelligence collection and analysis have become automated, especially in the technical collection disciplines. Computer systems are now able to automatically transcribe phone conversations, translate texts, or produce short summaries of texts. But despite these impressive advances, there is, at least currently, no way of replacing the human analyst and human judgments, as computers only process information without any deeper understanding of the what and the why. Traditional efforts for improving qualitative intelligence analysis seemed to have reached their limits, so it is not surprising the US intelligence community is now considering more innovative methods, including neuroscience-based approaches, which shall be outlined below.

4.2.1 Intelligence analysis and neuroscience

Neuroscience and the cognitive sciences can lead to new insights about why intelligence failures happen, which can help to minimize their occurrence. Intelligence failures are defined as 'a mismatch between the estimates and what later information reveals' (Jervis, 2010:2). According to CIA analyst Richards Heuer, '[m]ajor intelligence failures are usually caused by failures of analysis, not failures of collection. Relevant information is discounted, misinterpreted, ignored, rejected, or overlooked because it fails to fit a prevailing mental model or mind-set. The "signals" are lost in the "noise"' (Heuer, 1999: 65). One of the biggest challenges in this respect are cognitive biases that 'are mental errors caused by our simplified information processing strategies', which occur consistently and predictably (Heuer, 1999: 111). They affect both the producers and the consumers of intelligence. Cognitive biases result from selective perception and the way the brain processes information. They have nothing to do with any other forms of bias such as political bias, organizational bias, self-interest of the analyst and so on. The term 'cognitive bias' was introduced by the psychologists Amos Tversky and Daniel Kahneman in the early 1970s and their experimental work has since become the basis for various theories of human cognition and perception (Tversky and Kahneman, 1974; Heuer, 1999).

One of the most important discoveries in this field is the theory of 'cognitive dissonance' or 'cognitive closure', first formulated by Leon Festinger, which claims that individuals that are exposed to information that challenges their convictions or beliefs will experience psychological discomfort and will instinctively try to avoid it. According to Kjetil Anders Hatlebrekke and M.L.R. Smith, cognitive closure is 'a human defence mechanism, which attempts to assert that matters exist in a distinct order, structure and stability' (Hatlebrekke and Smith, 2010: 149). The sum total of an analyst's beliefs is called a 'mindset' and it is psychologically difficult to change, as the analyst instinctively seeks cognitive closure, which can result in wrong judgments when a situation changes but not the analyst's mindset. A derivative of cognitive dissonance is confirmation bias, which causes an analyst to seek out information that confirms the own assumptions, while ignoring or rejecting information that challenges these assumptions.

Although cognitive psychology has tremendously helped to improve intelligence analysis, there is still much that can be learned through neuroscience research that can lead to even greater insights into human cognition and how it is affected, for example, by the way the information is presented and how an analyst's brain makes use of information in terms of learning and reasoning. Important in this respect is the concept of neuroplasticity or the ability of the brain to reorganize itself by strengthening or weakening synaptic connections. Essentially, the brain reshapes itself according to how it is used (Landon-Murray, 2013: 74). This means that the way a person seeks out and assimilates information changes the brain in fundamental ways and causes or reinforces cognitive habits. Neuroscience research has shown that intellectual technologies such as the Internet impact on cognitive habits and mental capacity. Neuroscientist Gigi Vorgan conducted research on how daily Internet use affects the brain and found that it 'stimulates brain cell alteration and neurotransmitter release, gradually strengthening new neural pathways in our brains while weakening old ones' (quoted from Landon-Murray, 2013: 75). Media analyst Marshall McLuhan was therefore correct when he postulated that 'the medium is the message'.

As any teacher can tell, the cognitive habit of many students, to be constantly engaged in online activities, has substantially reduced their attention spans, memory and reasoning skills. Technology writer Nicholas Carr has argued in his recent book *The Shallows: What the Internet Is Doing to Our Brains* that Google makes us stupid as it creates the illusion that all answers can be found in a quick Internet search. This concept supports intellectual laziness, as nothing needs to be learned or memorized since it strengthens the wrong belief that all information can be retrieved at any time (Carr, 2011).

There are, of course, interesting consequences from this research for intelligence analysis since it concerns a profession built around cognitive processes (Landon-Murray, 2013: 73). Landon-Murray cautions that the Internet leads to a culture that emphasises quantity over quality, creating a heightened pressure to produce and thus reduces the ability for fact-checking, resulting in more errors. He also suggests that 'the nature and structure of their work assignments could in fact reinforce the disrupted modes of thinking that follow from heavy internet use. Thus, key skill sets that may already be flawed or lacking will be further compromised' (Landon-Murray, 2013: 78).

In order to address these issues, Landon-Murray suggests that intelligence analysts should spend at least part of their work-time away from their computers, avoiding distractive elements such as phones and e-mail. Special mental exercises and practices can also improve cognitive habits. Further research towards these ends is undertaken by IARPA. The agency has recently awarded $12 million for a three and a half year program called *Strengthening Human Adaptive Reasoning and Problem-solving* (SHARP) (Hurley, 2014). Amongst the approaches studied are meditation, games for strengthening memory and thinking and electrical brain stimulation. The purpose of SHARP is to 'rigorously assess whether innovative combinations of approaches can improve performance in high-performing adults such as those in the Intelligence Community...The researchers in this program will not only aim to improve adaptive reasoning and problem-solving, but also to understand how and why certain methods work' (Silva, 2014).

The NRC report makes some general suggestions for improving training and learning: neuroscience can help in: '[e]valuating the efficiency of training regimes and learning paradigms', measuring '[i]ndividual capability and response to training', '[m]onitoring and predicting changes in individual performance efficiency', selecting and assessing personnel and in '[m]onitoring and predicting social and group interactions' (NRC, 2009: 23). Neuroscience can improve the understanding of how the brain learns leading to more effective methods of teaching (Royal Society, 2012: 30). Neural imaging and monitoring also provide opportunities for an individualized approach to teaching and learning, where it can be checked how much an individual has learned and how fast the individual learns. Training could utilize virtual reality or holograms that create realistic environments and scenarios. Learning difficulties can be easily discovered and addressed through neural monitoring. Obviously, the government can use these approaches also for improving the selection of soldiers or other personnel for specific tasks and units. It would be easier to identify fast learners or to select individuals according to character traits or skills that are measurable through brain scans. The intelligence community could thereby identify individuals, who are creative thinkers and have more mental flexibility, which could greatly improve the quality of analysis.

4.2.2 *Neuro-cyber systems for intelligence analysis*

The performance of analysts can be enhanced through systems for augmented cognition. The analysts would wear a brain-monitoring device such as an EEG that can flag subconscious responses to stimuli. This helps the intelligence community to deal with the problem of information overload caused by vastly increased collection capabilities. Automated analysis is still inferior to human analysis, but could be eventually improved by neuro-cyber systems that train machines in cognitive tasks. James Giordano and Rachel Wurzman describe it in the following way:

> [i]nformation systems could conceivably be conjoined so that neural mechanisms for assigning and/or detecting salience (i.e. processes involving cortical and limbic networks) may be either augmented or modeled into neurotechnologic devices for rapid and accurate detection of valid (i.e. signal vs. noise) information within visual (e.g. field sensor, satellite and UAV-obtained images) and/or auditory aspects (e.g. narratives, codes) of human (HUMINT) or signal intelligence (SIGINT). Formulating and testing credible hypotheses while monitoring large amounts of information could be accomplished by computational cognitive frameworks that are capable of both self-instruction (e.g. using the internet as a 'training environment'), and learning from experience (e.g. via direct access to the operational environment).
>
> (Giordano and Wurzman, 2011: 57)

DARPA and IARPA are funding several research programs to this effect. In 2007, DARPA awarded a $4 million dollar contract to Honeywell to develop a technology that allows imagery analysts to quickly sift through vast amounts of imagery, while being able to spot important details (Hughes, 2007: 5). The program is called *Neurotechnology for Intelligence Analysts* (NIA), which monitors brain signals of analysts with an EEG. If there is a brainwave response to a particular image that indicates that the brain registered something of interest, it is flagged to the analyst so that the image in question will receive further review (Hughes, 2007: 5).

A similar effort is the IARPA project *Integrated Cognitive-Neuroscience Architectures for Understanding Sensemaking* (ICArUS). The project website states ICArUS' goal as 'the development of models that more accurately predict human sensemaking performance in both the cognitive and behavioral domains' (IARPA website). It is currently being developed in view of improving geospatial intelligence analysis, but it could also be adopted for any other big data analysis tasks. According to IARPA, 'ICArUS research will lay the groundwork for the development of a new generation of automated analysis tools that replicate the unique strengths of human sensemaking' (IARPA website). A program like ICArUS could sift through billions of communication transactions and other available data to predict the behaviour of millions of people, i.e. to find indications of a terrorist plot before it happened. This was actually the goal of the previous DARPA *Total Information Awareness* program, which seems to have now achieved some technological maturity and feasibility. Technology that can replicate aspects of human cognition and reasoning could one day lead to computers that can perform many, if not all, analytical and decision-making functions by themselves. Although strong AI still seems distant, a lot of progress has been made over the last ten years, especially with respect to quantum computing that could enable strong AI.

4.2.3 Neuroscience and AI

AI research and brain research have been intricately linked since the 1960s when computer scientists developed a brain-derived system known as 'neural networks' (Barrat, 2013: 214). It is software that simulates the behaviour of neurons by establishing and strengthening useful connections amongst neurons. In other words, neural networks can be taught to perform better over time – they learn, which is an important aspect of intelligence. So it is not surprising that both the BRAIN Initiative and the Human Brain Project explicitly include programs for AI research that aims to model the human brain at a hardware level in the hope that this could lead to computers that could perform difficult cognitive tasks at the same or above the performance level of humans. By mimicking the functioning of the brain in some key aspects computers would become a lot smarter than they currently are. The NRC report stated:

> it seems likely that increasingly sophisticated cognitive systems will be constructed in those two decades that, while not aiming processes in the human brain, could nonetheless perform similar tasks well enough to be useful, especially in constrained situations. In this case, success would not be determined by how closely the system resembled the brain's mechanisms for action, but how similar the performance of specific cognitive tasks was to a typical human user.
>
> (NRC, 2008: 95)

The main advantage of automating cognitive tasks is to reduce the need for human personnel and to potentially improve the quality of decision-making. Currently, computers are still struggling to perform cognitive tasks that are simple for a human such as recognising faces under varying light conditions and from different angles. The human brain manages to do this very efficiently using massive parallel processing. The human brain is extremely complex with its hundred billion neurons that constantly communicate with each other, but slowly, some of that complexity can be replicated in new 'neuromorphic' computer chips. The Human Brain Project subprogram in this area has produced a novel chip architecture that can simulate 200,000 neurons. The project website states: 'A single chip can simulate 16,000 neurons with eight million plastic synapses running in real time with an energy budget of 1W'. The distinctive advantage of neuromorphic computer architectures is indeed their extremely low power consumption compared to conventional computing. This would make the technology suitable for small mobile devices and applications. Neuromorphic systems would also have neural plasticity, which means that they can change through learning and can potentially improve themselves. Many AI researchers believe that brain-based approaches like neuromorphic computing are the most likely path towards achieving 'strong' or human-level AI in the future. Strong AI can lead to fully autonomous weapons systems that can learn and adapt to changing situations and to battlemanagement systems that effectively develop complex battle plans through extensive wargaming and then implement them by taking over many staff functions. At this point, there would be the danger that human decision-making in war is taken over by intelligent machines because they would be a lot faster and better able to deal with the growing complexity of war than humans (Adams, 2001). Opinion polls amongst computer scientists and IT professionals indicate that 10 per cent think strong AI will arrive before 2028, 50 per cent believe it may happen by 2050 and 90 per cent believe it will happen by the end of the twenty-first century (Barrat, 2013: 25).

4.3   Neuroscience and behavioural research

Neuroscience has already made numerous contributions to behavioural research, many of which are at least somewhat relevant to national security and intelligence. It is part of political intelligence to develop psychological profiles of leaders so that it becomes possible to more accurately predict their behaviour. Furthermore, behavioural research can lead to the development of analytical tools that enable the prediction of the behaviour of a collective. In other words, neuroscience could lead to more accurate models of individual and collective behaviour than can be achieved with standard psychological and social sciences methodology. Giordano and Wurzman suggested that:

> [t]he complex dynamics of political forces that contribute to such predictive difficulty are due, in part, to the numerous and varied agents involved, all of whose actions are individually determined. Thus, understanding the bio-psychosocial factors that influence individual and group dynamics, and being able to detect these variables with high ecological validity (i.e. "in the field", under real-world conditions) is important to both descriptive/analytic and predictive intelligence approaches.
>
> (Giordano and Wurzman, 2011: 58)

4.3.1   Neuroscience insights into individual behaviour

Some of the most important questions in the field of national security is why individuals become radicalized, turn to violence and commit acts of terrorism against innocent civilians. Despite very substantial efforts undertaken by psychologists there is to the present day no psychological profile of a 'terroristic personality' that could allow predicting whether an individual will become violent and threat to society (Borum, 2004: 3). Although there may not be much of a point in terms of using brain scans as a method for identifying who has a disposition towards radicalization, there are still insights to be gained from understanding the motivations of individuals using a neuroscientific approach.

The neuroscientist Gregory Berns, from Emory University, suggests that human behaviour is driven both by utility motivations and 'sacred values' to which humans remain committed. He tested this theory by checking via brain scans whether utility considerations and sacred values are indeed processed in the brain differently, which he found to be true (Berns et al., 2012). This means choices are not all considered within the brain in a uniform fashion and that whenever 'sacred values' are concerned people will not make decisions based on a utilitarian calculus. The finding has some very important implications for national security strategy, as current models of human decision-making need to be changed accordingly.

Current strategic thinking still emphasises the rational actor, which was originally developed by the think tank RAND in the 1950s. The basic assumption in this model is that an individual will consistently choose the option that promises the highest utility payoff. According RAND chronicler Alex Abella, Kenneth Arrow's work at RAND on these topics 'profoundly altered Western culture', as from now on all social processes were reduced to interactions of rational individuals (Abella, 2009: 51). These theories developed by RAND researchers were applied to the problem of winning the war in Vietnam, which led to disaster. The incremental approach to coercion based on game theory did not produce the desired result, as the North Vietnamese would not compromise (Freedman, 1996). Neuroscience can now make a compelling argument as to why rational individuals sometimes make choices that violate the principle of seeking the greatest utility. It seems whenever a decision-maker has to deal with an issue that concerns 'sacred values' any 'rational' utility considerations are swept aside, resulting in choices that the rational actor model cannot explain. The failure of relying on the rational actor model subsequently leads to highly flawed national security strategies.

Psychologist Eric Haseltine interprets the results of Berns' study in the way that we cannot change the behaviour of ideologically or religiously motivated terrorists through punishment. Painful military strikes could actually harden the resolve of the terrorists and radicalize individuals, who share their sacred values (Haseltine, 2015). A neuroscience approach can help understand when the decision-making of foreign actor is driven by utility calculation and when it is driven by sacred values, which can help to avoid tragic miscalculations as they happened during the Vietnam War. Of course, it is impractical to subject foreign leaders to a brain scan, but it may already help to be more conscious of the workings of the human mind. This means to discard the flawed rational actor model and replace it with a model that more accurately predicts human behaviour.

4.3.2 *Predicting social dynamics*

The behaviour of a collective is the sum-total of all individual behaviours in this collective. It follows that by understanding the motivations and behaviours of key or all individuals in that collective social dynamics can be identified, allowing for the accurate prediction of major societal upheavals such as uprisings and revolutions. The US military and the intelligence community have been, for a long time, in the business of predicting revolutions, although a number of failures in this respect indicates that their methodologies leave much to be desired.

In 1964, the US Army initiated a large-scale $6 million dollar social science research program called Project Camelot. It was the biggest grant that the US military had awarded to the behavioural sciences up to that time. The official purpose was to research the societies and cultures of particular target countries and regions with a focus on Latin America, but also in the Middle East, Far East and Europe, using established social science methodologies. According to General Dick, Project Camelot was supposed 'to help us to predict the potential use of the American Army in any number of cases where the situation might break out' (quoted from Nisbet, 1966: 46). The apparent intention was to use the knowledge gained from the research for COIN and psychological warfare.

But there was a darker side to Camelot. One of the countries studied was Chile, which happened to be a target for CIA regime change at the time. The CIA had spent $4 million between 1962 and 1964 to secure the election victory of Eduardo Frei against his socialist contender Salvador Allende, who later became president and was overthrown in a violent CIA-backed coup by General Pinochet in 1973 (Ranelagh, 1986: 514–520). Word got out in Chile about the Camelot behavioural study of their society by the US military, which led to the eviction of American Camelot scientists from the country (Nisbet, 1966: 49). The resulting controversy over the possible use of behavioural research for inciting and steering revolutions (instead of suppressing them) and other psychological manipulation of societies led to the early termination of the project in 1965 (Gusterson, 2009:5–6).

The current incarnation of Project Camelot is called Minerva Initiative, which is a collaboration between the Department of Defense and the National Science Foundation launched in 2008 that initially made $50 million in grant money available (Gusterson, 2008). According to the project website, the purpose of Minerva is to: '[l]everage and focus the resources of the Nation's top universities'; '[s]eek to define and develop foundational knowledge about sources of present and future conflict with an eye toward better understanding of the political trajectories of key regions of the world'; and '[i]mprove the ability of DoD to develop cutting-edge social science research, foreign area and interdisciplinary studies, that is developed and vetted by the best scholars in these fields'. The call for proposals mentions 'anthropology, economics, political science, sociology, social and cognitive psychology, and computational science' as relevant disciplines. Among the projects that have received funding are: 'New analytics for measuring and countering social influence and persuasion of extremist groups' by Arizona State University, 'Mobilizing media: a deep and comparative analysis of magazines, music, and videos in the age of terrorism' by Georgia State University, 'Tracking critical mass-outbreaks in social contagions' by Cornell University, 'Motivational, cognitive, and social elements of radicalization and deradicalization' by the University of Maryland.

At least some of the Minerva projects are clearly related to the problem of predicting revolutions/ radicalisation. Inevitably, there are also neuroscience-projects funded by the Minerva Initiative. For example, researchers from the University of Chicago received $3.4 million funding for the project 'Social and Neurological Construction of Martyrdom', which looks at how ISIS propaganda videos resonate with viewers. In a second step, 'the team will use functional magnetic resonance imaging (fMRI) to investigate the neural pathways through which martyrdom appeals evoke sympathy in the viewer. They aim to uncover exactly what is happening in the brain when an individual is persuaded to change their beliefs' (Basick, 2015).

4.3.3 The Sentient World Simulation

By far the most ambitious project for predicting social dynamics is the Sentient World Simulation, which the Pentagon announced in 2007. The project aims to create a synthetic virtual environment that mirrors the real world, including the simulation of billions of people or 'nodes' for the purpose of social wargaming. The Purdue University concept paper states in the summary that the goal

> is to build a synthetic mirror of the real world with automated continuous calibration with respect to current real-world information, such as major events, opinion polls, demographic statistics, economic reports, and shifts in trends. The ability of a synthetic model of the real world to sense, adapt, and react to real events distinguishes SWS from the traditional approach of constructing a simulation to illustrate a phenomena. Behaviors emerge in the SWS mirror world and are observed much as they are observed in the real world.
>
> (Cerri and Chaturvedi, 2006)

The paper also suggests 'SWS provides an environment for testing Psychological Operations (PSYOPS) and Civil Affairs activities, capable of illustrating the impact of these activities on populations' (Cerri and Chaturvedi, 2006). The Synthetic Environment for Analysis and Simulation developed at Purdue University is currently used by the DHS and DoD for simulating crises in the homeland (Baard, 2007). Theoretically one could use

SWS to reliably predict how a population would respond to a particular event before it happened to fine-tune PSYOPS and steer societies in the real world.

Of course, in order to make a SWS work accurately one would need enough data available on enough people so that the behaviour of the collective becomes predictable. DARPA has already the *Nexus 7* program, which can systematically monitor social media networks in a country such as Afghanistan to look for indications of instability or imminent attacks (Jacobson, 2015: 400). It is also not hard to see how the massive global NSA data collection effort comes in handy for creating files containing details of the every-day life of practically everybody within a given society. This would include data such as names, phone metadata, websites visited, social networks and more, allowing some prediction individual behaviour with predictive analytics (a statistical approach to big data) as used in the business world. NSA expert James Bamford suggested in 2012 that the NSA's new data warehouse in Bluffdale, Utah will have a capacity of 1 yottabyte (10^{24} bytes) and could store 500 quintillion pages of text (Bamford, 2012). Since the Snowden revelations it is known that the NSA collected in a single month 97 billion pieces of intelligence through its program *Boundless Informant* (Engelhardt, 2014: 10).

A few major cities in the US, among them Fresno, New York, Houston and Seattle already operate high-tech Real Time Crime Centers that assign every citizen and home a threat score (green, yellow, red), based on calculations with billions of data points such as 'arrest records, property records, commercial databases, deep Web searches and…social media postings' (Jouvenal, 2016). The software that does these calculations is called *Beware* and the threat score is used to assist police officers with understanding the threat level of the person before they intervene, which is controversial because it could shape their response towards a person with a high threat score. Furthermore, the algorithm that calculates threat scores is secret, which creates the concern that an innocent *Facebook* post containing some blacklisted keyword could raise somebody's threat score (Jouvenal, 2016).

The next step after giving everybody a threat score would be the prediction of everybody's behaviour in real-time. This kind of predictive technology is under development and is called *cognitive analytics*, which, according to a description from Deloitte University Press, is

> inspired by how the human brain processes information, draws conclusions, and codifies instincts and experience into learning. Instead of depending on predefined rules and structured queries to uncover answers, cognitive analytics relies on technology systems to generate hypotheses, drawing from a wide variety of potentially relevant information and connections. Possible answers are expressed as recommendations, along with the system's self-assessed ranking of how confident it is in the accuracy of the response. Unlike in traditional analysis, the more data fed to a machine learning system, the more it can learn, resulting in higher-quality insights.
> (Ronanki and Steier, 2014)

Another way of looking at cognitive analytics is to say that it is machine learning applied to big data. This is exactly the technology from which the NSA could benefit most: it collects moutains of data, but still lacks the technology and the resources to process and analyze all of that data, resulting in a 'bulk data failure' (Kamath, 2015). Without a sufficient analytical capability, almost all of the data would be collected in vain. The prospects for such an analytical capability to be developed in the future are good considering the rapid pace of advances in computer technology and AI.

4.4 Threat detection, identification and interrogation

Neuroscience could have a particularly strong impact in the homeland security and counterterrorism field. Since 9/11, governments have devoted a lot of attention and resources to fighting terrorism, which had been previously considered to be mostly a law enforcement issue. The attacks highlighted the great vulnerability of complex technological societies to new and ingenious ways of attack that could result in mass casualties. Considering the damage that can be done by a few terrorists with modern technology, threats need to be detected early. Neuroscience research could lead to new methods of threat detection, the biometric identification of individuals and deception detection.

4.4.1 Threat detection

There has also been a substantial growth in suicide bombings around the world since the 1980s, which are extremely rare in Western culture, but are used as a tactics by Islamic and Asian terrorist groups. Suicide bombings tend to result in the most casualties and are very difficult to prevent. According to sociologist Riaz Hassan, '[b]etween 1981 and 2006, 1,200 suicide attacks constituted 4 percent of all terrorist attacks in the world and killed 14,599 people or 32 percent of all terrorism related deaths' (Hassan, 2009). In the future it can be expected that terrorists will use WMD that could kill hundreds of thousands in one incident. The 'National Strategy to Combat Weapons of Mass Destruction' prepared by the NSC in 2002 argues that rogue states may use WMD terrorism 'to overcome our nation's advantages in conventional forces' and that there are terrorist groups that seek WMD, making WMD terrorism a serious threat to the US homeland (US NSC, 2002: 1).

Neuroscience can assist with developing methods for detecting terrorists before they can act. Since about 2005, the Department of Homeland Security has been funding research for the development of a non-invasive and real-time screening technology that can be deployed at airports and other public places. The new technology could assist security personnel to identify individuals with hostile intent. In 2007, DHS began funding the Screening Passenger by Observation Technique (SPOT) and has spent between 2007 and 2013 a total of $878 million dollars (US DHS, 2013: 1). The system is already deployed at 176 airports in the US (US DHS, 2013: 3). SPOT uses an analysis of facial expression for the detection of malintent and deception and its inventor claims an accuracy rate of 70 per cent. Individuals flagged by SPOT are subjected to a secondary security screening, which has led, according to DHS, to 1,700 arrests between 2006 and 2009 out of a total of 232,000 flagged (Weinberger, 2010: 414).

Another program that was started by DHS is the *Future Attribute Screening Program* (FAST) in 2008, which could be used at mass events to quickly screen a large number of people or to find a terrorist in a crowd. A program overview states: 'FAST seeks to improve the screening process at transportation and other critical checkpoints by developing behavior-based screening techniques that will provide additional indicators to screeners to enable them to make more informed decisions' (US DHS, 2008: 2). Amongst the non-intrusive sensors for real-time scanning individuals for malintent that are considered in this project are: a remote cardiovascular and respiratory sensor, a remote eye tracker, thermal cameras to assess electrodermal activity and eye movements, a high-resolution video for analysis of facial expressions and body movements, as well as other sensors such as an audio system for analysing the pitch of a voice. The analysis of this sensor data can reveal symptoms of high mental stress that could be expected in individuals, who are planning to detonate a bomb or have other malintent. The FAST system could be deployed at airports and other transportation hubs, at major sports events as mobile units, or for controlling entry to other secure areas. One experimental system has been deployed at an undisclosed public location for testing (Weinberger, 2011). The research is funded at around $10 million a year (Weinberger, 2010: 414).

The main criticism to a 'pre-crime' screening system, as envisioned by DHS, is that it would need to be extremely accurate. Otherwise DHS would end up detaining large numbers of innocent people. Similar to the polygraph, there is no scientific evidence that it could work with sufficient accuracy (Weinberger, 2010: 415). Steven Aftergood from the Federation of American Scientists is also sceptical: 'I believe that the premise of this approach – that there is an identifiable physiological signature uniquely associated with malicious intent – is mistaken. To my knowledge, it has not been demonstrated... Without it, the whole thing seems like a charade' (Weinberger, 2011). There are also concerns that such a system would also violate Fourth and Fifth Amendment rights (protections against unreasonable searches and seizures and the right not to be a witness against themselves).

4.4.2 Neuroscience and interrogation

Humans generally perform poorly in the task of identifying when another person is lying. As pointed out by a professional interrogator, 'even with training few people do much better than guesswork, and some do worse!' (Alder, 2007: 264). A behavioural study has shown that average people detect lies 54 per cent of the time or slightly above chance accuracy of 50 per cent and that trained interrogators perform only slightly better than average people (Costanzo and Gerrity, 2009: 185). So it is not surprising that governments have been looking for a technical solution to this dilemma, which eventually resulted in the invention of the polygraph by Cesare Lombroso in 1895. The CIA has used the polygraph in the personnel security process since 1948 (Sullivan, 2007: 15). Although the method has been controversial since the 1950s, the polygraph is now widely used in the US for vetting personnel. It is a requirement for employees of intelligence agencies to regularly undergo a polygraph test in order to obtain or keep a top secret security clearance. Unfortunately, even professional polygraph examiners admit that the polygraph is a very flawed instrument, which can produce very unreliable results. Citing a study by psychologist David Lykken, 'in real-life "field situations," when results were graded on the basis of the polygraph results alone, the innocent were called truthful only 53 percent of the time, which is to say, hardly better than guesswork' (Adler, 2007: XIV).

The only thing the polygraph does is to measure physiological responses to questions that indicate stress, which presumably occurs when a person is lying. Many honest people can get easily stressed when they are subjected to a polygraph test and thus fail the test without having lied. At the same time, experienced liars or psychopaths can beat the test by giving deceptive answers or by failing to produce any emotional response when lying. There are also the well-known techniques of causing oneself pain during the test or of taking calming drugs for distorting the test. Polygraph exams are now considered to be so unreliable that they are generally inadmissible as evidence in court (Frederickson, 2011). Despite of all these issues, there is currently not much of an alternative to the polygraph in the national security arena. This is the reason why it becomes necessary to look at other ways for detecting deception and neuroscience may offer better solutions to the problem.

4.4.3 'Brain fingerprinting'

One method invented by Lawrence Farwell is called 'brain fingerprinting' technique, which uses an EEG for brain monitoring. The CIA sponsored Farwell's research with $1 million in the early 1990s, but discontinued it out of concerns that the method could not be sufficiently evaluated (Alder, 2007: 264).

In theory, it works as follows: the brain produces a measurable P-300 brain-wave response in response to a stimulus that the brain recognises beyond the control of the individual tested. A suspect could be presented with information that only a guilty person would have and if there is a P-300 response it would indicate guilty knowledge. According to Farwell, 'brain fingerprinting' is one hundred per cent accurate in detecting guilty knowledge (Farwell et al., 2013). There have been questions raised with respect to the validity of Farewell's claims since the method uses proprietary technology that Farwell does not disclose in his research papers.

A Senate report investigating the technology found several general limitations of the method: it hinged on having specific details that can be tested, the test subject must be cooperative and sit still in front of a computer screen and the test does not work with tired persons (US Congress, 2007: 2). Canli et al. also point out that the approach 'requires the interrogator to have extensive knowledge of the crime...and a stimulus set that is uniquely familiar to the perpetrator. This approach may therefore not be useful for scenarios in which the interrogator does not have unique attack information or in which the suspect may have an innocent reason for having the knowledge' (Canli et al., 2007: 6).

Furthermore, brain fingerprinting can also be useful for the biometric identification of individuals. Research at the Basque Center on Cognition, Brain, and Language aims to identify subjects based on their unique brainwave response to particular stimuli such as particular words. The Center claims that they have achieved 94 per cent accuracy (Warmflash, 2015). This could be used as an anti-theft system in cars or for authorising bank transactions instead of keys or passwords, for example. Of course, individuals could be identified using brain scans as everybody has a unique brain, although accuracy may not be good thanks to brain plasticity (Ehrenberg, 2015).

4.4.4 MRI scans

Magnetic Resonance Imaging (MRI) machines use electromagnetic radiation that passes through the skull in order to create three-dimensional images of the brain. An MRI has huge cylindrical magnetic coils in which the head is placed that produce a very strong magnetic field. The magnetic field changes the alignment of nuclei inside the body and when reverting back they produce an echo that can be analyzed (Kaku, 2014: 22). MRIs have a spatial resolution of 0.1 mm, which is still not yet good enough for monitoring a single neuron (Binhi, 2009: 30). Functional MRIs (fMRIs) work differently: they detect the presence of oxygen in the blood, which is indicative of neural activity. fMRIs have a lower spatial resolution of 1 mm, but it has a higher temporal resolution of several milliseconds, which is important for studying specific brain functions (Binhi, 2009: 30). fMRIs are considered to be less harmful than X-rays and they have been used for research into mind-reading and lie detection.

In fact, there is already a small industry that has been built around the presumption that fMRI brain scans can enable deception detection. The Royal Society report mentions specifically *Cephos Corporation* and *No Lie MRI* (Royal Society, 2012: 17). The California-based company *No Lie MRI* claims that it can determine, with 90 per cent accuracy, whether a person is lying, based on fMRI brain scans. fMRIs measure blood flows in the brain that are indicative of neural activity. The main assumption is that brain scans can show increased brain activity, which would indicate a lie, since the brain would need to make a greater effort to invent some untruth

(Stix, 2008). Activity in particular parts of the brain could also indicate a specific cognitive task such as the retrieval of an existing memory. However, historian Ken Alder has pointed out some of the fundamental flaws in using brain scans for lie detection: people could have 'false memories', or just memorize lies in advance well, or in other ways refuse to play along with the examiners (Alder, 2007: 265).

Up to now, fMRI lie detection is highly controversial and may be prone to error, but many researchers are enthusiastic and several companies are moving forward in using fMRI scans for deception detection and indeed 'mind reading'. For example, researchers from the University of California, Berkeley have used an fMRI to translate an image seen by a test subjects into words that describe what they see (Smith, 2013). When test subjects placed in an fMRI were shown a scene from the movie *Bride Wars*, when Anne Hathaway is in a conversation with Kate Hudson, the program correctly describes it as 'woman' and 'talk' (Smith, 2013). Similarly, researchers from the University of Washington managed to determine in real-time whether a test subject was looking at houses or faces, using an invasive electrocorticography (ECoG) (Miller et al., 2016).

These techniques work through an analysis of how visual perceptions are encoded in the brain, but researchers have pointed out that visual perception is easy compared to the encoding of other information in the brain. The decoding of actual thoughts with a MRI is also possible at a rudimentary level. However, physicist V.N. Binhi has argued that up to now researchers with fMRIs could only identify nouns that test subjects were thinking and that it would be impossible to decipher complex thoughts using this technology (Binhi, 2009: 33).

From a practical point of view, using an fMRI in threat detection and interrogation is anything but ideal. fMRIs are currently very large and very expensive machines that are not mobile. However, physicist Michio Kaku envisions that future MRI's could be handheld and as small as a cell phone with the ability of producing 3-D pictures of the brain (Kaku, 2014: 72). Kaku also cautions that '[f]or the foreseeable future brain scans will continue to require direct access to the human brain in laboratory conditions' (Kaku, 2014: 76).

4.4.5 Russian approaches to mind-reading

Russian researchers have long relied on less sophisticated, but more ingenious approaches to mind-reading, which has resulted in some interest on part of the US national security establishment. The two US military analysts, Janet and Chris Morris, have travelled to Russia in the early 1990s to take advantage of the new openness and learn about Russian NLW technologies (Shukman, 1996: 213). They met with the renowned Russian inventor and researcher Igor Smirnov, who demonstrated his technology for creating a model of a person's mind. Chris Morris was attached to an EEG while looking at a computer screen. It displayed in rapid succession apparently nonsensical words that were at least six letters long and that Morris could hardly read. After twenty minutes the procedure was over and the data collected on Morris' mental responses was then analyzed by a software and then arranged in a three-dimensional graph, which 'showed the relative importance in Chris' mind of ten different factors including family, sex, job, father, health and anxiety', which 'rang true' (Shukman, 1996: 225). Smirnov used the technology for a treatment that he called 'psycho correction' that used subliminal messages and electromagnetic fields to heal the mind of addiction, trauma, etc.

In 2007, Smirnov's company Psychotechnology Research Institute, headed by his wife Rusalkina since Smirnov's death in 2004, was considered for a multi-million dollar contract by DHS in the context of its Project Hostile Intent. The company offered its mind-reading technology Semantic Stimulus Response Measurements Technology or SSRM Tek, which tests individual brain responses to subliminal stimuli such as a picture of Osama bin Laden (Weinberger, 2007c). Since Smirnov's company refused to hand over the software for testing, it cannot be evaluated whether it actually works (Weinberger, 2007c).

4.4.6 Synthetic telepathy as interrogation tool?

In the previous chapter, it was mentioned that the US Army sponsors 'synthetic telepathy' as a clandestine communication system, which translates thoughts into text. It seems obvious enough that a technology such as synthetic telepathy could be utilized in interrogations and for threat detection. Apparently Lockheed Martin was/ is working on an EEG-based mind reading device called 'BioFusion', which was thought up by neuroengineer John Norseen and is apparently based in part on the Russian technology described above. According to journalist Douglas Pasternak,

> BioFusion would be able to convert thoughts into computer commands, predicts Norseen, by deciphering the brain's electrical activity. Electromagnetic pulses would trigger the release of the brain's own neurotransmitters to fight off disease, enhance learning, or alter the mind's visual images, creating what Norseen has dubbed 'synthetic reality'. The key is finding 'brain prints'. 'Think of your hand touching a mirror,' explains Norseen. 'It leaves a fingerprint.' BioFusion would reveal the fingerprints of the brain by using mathematical models. 'Just like you can find one person in a million through fingerprints,' he says, 'you can find one thought in a million'.
>
> (Pasternak, 2000)

Norseen stated that this electromagnetic manipulation can allow in principle 'to manipulate what a person is thinking', which was also a claim made by Igor Smirnov (Pasternak, 2000). An article by Sharon Berry in the military *SIGNAL Magazine* describes BioFusion in the following way:

> it combines sensors to examine biological systems to understand how information and neural structures produce thought and to display the thought in mathematical terms. By creating an advanced database containing these terms, researchers now can look at brain activity and determine if a person is lying, receiving instructions incorrectly or concentrating on certain thought types that may indicate aggression.
>
> (Berry, 2001)

In 2007, Norseen passed away and not much has been heard about BioFusion since 2001. So it is unclear whether the research was successful or not or whether it continues as a classified project. There are several reasons to be sceptical about a BioFusion type device or synthetic telepathy in general. Brain plasticity is an obstacle to creating a thought database. Since the brain not only uses digital electrical signals, but also analogue chemical signals, important parts of a person's mind may be unrecoverable by such a device. Finally, there is also the possibility that people could think in a way that is not understood by the machine to protect their privacy or deliberately change their brains through mental exercise or more sophisticated ways. This would make a BioFusion type device unsuitable as an interrogation tool since individuals could deliberately think in a different language or in an erratic pattern to fool the device. The obstacles to true mind-reading have also been stated in the Royal Society report: 'The technology is not advanced enough to distinguish the subtle differences between the vast numbers of brain states... There are very limited prospects for a universal thought reading machine' (Royal Society, 2012: 16). At the same time, in a few decades when sufficient computing power and better brain measuring instruments will be available this technology could come to fruition.

4.5 Conclusion

This chapter has given an overview of neuroscience applications in the field of intelligence and prediction in four areas: (1) strategic intelligence, (2) intelligence analysis and decision-making, (3) neuroscience and behavioural research and (4) threat detection, identification and interrogation. Neuroscience can contribute to strategic intelligence by improving the understanding of foreign cultures, which can help avoiding political and strategic miscalculations. Cultural neuroscience claims that people from different cultures have different cognitive habits and process information differently. This ties in with intelligence analysis and decision-making. Since most intelligence failures are failures of analysis due to cognitive biases, a lot could be achieved by better understanding the causes and dynamics of cognitive biases, which will help to avoid them. Intelligence analysis can be also improved through the development of neuro-cyber systems that assist analysts and that can learn from analysts, which will eventually lead to machines that can replicate human sensemaking and reasoning. Furthermore, neuroscience is contributing much to the behavioural sciences in view of understanding individual behaviour and group dynamics. Very powerful predictive technology is being developed that can be used for predicting social upheaval or for wargaming PSYOPS, some of which can be assumed to already exist. Finally, there are substantial investments by the US government to develop neuroscience-based technologies that can assist in threat detection, biometric identification and interrogation. The problem of identifying individuals with malintent is a complex one and critics are concerned that there would be far too many false positives and still no guarantee of no false negatives. Similarly, neuroscience-based technology for deception detection does not yet seem ready for a national security application, but this could change in a few years when the technology for 'mind-reading' will be more advanced.

5   The degradation technologies I

Drugs and bugs

This chapter will discuss neuroscience and biotechnology-based biochemical and biological approaches that can be used for degrading human performance, or in other words, neuroweapons. As suggested in the introduction, neuroweapons are weapons that specifically target the brain or the central nervous system in order to affect the targeted person's mental state, mental capacity and ultimately the person's behaviour in a specific and predictable way. Such weapons fall broadly into the 'nonlethal' category, although it has to be stressed here that some of them could produce permanent and severe damage to people. In principle, there are four different approaches for creating weapons that can produce such affects: (1) *drugs*, or chemical and biochemical incapacitants, (2) *bugs*, or biological incapacitants (microorganisms and biological toxins), (3) *waves*, or directed energy weapons and (4) *bytes*, or cyber and informational weapons (the latter two types will be discussed in the next chapter). These weapons can have mild or merely incapacitating effects, in-between behavioural effects (e.g. by inducing emotions) or may have extreme effects (mental coercion) (Sirén, 2013, 86). Publicly acknowledged are only NLW with mild or incapacitating effects. However, there are many reasons to believe that more specific behavioural effects using biochemical or biological agents may be possible in the future, as will be shown below. It will be argued that the use of chemical and biochemical agents for affecting behaviour is currently the most promising approach for a neuroweapon and there is a good amount of scientific literature that indicates a lot of potential in this area (Pearson et al., 2007). Currently less understood are the biological and genetic roots of behaviour. Nevertheless, new bioweapons could be engineered to have some targeted behavioural effects (Moreno, 2012: 200–201).

5.1 Drugs

The type of neuroweapon that is currently most discussed in the academic literature are chemical and biochemical incapacitants, which can sedate, disorient, or produce other fairly specific behavioural effects. Biochemistry is the chemistry of life and deals with biologically active chemicals of which bioregulators that control specific biological functions are a subset (Pearson,

2006: 153). Neuroscience is increasingly uncovering the link between brain and behaviour, especially with respect to the chemical processes within the brain that are responsible for memory, emotion, perception, mental state and behaviour. It is understood that this new knowledge of the brain, e.g. how neurons communicate via biochemicals and how neuroreceptors work, will lead to new treatments for mental disorders, but could also be abused or weaponized (Dando, 2005: 18; Royal Society, 2012: 43). The new biochemical weapons that are believed to be under development in many countries would effectively bridge the distinction between the chemical and the biological, as they are chemical compounds that are produced by biological processes (Pearson, 2006: 153). They are, therefore, covered by both the CWC and the BWC (Royal Society, 2012: 8). Nevertheless, the American Joint Nonlethal Weapons Directorate lists biochemical calmatives as number six on its priority list for NLW development and the National Institute of Justice is sponsoring research at Pennsylvania State University to 'explore the potential of operationalizing calmatives and to examine possible pharmaceuticals, technologies, and legal issues' (Davison, 2009: 81; 125). Calmatives were defined in a US Air Force research paper as 'biotechnical agents which are sedatives or sleep-inducing drugs, [including] alfentanil, fentanyls, ketamine, and BZ' (Bunker, 1996: 10). In addition to calmatives, there are also hallucinogens that, instead of inducing relaxation and sleep, could severely disorient and confuse people to the point of causing psychotic behaviour. There are also chemicals that induce hypnotic states and cause hypersuggestability, which could be useful for interrogation and mental coercion. Finally, there are other relevant bioregulators that can induce trust or manipulate memory, which may have some intelligence or military applications.

According to Alan Pearson, incapacitating agents should have the following characteristics to be operationally useful:

1. 'Be highly potent (micrograms/kilogram [m/kg] body weight, or less)
2. Have a rapid onset (minutes)
3. Have a defined, short (minutes-to-hours) duration
4. Have effects that are reversible
5. Be stable in storage and delivery
6. Have a significant and predictable effect(s) at a given dose or dose range
7. Be capable of rapid, often covert, dissemination in defined, controllable, and appropriate amounts
8. Have a high safety margin' (Pearson, 2006: 159).

It is technologically very difficult to create incapacitants with these charatceristics, which means that in practice, chemical agents will not be useful in many operational settings, especially when multiple people are targeted at once. However, there could be niche applications such as hostage liberation, interrogation and terrorist use.

5.1.1 Calmatives

Calmatives are drugs that have a calming effect and that can make people sleepy or distract/ disorient them by depressing the CNS. Many drugs would fall into this loose category of calmatives, which is more of a description of intent rather than a medical term. According to a report on calmatives by Pennsylvania State University, '[p]harmaceutical agents considered under the topic of "calmatives" include compounds known to depress or inhibit the function of the central nervous system', such as: 'sedative-hypnotic agents, anesthetic agents, skeletal muscle relaxants, opioid analgesics, anxiolytics, antipsychotics, antidepressants and selected drugs of abuse' (Lakoski et al., 2000: 2).

Calmatives could be used in hostage situations for incapacitating the hostage-takers without harming the hostages, in civil unrest situations to calm an angry mob, or for the management of prisoners. The report is generally enthusiastic about new chemical capabilities that would be useful in the War on Terror, suggesting that 'drugs can be tailored to be highly selective and specific for known receptor (protein) targets in the nervous system with unique profiles of biological effects on consciousness, motor activity and psychiatric impact' (Lakoski et al., 2000: 3).

Calmatives typically interact with neurotransmitters in the brain, which transmit messages between nerve cells (dopamine) and that control cognition, emotion and motor functions. By targeting specific neurotransmitter receptors such as the dopamine D-3 receptors, specific effects on the cognitive system can be achieved. For example, Serotonin Reuptake Inhibitors (SSRI) block the reuptake of serotonin in the brain and lead to increased serotonin production, which has a calming effect and enables better control over behaviour (Lakoski et al., 2003: 26). SSRI are widely used as anti-depressants and can be also used for the management of behavioural disorders.

A class of drugs with calmative qualities, which has received most attention because of the Moscow theatre siege, are opioids. They are non-opium-based narcotics with strong analgesic properties that are typically used for pain management, but that may also produce other effects, including sedation, unconsciousness, hypotension, bradycardia, muscle rigidity and severe respiratory depression (Pearson, 2007: 72). Although the exact compound used by the Russians remains a state secret, it is believed to have been a derivative of fentanyl (Crowley, 2009: 68). According to the CDC, fentanyl is a rapid-acting odourless opioid that is 80 times more potent than morphine (CDC 2011). But it still took 30 minutes from the release of the gas to storming the theatre by Russian Special Forces (Crowley, 2009: 68), indicating that the onset of the effect takes a considerable amount of time.

Various analgesics and anaesthetics drugs may be suitable as biochemical incapacitants, including barbiturates, neuroleptics and benzodiazepines. Suggested delivery mechanisms for calmatives are 'application to drinking water, topical administration to the skin, an aerosol spray inhalation route, or a drug-filled rubber bullet' (Lakoski et al, 2000: 10). In a military context, they

Table 5.1 Calmatives

Class of drug	Mechanism of action	Effects	Risks
Benzodiazepines	Stimulates GABA receptors in the brain	Can have rapid onset; sedates; decreases arousal; decreases reaction time; reduces anxiety; produces anaesthetic effect; produces motoric lethargy; produces amnesic effects	Respiratory and cardiovascular depression; memory loss; can induce mania, anger and hostility in some people
Alpha₂ Adrenergic Receptor Agonists	Stimulates α-adrenergic receptors	Sedates; reduces heart rate; relaxes blood vessels; produces anaesthetic effects	Increasing blood pressure; anxiety; dizziness; headache; nausea
Dopamine D3 Receptor Agonists	Blocks dopamine receptors that regulate communication between nerve cells	Antipsychotics; reduces agitation; sedates; blocks convulsant and lethal effects of cocaine	High blood pressure; dizziness; drowsiness; heart rate changes
Selective Serotonin Reuptake Inhibitors	Blocks one type of receptor of serotonin, specifically the serotonin reuptake sites	Antidepressants; treats obsessive compulsive disorders; enables greater control of behaviour	Requires repeated administration for calming effect; may increase aggression; suicidal thoughts
Serotonin 5-HT₁ₐ Receptor Agonists	Blocks one type of receptor of serotonin associated with behavioural and physiological functions	Relieves anxiety without sedation	Tachycardia; may cause palpitations, nervousness and gastrointestinal distress
Opioid Receptors and Mu Agonists	Inhibits pain neurotransmission	Analgesic for treating severe pain; produces euphoria, indifference to the surroundings, sedation, depressed respiration	Potentially fatal respiratory depression; addictiveness; miosis
Neurolept Anaesthetics (propofol)	Inhibits the GABA receptors by stimulating the chloride channel	Hypnotic agent that produces rapid induction of anaesthesia; produces conscious sedation; minimal side effects	Low blood pressure; change of heart rate
Corticotropin-releasing Factor Receptor Antagonists	Blocks receptors for stress induced hormones	Sedates; treats depression and affective disorders	Experimental; no human testing yet
Cholecystokinin B Receptor Antagonists	Blocks cholecystokinin (CCK) receptors	Cholecystokinin is responsible for panic attacks – blocking it reduces anxiety and thereby, behaviours motivated by anxiety	Stage of clinical trials

Source: Compiled from Joan M. Lakoski, W. Bosseau Murray & John M. Kenny (2000), 'The Advantages and Limitations of Calmatives for Use as a Non-Lethal Technique', Pennsylvania State University, pp. 16–45.

could be delivered by mortar or artillery shell, or by a grenade launcher as an airburst munition, some of which was at an advanced stage of development in the 1990s (Pearson, 2007: 80–87). The JNLWD apparently also considered 'microencapsulation' of psychoactive drugs that could be binary systems and be delivered by shot gun, airburst munitions and UAVs (Davison, 2009: 133).

5.1.2 Hallucinogens

As discussed in the first chapter, the US Army was pursuing the concept of 'psychochemical warfare' in the 1950s up to the mid-1970s, during which they were looking for incapacitants that were divided into 'on the floor agents' (calmatives) and 'off the rocker agents' (hallucinogens) (Davison, 2009: 107). In the end, the Army considered calmatives less safe and began to actually develop nonlethal CW along the lines of hallucinogens.

In the 1950s, there was a strong interest of the CIA and the US Army to develop LSD as a NLW since even very small dosages would have a very strong effect. According to the original paper by L. Wilson Greene, '[t]he symptoms which are considered to be of value in strategic and tactical operations include the following: fits or seizures, dizziness, fear, panic, hysteria, hallucinations, migraine, delirium, extreme depression, notions of hopelessness, lack of initiative to do even simple things, suicidal mania' (Khatchadourian, 2012). LSD seemed to have some of the desirable properties, but plans for weaponizing the drug were given up because of 'its high production costs and side effects' (Davison, 2008: 108).

A different agent that peaked the military's interest was the hallucinogen 3-quinuclidinyl benzilate (BZ), which was discovered in the late 1950s and considered for use in Vietnam in the 1960s. It is even more potent than LSD and also cheaper to produce.

> BZ would inhibit the production of a chemical substance that facilitates the transfer of messages along the nerve endings, thereby disrupting a person's perceptual patterns. The effects would generally last three days, although some symptoms – headaches, giddiness, disorientation, auditory and visual hallucinations and maniacal behavior – have been known to persist for several weeks.
>
> (Lee and Shlain, 1982: 20)

By 1962, the US Army was producing 750 lb M 43 cluster bombs and 755 lb cluster generator ammunitions containing BZ between 1962 and 1965, although only 1,500 BZ munitions were stockpiled (Davison, 2009: 108). Overall 50 tons of BZ were produced, which was enough to incapacitate every person living on earth in the 1960s (Lee and Shlaim, 1982: 22). The US Army was apparently mostly interested in the calming effect of BZ, but found out that sometimes, rather than calming an enemy, it could turn them just as well into raving lunatics (Lee and Shlaim, 1982). Pearson suggested that the unpredictability of BZ was due to its characteristic of affecting five subtypes of muscarinic acetylcholine receptors, all of which would regulate different functions in the brain (Pearson, 2006: 159).

In addition to unpredictability, BZ has numerous other characteristics that makes it little attractive as an incapacitant: onset can take eight hours, its dissemination generates a visible cloud, eliminating surprise, and it has a high lethality rate (Pearson, 2006: 156). By 1976 BZ was declared obsolete by the US Army and its destruction began in the 1980s (Davison, 2009: 36). No other psychochemical incapacitant replaced BZ, at least officially. By now BZ is almost forgotten and is little more than a footnote in today's literature on chemical incapacitants.

Far less is known about Soviet research into psychoactive substances that may be useful in psychochemical warfare. Soviet defector Ken Alibek mentioned in his book *Biohazard* that the Soviet Union was interested in the development of behaviour-modifying drugs in a domestic context and that there was an active program code-named *Flute* (Alibekov and Handelman, 1999: 171–172). Another program revealed by the Czech defector General Jan Sejna was called *People's Friendship* and 'aimed at large-scale drug trafficking with the purpose to destroy the moral coherence of western societies at their own expenses' (Rosza, 2009: 173). Recent research has uncovered a 'drug weapons' program of the Hungarian People's Republic, which existed between 1962 and 1972 (Rosza, 2009: 174–175). The hallucinogen methylamphetamine was used as an experimental model for a psychochemical weapon. The program was apparently closed down because of lacking financial and intellectual resources, which made success unlikely (Rosza, 2009: 175).

5.1.3   Hypnosis-inducing incapacitants

Apart from calming and disorienting effects of biochemicals, the CIA and the US military wanted a drug that put people into a very suggestible state of mind that would make it easy to manipulate them psychologically. Under the influence of the drug, enemy soldiers could be made to surrender or to give up secret information during an interrogation. The CIA investigated a range of 'truth drugs' in the 1940s and 1950s of which barbiturates (sodium amytal, sodium pentothal, seconal) and scopolamine were considered most promising at the time (Bimmerle, 1993). Barbiturates are sedatives that can render people unconscious and at lower doses make people lose control of themselves.

Similar in some respects is the hallucinogen scopolamine that has been researched intensively since the 1940s by the US security establishment. It is an alkaloid drug, which is chemically related to BZ (Khatchadourian, 2012). The drug is derived from the burrachero tree that grows in Colombia, where it is also known as *burrundanga*. In modern medicine scopolamine is used for treating motion sickness and gastrointestinal disorders. A CIA publication comments on scopolamine: 'Among the most disabling of the side effects are hallucinations, disturbed perception, somnolence, and physiological phenomena such as headache, rapid heart, and blurred vision, which distract the subject from the central purpose of the interview. Furthermore, the physical action is long, far outlasting the psychological effects' (Bimmerle, 1993).

Although unsuitable as a truth drug, scopolamine has some pretty remarkable effects that can make it a potential psychochemical weapon. According to an article by Alfredo Ardila and Carlos Moreno, '[t]he clinical picture of intoxication with Burundanga is marked by two main characteristics: (1) the severe anterograde amnesia, and (2) the submissive behavior of the subject' (Ardila and Moreno, 1991: 241). The drug prevents the formation of memory by blocking the neurotransmitter acetylcholine, effectively causing amnesia (Ardila and Moreno, 1991: 238). Secondly, it causes the 'submissive and obedient behavior of the victim' to the point that the 'person follows any command, does not present any resistance, and offers money and belongings to the offender' (Ardila and Moreno, 1991: 242).

In recent years, a lot has been written about the alleged property of scopolamine to turn people highly suggestible 'zombies', which is exploited by criminals across Latin America with an estimated number of 50,000 incidents per year (US DoS, 2011). One of the world's experts on scopolamine, Camilo Uribe, has claimed: 'The victim can't say no; he has no will and becomes very open to suggestion. It's like a chemical hypnotism' (quoted from De Cordiba, 1995: A1). In other words, the claim made is that the drug can disable free will and turn people into mere puppets under external control. Despite its known inadequacy as a truth drug, scopolamine and other psychotropic drugs were apparently administered to Guantanamo Bay detainees for undetermined reasons (O'Brien, 2012). Most likely the intention was to make the prisoners more compliant. However, scopolamine can produce unpredictable results. In larger doses it can cause respiratory failure and death, which makes it a drug that is difficult to weaponize in an effective way (US DoS, 2011).

5.1.4 Biochemicals that produce other effects

Bioregulators are drugs produced by the body that regulate body functions. They can directly affect human behaviour and could be suitable for weaponization, which was already discussed back in 1972 in a DIA report (US DoD, 1972: 71–72). According to Robert Tucker:

> Of particular concern is the possible development of a new generation of biochemical 'calmative' agents that would act on the central nervous system in highly specific ways. Pharmaceutical companies are currently developing new therapeutic drugs modeled on natural body chemicals called 'bioregulators,' many of them peptides, that control vital homeostatic systems such as temperature, sleep, water balance, and blood pressure. In the brain, a large class of bioregulators act on neural circuits

to modulate awareness, cognition, and mood. Based on this research, it may eventually become possible to develop modified bioregulator molecules called analogues that can cross the blood-brain barrier and induce a state of sleep, confusion, or placidity, with potential applications in law enforcement, counterterrorism, and urban warfare.

(Tucker, 2010: 20–21)

It has been established that humans emotions are biochemically encoded and can be therefore biochemically manipulated. A particular bioregulator that has been discussed for military use is the neurohormone oxytocin, which is linked to sexuality and is released after child birth to strengthen the bonding of mother and child. Oxytocin acts both as a hormone in the bloodstream and as a neurotransmitter in the brain and is responsible for promoting prosocial behaviour (Zik and Roberts, 2014: 31). In particular, it is believed that oxytocin creates trust among people, which could be exploited in negotiations, interrogations, or for making psychological operations more effective. Jonathan Moreno writes: 'Exposing enemy troops to oxytocin could undermine their ability to maintain the "us versus them" mentality underlying the capacity to kill, ultimately enhancing the possibility of surrender without bloodshed' (Moreno, 2012: 90). The bioregulator would decrease the desire of the enemy to fight and thus make them more amenable to a negotiated settlement.

On the other hand, oxytocin has been also described as a social boundary regulator, which also has the defensive function of protecting the own in-group against harm (Zig and Roberts, 2014: 32). The military use of oxytocin may therefore not be without the peril that it could actually strengthen the determination of a group to resist. Oxytocin is already commercially marketed as an aphrodisiac under the brand name 'liquid trust'.

Previous research in this area aimed at exploiting pheromones, which are biochemicals that are secreted for social signalling such as readiness for sex (Cutler, 2000: 65). The US military thought of exploiting the behavioural effects of pheromones by developing a 'gay bomb', which would make enemy soldiers 'sexually irresistible' to each other, thereby undermining morale, but the concept was never funded (BBC, 2005).

5.1.5 Problems with the concept of incapacitating chemicals

The CWC of 1992 (in effect 1997) made the use of any chemicals as 'a method of warfare' illegal regardless of whether the intention is lethality or nonlethality (Pearson, 2007: 106). The CWC comprehensively forbids the development, production, stockpiling and use of CW. In particular, the Convention prohibits any 'toxic chemical', which is defined as: 'Any chemical which through its chemical action on life processes can cause death, temporary incapacitation or permanent harm to humans and animals' (Art. II.2). This would suggest that new incapacitating CW would fall within the scope of the CWC. However, the CWC exempts the use of 'toxic chemicals' for 'purposes not prohibited under this convention' such as '[l]aw enforcement including riot control purposes' (quoted from Koplow, 2006: 37). The use of riot control agents (RCA) that only produce a temporary irritant effect on eyes, nose and skin, used in domestic law enforcement is legal. Other nonlethal chemicals that are considered for law enforcement are calmatives and malodorants (Pearson, 2007: 109).

Since the threat of terrorism in particular has eroded the clear distinction between warfare and law enforcement, it creates a loophole that may allow the military to use nonlethal CW in specific situations, especially since new biochemicals are different from RCA and are not clearly covered under the CWC. According to the Royal Society report, this would create legal uncertainty over what chemical incapacitants may be used in what situations. The report expressed the concern that 'law enforcement purposes could potentially provide camouflage for military interest in incapacitating chemical agents or other chemical weapons', effectively eroding the chemical weapons taboo (Royal Society, 2012: 10). Furthermore, the use of biochemical incapacitants also conflicts with the BWC, which comprehensively prohibits 'microbial and other biological agents, or toxins whatever their origin or method of production, of types and in quantities that have no justification for prophylactic, protective or other peaceful purposes' (Article 1). As pointed out by the Royal Society, bioregulators would be also covered by the BWC (Royal Society, 2012: 59).

Apart from the legal issues related to biochemical incapacitants, there are substantial obstacles in terms of practicality that make them difficult to use in operational settings. The biggest obstacle is the 'dose response problem', which means that any drug can have a range of different effects based on the dosage that an individual receives. According to the UK Defence Advisory Board, '...any chemical agent, a small dose of which is capable of profound disturbance of bodily or mental function, is certain to be able to cause death in large dose ... and no attack with a chemical warfare agent is likely to be designed with the primary objective of avoiding overhitting' (quoted from Pearson, 2006: 156). This means that biochemical incapacitants can easily kill if the dosage that any individual receives in an attack cannot be precisely controlled. For example, during the Moscow theatre siege of October 2002 the FSB used the opioid fentanyl to render hostage takers and hostages unconscious before the theatre was stormed. The gas also accidentally killed 127 hostages (out of over eight hundred) because of a delayed and wrong medical emergency response and altogether 650 people had to be hospitalized (Pearson, 2007: 155).

The question then becomes: what is an acceptable lethality rate for a biochemical incapacitant? The 'nonlethal' fentanyl killed around 16 per cent in Moscow, which is more than twice the lethality rate of the 'lethal' chemicals used during the First World War (Fidler, 2005: 532–533). The Russians have not registered fentanyl as required under the CWC and it was arguably used as a 'method of warfare' against armed rebels (Koplow, 2006: 108–109). It is therefore notable that following the incident there was no international outcry regarding this potential breach of the CWC, which may suggest that governments considered the use of fentanyl in this case to be lawful.

Michael Crowley, the coordinator of the University of Bradford non-lethal weapon research project, has warned that future advances in the field of biochemicals may allow for more targeted effects with respect to altering human behaviour:

> Incapacitant research, if allowed to continue unchecked and in secret, may lead to even more dangerous developments. If the revolutionary advances in genomics, biotechnology and neuroscience are put to use by weapons designers then tomorrow's arsenals may well hold weapons that alter your thoughts and emotions, weaken your immune system, target your sight or leave you sterile.
>
> (Crowley, 2009b)

Biochemical incapacitants may be used in insidious ways such by introducing them to a population through the water and food supply. Authoritarian regimes may forcibly medicate their citizens to make them more compliant. Biochemical drugs could be administered via dart guns or by an injector equipped micro UAVs, which could apply a precise dose and defeat protective suits (Hauck, 2005: 253). As a result, chemical weapons will continue to have their niche in future warfare. It will be possible to use them discriminately and in a nonlethal manner, which will make governments that develop such weapons, likely push for a modification of the CWC to accommodate nonlethal chemical options.

5.2 Bugs

There have been concerns about the possibility of developing neuroweapons as bioweapons. These new envisioned new bioweapons could specifically attack the CNS and brain and may have the capability not only to incapacitate, but to produce more specific behavioural effects. Biological weapons (BW) are biological organisms or biological toxins that are intended to injure, damage, or kill humans, animals and plants. They can be grouped into viruses, bacteria and fungi (Croody et al., 2002: 196). Viruses are more contagious than bacteria and they replicate by inserting their own DNA into a cell so that it will produce copies of the virus, while bacteria are larger microorganisms that can survive outside of a host. Fungi are typically used for producing biological toxins such as ricin or botulinum toxin that kill in concentrations as small as 0.1 micrograms (Croody et al., 2002: 215).

A biological weapon consists of a method for isolating/ growing/ preserving a disease agent, a vector for effectively disseminating the agent and a treatment, vaccine, or antidote that protects own personnel (van Aken and

Hammond, 2003). Like CW, many BW tend to incapacitate rather than kill with the exception of a few highly lethal agents such as anthrax, smallpox, hemorrhagic fevers and some others. Although biological weapons have been comprehensively outlawed by the BWC in 1972, concerns remain about bio-terrorism by state and nonstate actors (US NSC, 2009: 2). Next generation bioweapons are likely going to be very different from the lethal bioweapons threat of the twentieth century.

They could be genetic bioweapons that are genetically targetable or that are genetically altered pathogens to produce very specific effects and may be engineered in a way to be not recognizable to expert communities (Petro et al., 2003: 164). This section will review some known incapacitating BW and will explore the possibility of pathogens that directly target the brain and alter human behaviour.

5.2.1 *Biological warfare*

Some general discussion of biological warfare is necessary to illustrate its nature before the new 'advanced biological warfare' threat can be discussed. In practice, biological weapons are quite difficult to develop and to employ, which is the main reason why they have been rarely used in warfare. Even the superpowers of the Cold War, despite working for decades and spending billions of dollars on their development, were not able to produce many bioweapons that could be actually operationally used (Ben Ouargham-Gormley, 2013: 482–483).

Famous historical cases of biological warfare include the siege of the Black Sea city Kaffa in 1347–48, where the Mongols catapulted plague-infected corpses over the city walls (Croody et al., 2002: 219); during the First World War German agents infected horses and cattle with glanders and anthrax to disrupt allied supply lines (Croody et al., 2002: 222); biological weapons (BW) have been used on a large scale by Japanese forces by the infamous Unit 731 in China, which killed at least 270,000 Chinese soldiers and civilians (Croody et al., 2002: 225); the Japanese attacked at least 11 Chinese cities over which they released over 15 million fleas infected with plague to create an epidemic (Christopher et al., 1997: 413).

During the Second World War, Britain and the US were also actively researching BW in view of using them against the Axis powers. When the war was over, the US military continued BW research at Fort Detrick (at that time: Camp Detrick) with open-air testing that was conducted at the Dugway Proving Grounds in Utah (Mangold and Goldberg, 1999: 39). The US Army was considering various pathogens that could be released on the enemy, among them anthrax, smallpox, hemorrhagic fevers, brucellosis, tularemia, the plague, Venezuelan equine encephalitis (VEE), cholera, botulinum and dysentery (Mangold and Goldberg, 1999: 34–35).

From 1951 to 1969, the US military perfected the delivery of bioweapons for seven standardized biological weapons that could be used in the event of war (Croody et al., 2002: 232). In a surprising move, President Nixon ordered all offensive BW research to be stopped and all bioweapons stockpiles to be destroyed in 1969, which created the foundation for the BWC of 1972. The convention states: 'Each State Party to this Convention undertakes never in any circumstances to develop, produce, stockpile or otherwise acquire or retain…Microbial or other biological agents, or toxins whatever their origin or method of production, of types and in quantities that have no justification for prophylactic, protective of other peaceful purposes' (quoted from Koplow, 2006: 42). This comprehensively outlaws the use of any kind of biological agents or their toxic residues for anything other than peaceful purposes.

The Soviet Union violated the BWC on a massive scale by hiding its biowarfare research within a civilian medical research program called *Biopreparat*, which employed up to 60,000 workers at 18 biowarfare labs and facilities across the country funded at $1 billion a year (Ainscough, 2002: 3). They worked on weaponizing anthrax, tularemia, plague, smallpox, glanders, Venezuelan equine encephalitis (VEE) and botulinum toxin (Croody et al., 2002: 234–235). The Soviet Union stockpiled tons of bioweapons (20 tons alone of plague) and developed plans for integrating biowarfare attacks into their overall war plans, which included plans for ICBMs and strategic bombers loaded with anthrax and smallpox (Ainscough, 2002: 4). The Soviets even kept bioweapons in their embassies around the world and planned to spread them by human agents in the event of war or for biological terrorism and sabotage (Kouzminov, 2005: 32).

The West was largely unaware of these schemes until Soviet microbiologist Vladimir Pasechnik defected to England in 1989. Pasechnik confirmed that the USSR was conducting offensive biowarfare research and described these efforts in great detail, including the BW-carrying ICBMs for 'deep targets' destined for London, New York, Washington, Chicago, Seattle and Los Angeles (Mangold and Goldberg, 1999: 95). Ken Alibek, the deputy director of *Biopreparat*, who defected to the US in 1992, revealed many more details about the grand scale of the Soviet biowarfare efforts. By 1990, the Soviets had weaponized no fewer than 52 diseases and had genetically engineered smallpox for greater virulence and resistance to antibiotics (Ainscough, 2002: 6). The biowarfare research was so advanced that the Soviet bioweaponeers were even creating entirely new life forms through genetic engineering that combined genes from one virus into another to make a pathogen more lethal (Ainscough, 2002: 6–7). President Yeltsin closed down the program in 1992, but Alibek has argued that the biowarfare research continues in today's Russia (Croody et al., 2002: 236).

5.2.2 Incapacitating bioweapons

Many biological weapons have a relatively low lethality rate, which means that their overall effect is one of mass incapacitation rather than mass killing.

Even the Bubonic plague, which was used by Japanese germ warriors, only has a lethality rate of 50 per cent (Croody et al., 2002: 207). Very much like CW, the distinction between a lethal and a nonlethal or incapacitating bioweapon is not absolute (Beckett, 1983: 105). However, both the US and the Soviet bioweapons programs aimed to specifically develop bioweapons that had a very low lethality rate for various reasons.

The US military wanted nonlethal microbial organisms, mostly for biowarfare simulations that can be conducted safely in civilian areas (Cole, 1988: 45-45). They also considered nonlethal bioweapons for use against enemy populations. Neil Davison has found documents that explain the rationale behind the US military's interest in incapacitating BW. One document states:

> The biological agents, while having much of the versatility of chemicals, lack a rapid onset of effect. Their tactical incisiveness is severely limited so they are less applicable to the class of conflict discussed in this paper [limited and urban warfare]. They may, however, have a substantial application in capturing and neutralizing hostile cities at highly intense levels of limited warfare.
>
> (Davison, 2006: 17)

Davison quotes from a 1973 SIPRI study on NLW that suggests that a desired low mortality rate of 1–2 per cent would allow for greater flexibility, as such bioweapons could be used in situations where civilians and own personnel might be exposed. Mass incapacitation would overwhelm emergency and logistical services (Davison, 2006: 17).

According to Ken Alibek, the Soviets intended to release nonlethal 'operational bioweapons' about 100 km to 150 km behind enemy lines to affect 'rear services and reinforcements. These agents, such as tularemia, brucellosis, glanders, and Venezuelan equine encephalomyelitis [VEE], would not generally kill soldiers, but would incapacitate them and thereby make it easier to destroy an enemy's defenses' (Alibek, 1999: 2). It is possible that the Soviets operationally tested glanders during the 1980s Afghanistan war on a small scale (Alibekov, 1999: 5).

Brucellosis was seen as particularly suitable as a nonlethal bioweapon. According to information from the Federation of American Scientists, brucellosis has the following symptoms, including 'fever, sweats, anorexia, fatigue, malaise, weight loss, and depression. Localized complications may involve the cardiovascular, gastrointestinal, genitourinary, hepatobiliary, osteoarticular, pulmonary and nervous systems. Without adequate and prompt antibiotic treatment, some patients develop a chronic brucellosis syndrome with many features of the chronic fatigue syndrome' (FAS, 2007). The disease agent can be delivered as an aerosol and the fatality rate of brucellosis is less than 0.5 per cent. In 1954, the US Army filled 500 pound cluster bombs with brucellosis at the Pine Bluff Arsenal in Arkansas (G. Thomas, 2008: 106–107).

VEE is another incapacitating bioweapon that was developed by both the US and the Soviet biowarfare program. In nature it is transmitted by mosquitoes, but other disease vectors, including sexual transmission, are possible. 'Symptoms are exhibited in nearly 100 percent of those infected, including inflammation of brain meninges. After a 1 – to – 5 – day incubation period, symptoms include severe headache and fever along with nausea, vomiting and diarrhea in some cases' (Croody et al., 2002: 212). It causes fever, lethargy and fatigue that lasts up to two weeks with no cure and has only a very low lethality rate of 1 per cent (Mangold and Goldberg, 1999: 386).

A contagious disease that attacks CNS and brain and that has yet to be weaponized is Lyme disease, which was first discovered in 1975 after an outbreak in Old Lyme, Connecticut. It has been rumoured to have been accidentally released from the Plum Island Animal Research Center near Long Island, New York, which is just eight miles away from Old Lyme. The bacteria *burrelia burgdorferi*, which causes the disease, was possibly contracted by birds and deer that spread it to the human population across the water to Connecticut and Long Island (Carroll, 2005: 26). The disease is

> [c]haracterized by symptoms such as facial paralysis and stiff swelling in the neck and joints, Bb also causes maladies like meningitis and encephilities – both swellings of the brain – and cardiac problems, including atrioventricular block, myopericarditis, and cardiomegaly. Because it attacks the body's central nervous system, additional symptoms of Lyme disease include acute headaches, general fatigue, fever, moodiness, and depression.
>
> (Carroll, 2005: 5)

It is a disease that infects more than 300,000 people a year in the US, using ticks as a vector with currently no vaccine or cure available. If left untreated Lyme disease is slowly debilitating victims without directly killing them. Since it is not a known bioweapons agent a deliberate biowarfare attack with a modified Lyme disease might thus go unnoticed, thereby putting a larger number of people at risk.

5.2.3 Next generation bioweapons

The bioweapons research undertaken during the Cold War has clearly resulted in the development of some very dangerous pathogens. But as dangerous as weaponized anthrax and smallpox may be, they can hardly compare to the much more sophisticated future bioweapons that are enabled by today's biotechnology. James Petro from the DIA has argued that advances in biological research will enable a new class of biological weapons that will 'elicit novel effects' (Petro et al., 2003: 161). 'Emerging biotechnologies likely will lead to a paradigm shift in BW agent development; future biological agents could be rationally engineered to target specific human biological systems at the molecular level' (Petro et al., 2003: 161–162). Furthermore, they suggest that pathogens can be made more dangerous with recombinant DNA technology; an organism's genetic makeup could be changed for 'antibiotic resistance, increase aerosol stability, or heightened pathogenesis' (Petro et al., 2003: 162).

In the 1990s, after the revelations of the Soviet biowarfare defectors, DARPA was investigating the future bioweapons threat (Jacobson, 2015: 295). In 1997, they commissioned the defence advisory group Jason to produce a still classified report on the future threat of bioweapons that predicted several new types of bioweapons (Ainscough, 2002: 17):

- *Binary biological weapons* are analogue to binary chemicals: they consist of two innocuous parts that when mixed form a pathogen. This would make the handling of the BW agent safer. 'To produce a binary weapon, a host bacteria and a virulent plasmid could be independently isolated and produced in the required quantities. Just before the bioweapon was deployed, the two components would be mixed together' (Ainscough, 2002: 18).
- *Designer genes* refers to the creation of pathogens at a molecular level to enhance desirable features such as resistance to antibiotics. Different viruses could be genetically combined to create 'chimeras' that are more potent. For example, a new strain of influenza could be created by hybridizing an influenza strain with another virus (Ainscough, 2002: 19).
- *Gene therapy* could be weaponized in the sense that pathogenic genes could be inserted into human DNA or specific gene expression could be manipulated in some nefarious way, e.g. triggering cancer. It might be possible to deliver DNA in a 'microbullet' to 'cause disease or injury by controlling genes' (Guo, 2005: 77)
- *Stealth viruses* could covertly infect a population and remain dormant until they are triggered by some internal or external condition, e.g. sun burns or high amounts of stress. '[A]n unwitting population could be "slowly pre-infected with a stealth virus over an extended period, possibly years, and then synchronously triggered"' (Jacobson, 2015: 298). For example, by sun burns or other environmental conditions. This could be used for blackmailing individuals or states (Ainscough, 2002: 21).
- *Host-swapping diseases* are generally animal diseases that are engineered to infect humans. An example of a disease that 'jumped' species is HIV that was first present in chimpanzees and then spread to humans (Ainscough, 2002: 22). Rabies could be engineered to infect humans to drive them crazy.
- *Designer diseases* would be enabled by futuristic biotechnology that makes it possible to determine the desirable characteristics of a pathogen and then create it from scratch. 'Our understanding of cellular and molecular biology has advanced nearly to the point where it might be possible to propose the symptoms of a hypothetical disease and then design or create the pathogen to produce the desired disease complex' (Ainscough, 2002: 18–22).

5.2.4 *'Gene drives' and transgenic animals as bioweapons*

Scientists have been working on genetically modified mosquitoes for many years for the purpose of suppressing mosquitoes-borne diseases such as malaria and dengue fever. The goal is typically to suppress the problematic species of mosquitoes and then replace it with a new GMO species. The GM species *Aedes Aegypti*, which was engineered to have a shortened life expectancy, was released in field trials in Malaysia in 2010 and Brazil in 2011, resulting in an 85 per cent reduction in *Aedes Aegypti* population density in the test areas (Araujo et al., 2015: 584). In 2015, six million transgenic GM mosquitoes have been released in a Brazilian city. The mosquitoes were created with the help of germ-line transformation, where transgenes are introduced in mosquito embryos through microinjection (Lobo et al., 2006: 1312).

A faster and much cheaper method for producing transgenic species could be 'gene drives' that can spread a particular gene across a wild population. While a sexually reproducing organism has a 50 per cent chance to pass on a gene to the next generation, gene drive technology can increase the chance to one hundred per cent, which produces something like a genetic chain reaction across a population (Fernandez, 2015). There are certain limitations to future gene drives: they only work with sexually reproducing species and they require dozens of generations for spreading across a species population, which makes it suitable for altering short-lived species, but unsuitable for the human species (it would take hundreds of years to alter the human species with a gene drive) (Oye et al., 2015: 626).

'Potential beneficial uses of such "gene drives" include reprogramming mosquito genomes to eliminate malaria, reversing the development of pesticide and herbicide resistance, and locally eradicating invasive species' (Oye et al., 2015: 626). However, scientists have become concerned that 'gene drives' that modify wild species could also affect non-targeted species and could be used as future bioweapons. For example, David Gurwitz, a microbiologist from the University of Tel Aviv, argued: 'Just as gene drives can make mosquitoes unfit for hosting and spreading the malaria parasite, they could conceivably be designed with gene drives carrying cargo for delivering lethal bacterial toxins to humans' (Fernandez, 2015).

It has been suggested that transgenic species could be created 'to produce large quantities of bioregulatory or toxic proteins':

> Transgenic insects, such as bees, wasps, or mosquitoes, could be developed to produce and deliver protein-based biological warfare agents. By employing future discoveries related to insect ontogeny and genetic manipulation, a mosquito potentially could be genetically altered to produce and secrete a highly potent bioregulator or toxin in its saliva. The insect would then intoxicate people with the protein by inoculation during its feeding process. Because many bioregulators and toxins are thought to be effective at exceedingly low doses, an individual

may succumb to infection after having been bitten by a few transgenic mosquitoes.

(Petro et al., 2003: 163–164)

Many experts think that gene drives present an unprecedented biosecurity threat since they could enable future bioterrorists to genetically modify wild species, such as mosquitoes to infect humans with diseases or to inject them with bioregulators. As discussed above, bioregulators control important functions in the human body and brain, including mental states and behaviour. Transgenic mosquitoes could be designed to release toxins that tranquilize humans or otherwise debilitate humans, e.g. to temporarily incapacitate a population. 'Entomological warfare' (warfare with insects as weapons) would acquire an entirely new and frightening dimension.

5.2.5 Brain-targeted bioweapons

Some existing disease agents may be genetically engineered to attack the brain and CNS to debilitate the enemy or to change the enemy's behaviour in a desirable manner. Jonathan Moreno quotes a biodefence expert, who claimed: 'rapid-onset, brain-targeted biological weapons are something we need to worry about now, not some time in the future' (Moreno, 2012: 200–201). Moreno suggested:

> Inside the central nervous system, the technological surprise stems from designer peptides produced from synthetic genes that have effects quite distinct from those normally associated with the pathogen. For example, when produced in the brain, they could function as malign neuromodulators, disabling brain functions by modifying the relationships and communications between neurons. In such advanced neuroweapons, the infectious pathogen is really just a Trojan horse, selected for its ability to get the synthetic gene quickly into a target it cannot otherwise reach...Various kinds of disabling reactions, from intense fatigue or confusion to the loss of sensation, could be attempted that would neutralize enemy forces.
>
> (Moreno, 2012: 200–201)

The Soviets were apparently already looking into designer peptides and bioregulators as biological weapons. According to defector Alexander Kouzminov, 'we sought information about...human hormones, particularly peptides or, more precisely, neuropeptides (regulators of memory) which control human emotions and which could be used as biochemical weapons... its overt use would result in short-term or long-term corruption of human mental processes, and induce uncontrollable feelings of fear and panic, and could even lead to death' (Kouzminov, 2005: 83–84). It remains unclear to what extent the Soviets were successful in their efforts.

The development of behaviour-modifying microorganisms for 'mind control' is also less absurd than it initially sounds. There are numerous examples in nature of parasites that can alter the behaviour of their host to their own benefit: wasps that insert their eggs into a maggot may get infected by the *L. boulardi filamentous* virus, which can change the wasp's behaviour in a very specific way to spread itself (Yong, 2012). *Toxoplasma gondii* (T. gondii), which is a parasite contained in cat faeces, can manipulate the behaviour of rodents that come into contact with it by changing their brain. *T. gondii* is listed as a potential microbial neuroweapon in James Giordano's and Rachel Wurtzman's *Synesis* article. Its effects are described as 'loss of impulse control; agitation, confusion' (Giordanao and Wurtzman, 2011: 65).

The parasite alters the instinctive response of rats and mice to cat smells so that they lose their natural fear of the scent by altering the genes that affect how the neurons in a rat's brain process smell (Bast, 2015: 894). The disease caused by the parasite is known as toxoplasmosis and can occur in humans, who have come in contact with cat faeces. A study indicates that 'men carrying the *T. gondii* latent infection are more likely to disregard rules and to become excessively jealous and suspicious' and that 'cognitive response times become longer (responses become delayed) for those who are affected by this pathogen' (Bast, 2015: 896). Other research shows that toxoplasmosis may even cause mental illness, as well as mood and neurological disorders (Browne, 2012). With modern biotechnology it would be possible to design a highly contagious pathogen that causes mental illness and literally drives people crazy. More nuanced effects such as altering moods or causing passivity may also be possible.

5.2.6 Biological incapacitants as neuroweapons

Biological weapons have certain advantages over other kinds of weapons that make them suitable not just as WMD, but more importantly, as weapons of effective mass incapacitation, massively degrading an enemy's ability (and possibly will) to resist. They could be designed to spread from person to person or they could be delivered by biological disease vectors such as genetically engineered mosquitoes, ticks or other insects, which makes their distribution amongst a target population comparatively cheap and easy. Bioweapons could be spread covertly amongst a larger population and may remain dormant until activated by some external trigger (Jacobson, 2015: 298). Low lethality rates would be particularly advantageous for stealth attacks, as attacks could be disguised as natural epidemics and would not alert biosecurity experts in the same way as a lethal pathogen would. Bioweapons that target CNS and brain would have behavioural effects of some sort, ranging from mere mental incapacitation and lethargy to causing psychotic, abnormal, or other irrational behaviour. Bioregulators produced and transmitted by transgenic insects could cause a wide range of specific behavioural effects. Furthermore, genetic bioweapons could 'target a specific population based on genetic or cultural traits' (Petro et al., 2003: 164). Advanced delivery systems such as microbullets and micro UAVs could also result in bioweapons attacks that are highly targeted and that have very precise effects (Guo and Yang, 2005). Considering all these possibilities, it may well be that advanced biological weapons that are highly contagious and resilient to antibiotics and other treatment will become the most serious threat to global security.

5.3 Conclusion

Chemicals are the most obvious approach for the development of neuroweapons. The whole discipline of psychiatry is based on the assumption that mental states and behaviour can be regulated or controlled with pharmaceuticals. Furthermore, there is more than a century of medical research into analgesics and anaesthetics, which can calm, tranquilize and put people to sleep. The possibilities for weaponisation are fairly obvious. The US military already became interested in 'psychochemical warfare' back in the late 1940s and that interest never fully went away. However, the obstacles to creating an effective psychochemical weapon have been very substantive in the past. The biggest challenge is the so-called 'dose-response problem': in a combat situation it is difficult to control the dosage any one person attacked receives, resulting in either under- or overdosing, which means that some people may experience no effect while others may suffer lethal effects or permanent injury. The only way around this dilemma is to limit the use of psychoactive chemicals to individual targeting, e.g. through microencapsulation of the agents and targeted delivery by dart guns or micro-UAVs. The use of nonlethal biological agents is even more problematic in terms of targeting, as their spread tends to be less controllable. Incapacitating biological agents such as brucellosis and VEE have been part of the US and Soviet bioweapons arsenals during the Cold War and could be used for weakening enemy forces and populations. In the future, biotechnology will provide the tools for engineering new diseases that can have much more specific and behavioural effects. Chemical and biological weapons will increasingly overlap in the form of bioregulators that are chemical compounds created by biological processes, which could be spread by biological vectors such as mosquitoes or other insects. Regulatory mechanisms that are currently in place will need to be strengthened in order to prevent an erosion of the chemical and biological weapons taboo. Legal loopholes need to be closed and monitoring mechanisms need to be improved. Unfortunately, it seems likely that biochemical and biological incapacitants could be used for high-tech repression by authoritarian governments or for bioterrorism by 'rogue states' and by other illegitimate actors.

6   The degradation technologies II

Waves and bytes

This chapter will look at two very controversial areas of possible neuroweapons development: nonlethal directed energy weapons or 'electromagnetic mind control' and cyber or 'informational' weapons that 'hack' the human mind, either through a BCI or through perception, e.g. subliminals. A general problem with respect to reviewing these technologies, which has been acknowledged by many NLW experts, is the fact that much of this kind of NLW research is highly classified. Neil Davison's excellent book on NLW mentions 'secrecy' more than twenty times and he devotes three long paragraphs to it (Davison, 2009). The NRC report on 'Emerging Technologies' only has some small sections dealing with neuroweapons (NRC, 2008: 133). At least it acknowledges that the 'neurotechnology degradation market segment is completely underground with only speculative information available. This cognitive weapons market does exist' (NRC, 2008: 129). The Royal Society report only discussed neuropharmaceuticals and bioregulators as potential neuroweapons, mostly disregarding other approaches (Royal Society, 2012: 43–59).

A researcher of neuroweapons not only has to deal with the issue that official information is unavailable, but furthermore with the problem of outright disinformation. Steven Aftergood has pointed out that Cold War secrecy systems such as Special Access Programs (SAPs) 'authorizes defense contractors to employ "cover stories" to disguise their activities. The only condition is that "cover stories must be believable"' (Aftergood, 1994: 44). Some researchers such as Robert O. Becker have suggested that governments have purposefully disseminated disinformation to the public to muddy the waters (Binhi, 2009: XI). This claim begins to make sense if one scans through some of the popular electromagnetic mind control literature often found on the Internet, which alleges the existence of technology far beyond published science, but does so with such detail and consistency that a coordinated disinformation effort seems plausible. Aftergood has suggested with some irony: '[i]f nonlethal technology is so fragile that simply acknowledging it would negate its effectiveness, then it probably isn't worth much' (Aftergood, 1994: 45).

Taking into account the obstacle of official secrecy and the vast amount of misinformation and disinformation regarding 'mind control' weapons, the chapter will focus on the known principles and technologies, as well as known NLW systems and research. It is argued that acoustic and electromagnetic weapons can have behavioural effects, although extreme effects or actual 'mind control' seem less likely. The second section of the chapter will investigate approaches like subliminals, 'narrative networks' and 'mind viruses' that attack neuro-cyber systems, which could potentially produce anything from mild to extreme effects. The chapter concludes that NLW are already reaching some maturity, where they can be useful on the battlefield and that behavioural effects, within limitations, are certainly possible.

6.1 Waves

Since Project Pandora, which investigated possible health damage and behavioural effects of low-energy EMF in the 1960s, there has been a lot of speculation about 'psychotronic' or electromagnetic mind control weapons (Davison, 2009: 163). While very little tangible proof for the existence of such weapons has materialized after more than half a century, there is still mounting scientific evidence that electromagnetic mind control using directed-energy weapons (DEW) could be possible at some point in the future. This section will discuss DEW as potential neuroweapons in the light of weapons, technologies and scientific research that is in the public domain.

The Pentagon defines DEW as: 'A weapon or system that uses directed energy to incapacitate, damage, or destroy enemy equipment, facilities, and/or personnel' (US DoD, 2007: I-8). This includes all weapons that use energy for producing a weapons effect, such as lasers, masers, high energy radio frequency (HERF) weapons, high-powered microwaves (HPM) and acoustic weapons. The US military is already investing heavily into DEW with the DEW research budget growing from $7.72 billion in 2010 to $12.15 billion in 2014 since they are believed to be a 'game changer' (Ellis, 2015: 10). Some of these weapons are anti-personnel weapons that can be used for crowd control and for incapacitating potential attackers. Relevant with respect to neuroweapons are: (1) stun weapons that jam the CNS or brain, (2) high power RF, microwave and millimetre wave weapons that cause pain or debilitate, (3) acoustic weapons that cause pain, debilitate, or otherwise incapacitate, (4) 'voice-of-God'-weapons that may be used in PSYOPS attacks and (5) low power EMF that may be able to interfere with biochemical processes und functions in the brain and other organs.

6.1.1 Stun weapons

Stun weapons are NLW that have an immediate incapacitating effect on the target by causing the loss of control of a victim's body or by rendering the victim unconscious. To achieve this, stun weapons attack the CNS and the brain. Currently, stun weapons tend to use electricity and rely on direct contact to incapacitate an individual and are called Tasers. The Taser was first invented by Thomas A. Swift in 1970. It fires either a wired projectile for establishing direct contact, or may use a self-contained 'Taser round' (Davison, 2009: 22). A more advanced stun weapon is the myotron that through touch scrambles brain signals in the motor cortex by emitting rapid pulses of electromagnetic radiation, resulting in the victim losing control over all voluntary muscle movements for up to 30 minutes (Pasternak, 1997). Although the myotron is not a DEW, it demonstrates the potential for a stun weapon out of *Star Trek* that paralyzes a victim for an extended period of time.

According to military analyst James Dunnigan, RF signals can be used for hacking the CNS and causing a stun effect. He writes in his book: 'Radio frequency (RF) systems. These are radio transmitters that jam and short-circuit the human nervous system. This temporarily disables the people the radio beams are aimed at. Think "Phasers on stun" and you'll get the idea. A tricky bit of business, this, for it can also cause long-term effects. It works on animals; no one is admitting to human experiments' (Dunnigan, 1996: 223). Although no RF anti-personnel weapons have surfaced yet, there are stun weapons openly under development that rely on light and sound to hack a person's brain.

Since 2001, the Pentagon's Joint Non-Lethal Weapons Program (JNLWP) has been working on a distance weapon called *Personnel Halting and Stimulation Response Rifle* (PHASR) – previously known as *Portable Efficient Laser Testbed* (PELT), which was originally described as a nonlethal laser dazzler for temporarily blinding and disorienting enemies. According to Davison, the PHaSR is being designed to employ a two-wavelength laser system, one to heat the skin of the target person and the other as a 'dazzling' weapon against the eyes' (Davison, 2009: 76). A JNLWP fact sheet summarises:

> PHaSR achieves the desired degree of protection through the synergistic application of two non-lethal laser wavelengths during the course of protection activities that will deter, prevent, or mitigate an adversary's effectiveness. The laser light from PHaSR temporarily impairs aggressors by 'dazzling' them with one wavelength. The second wavelength causes a repel effect that discourages advancing aggressors.
>
> (quoted from Hambling, 2008)

The military stopped funding for PHASR in 2008 after constructing a prototype, but research on a long-distance stun weapon continues. The company Genesis Illumination patented in 2011 the StunRay light dazzler, which is ten times more powerful than an aircraft landing light and which works by overloading the neural circuits connected to the retina. The company claims that the weapon is effective up to 150 feet and can incapacitate a person for up to 20 minutes (V. Ross, 2011).

A future dazzling laser could take advantage of the so-called 'Bucha effect' in order to stun or disorient an adversary. The 'Bucha effect' was discovered in the 1950s after several helicopters crashed for unknown reasons. The rotor blades caused a flicker effect that disoriented the pilots thereby causing the crashes (Ronson, 2002: 147). An Air Force research paper on NLW describes the Bucha effect as: 'High intensity strobe lights which flash at near human brain wave frequency causing vertigo, disorientation, and vomiting' (Bunker, 1997: 16). Persons with epilepsy may suffer seizures as a result of exposure to incoherent light sources (Bulletin of Atomic Scientists, 1994).

The Soviets were apparently first to weaponize the Bucha effect. During the Vietnam War the American military recovered a so-called LIDA machine that had been used by Soviet interrogators on American PoWs (Begich, 2006: 98). The machine can produce pulsed electromagnetic 40 MHz signals, stroboscopic light, heat and auditory signals for brain entrainment. The device is rumoured to have such a strong tranquilizing effect to the point of putting people to sleep almost instantly. The captured LIDA machine was tested in the US by physicist Ross Adey, who found that it could indeed prolong sleep states or shift sleep patterns to deeper levels (Tyler, 1986: 254).

The previously cited DIA report from 1972 discusses Soviet light and sound weapons in the context of behavioural control in some detail. It states that there is 'general agreement that flicker has the potentiality of causing considerable interruption of the normal functions of the human nervous system', suggesting that flicker rates in the alpha rhythm of the EEG interfere with consciousness and that 'visual illusions appear to be produced by frequencies above 10 – 12 Hz' (US DIA, 1972: 79)

After 40 years of investigating the concept, the US military is reportedly still interested in a pulsed light- and sound-based stun weapon. The *Distributed Sound and Light Array Debilitator* (DSLAD) will, according to Sharon Weinberger, 'use essentially off the shelf technology to see if combining aversive noises with light produce some special debilitating effects. Anecdotal effects include dizziness and loss of balance, and of course, nausea' (Weinberger, 2007d). The DSLAD was tested by the JNLWP engineers in 2009, who claimed that 'preliminary results indicate that non-lethal sound and light at moderate intensities are effective at producing measured effects in targeted individuals' (Bowen, 2009).

Former head of the Los Alamos NLW program, John Alexander, suggested in his book the use of Pulsed Periodic Stimuli (PPS) in the form of acoustic waves for achieving a stun effect: 'The technique [PPS] can be applied to situations where it is desirable to cause perceptual disorientation in targeted individuals. This is important, as it is the first acoustic weapon that does not rely on high intensity to cause the desired effects. Rather low-intensity, pulsed, acoustic energy can induce fairly strong effects in humans' (Alexander, 1999: 102). The technique referred to by Alexander does not appear anywhere else in the NLW literature, but may exist within classified programs.

6.1.2 *High-power microwave and millimetre wave weapons*

The Soviets had been working on potentially lethal high-power microwave weapons since the 1960s (Gallagher, 1987: 23). In the US, interest in HPM weapons grew in the 1980s due to discoveries related to generating microwaves with higher powers and at higher frequencies (Davison, 2009: 164). These weapons are generally shrouded in secrecy. However, there were unconfirmed reports that Britain was working on a secret radiofrequency/ microwave weapon and that such a weapon may have been used on peace activists at Greenham Commons in 1984–85 (Kattenberg, 1987: 9). According to an article by international lawyer Louise Doswald-Beck and electronics expert Gerald Cauderay,

> [r]esearch work in this field [electromagnetics] has been carried out in almost all industrialized countries, with a view of using these phenomena for anti-material or anti-personnel weapons. Tests have demonstrated that powerful microwave pulses could be used as a weapon to put an adversary hors de combat or even kill him. It is possible today to generate a very powerful pulse (e.g. between 150 and 3,000 MHz), with an energy level of several hundreds of megawatts. Using specially adapted antenna systems, these generators could in principle transmit over hundreds of meters sufficient energy to cook a meal.
>
> (Doswald-Beck and Cauderay, 1990)

So far, only two anti-personnel microwave weapons systems have been publicly acknowledged: the Active Denial System (ADS) and the MEDUSA system.

The development of the ADS began probably in the 1990s and the program was declassified in December 2000 (Davison, 2009: 166–167). The technologies used in the ADS are similar to high-powered microwaves (HPM) and are being developed as anti-material NLW that target electronics (Davison, 2009: 167). The ADS device can be mounted on a HUMVEE or a similar vehicle and it uses 95 GHz millimetre waves. The microwaves only penetrate 0.4 mm of the skin and cause a strong burning sensation by heating the skin to 44 to 55 degrees centigrade (Davison, 2009: 167). After intensive testing on 700 human volunteers, the Pentagon determined that the ADS is safe for use as a nonlethal weapons system and decided to deploy it in Afghanistan in 2010. The system was recalled after a month without ever having been used because of the controversial nature of the weapon (Shachtman, 2010). Another factor may have been the complexity of the system that makes it difficult to deploy, including the need to cool down the pulse generators for 16 hours before it can be fired (Weinberger, 2012).

The other microwave-based weapon that has been openly announced in 2003 is the *Mob Excess Deterrent Using Silent Audio* (MEDUSA) system that was under development by the Sierra Nevada Corporation. The main idea behind MEDUSA was to use the Frey or microwave auditory effect to beam noise into the heads of adversaries to cause discomfort and disorientation.

> Potential applications of the MEDUSA system are as a perimeter protection sensor in deterrence systems for industrial and national sites, for use in systems to assist communication with hearing impaired persons, use by law enforcement and military personnel for crowd control and asset protection. The system will: be portable, require low power, have a controllable radius of coverage, be able to switch from crowd to individual coverage, cause a temporarily incapacitating effect, have a low probability of fatality or permanent injury, cause no damage to property, and have a low probability of affecting friendly personnel.
>
> (US Navy, 2004)

Although the initial Navy evaluation of MEDUSA was positive, Sierra Nevada Corporation has discontinued the project, possibly because it may have shown to permanently damage human brain tissue (Hambling, 2008).

Many NLW researchers believe that there is more to microwave weapons systems like ADS and MEDUSA than meets the eye. An article in *Nature*, written by an anonymous scientist, argues:

> Many records related to the Active Denial System remain classified and inaccessible to the public and the scientific community. The US Air Force's unwillingness to reveal the full scope of its research into the biological effects of high-power microwaves in the 1990s, which included work on their auditory and lethal effects, flies in the face of the defence department's claims that it is interested in classifying only weapons technology, and not science. If, as the Air Force says, the biological research never led to weapons, then there is no reason not to release it.
>
> (Nature, 2012: 178)

It seems probable enough that high power electromagnetic waves and fields could have an influence on the brain and affect mental capacity and behaviour. Physicist Vladimir Binhi stated: 'The brain and nervous system are considered particularly vulnerable to EMF [electromagnetic fields] because their functioning is based on transfer of nerve impulses, which involve electrical processes within and in between nerve cell membranes' (Binhi, 2009: 61). It is therefore theoretically possible that high-powered microwave (HPM) weapons could disrupt or interfere with brain functions.

Even based on the acknowledged thermal bioeffects of microwaves it would be possible to build an effective crowd control weapon that can severely disorient people. A DoD study on 'Bioeffects of Certain Non-Lethal Weapons' suggests that heating the brain with microwaves could affect mood and behaviour. The report states:

Body heating to mimic a fever is the nature of the RF incapacitation. The objective is to provide heating in a very controlled way so that the body receives nearly uniform heating and no organs are damaged. Core temperatures approximately 41°C are considered to be adequate. At such temperature a considerably changed demeanor will take place with the individual. Most people, under fever conditions, become much less aggressive; some people may become more irritable. The subjective sensations produced by this buildup of heat are far more unpleasant than those accompanying fever. In hyperthermia all the effector processes are strained to the utmost, whereas in fever they are not. It is also possible that microwave hyperthermia (even with only a 1°C increase in brain temperature may disrupt working memory, thus resulting in disorientation.

(US DoD, 1998)

According to a US Army intelligence report summarizing Chinese research into the bioeffects of HPM (that produce an EMP), even lethal effects are possible. The report states: 'The high mortality rates of animals (especially for primates) exposed to EMP radiation in the recent Chinese experiments are in graphic contrast to the lack of reported bio-effects associated with EMP exposures during the period of atmospheric testing (during the 1950s and 1960s) by the United States and other nations' (US Army, 2005: 134). The report suggests that this may be explained by the very high energy levels used by the Chinese researchers. The Soviets were reportedly able to kill goats from 1 km distance with an EMP device and cause an incapacitating/ disorienting effect over 2 km distance (De Caro, 1987: B03).

A report by the European Parliament claims that '[s]ome microwave systems have been proposed, which can raise body temperature to between 105 to 107 degrees F, to provide a disabling effect in a manner based on the microwave cooker principle. However, the greatest concern is with systems which can directly interact with the human nervous system'. The report indicates that secret research in the US and Russia in this field 'can be divided into two related areas: (i) individual mind control and (ii) crowd control' (European Parliament, 1999: LIII).

While behavioural effects are clearly possible using high powered RF and microwaves, there are also some fairly obvious limitations that have likely hampered the development of anti-personnel microwave weapons: (1) the effective range of these weapons is fairly limited because of 'atmospheric breakdown' that absorbs more of the energy over longer distance (Brunderman, 1999: 19); (2) there is a high risk of fratricide, as anything within the beam of the weapon and close to the antenna will be radiated and affected (Brunderman, 1999: 20); (3) adverse weather conditions such as strong humidity or rain can greatly impact on the effectiveness of a RF/ microwave weapon (Weinberger, 2012); and (4) high powered RF and microwave weapons could easily produce permanent injury such as brain damage. Taken together, this might suggest that HPM is a technological dead-end (Weinberger, 2012).

Table 6.1 The electromagnetic spectrum

Wave type	Frequency range	Frequencies	Wavelengths
Radio Waves	Extremely Low Frequency (ELF)	3 Hz – 30 Hz	10 Mm – 100 Mm
	Super Low Frequency (SLF)	30 Hz – 300 Hz	1 Mm – 10 Mm
	Ultra Low Frequency (ULF)	0.3 kHz – 3 kHz	100 km – 1000 km
	Very Low Frequency (VLF)	3 kHz – 30 kHz	10 km – 100 km
Radio Waves	Low Frequency (LF)	30 kHz – 300 kHz	10 km
	Medium Frequency (MF)	300 kHz – 1,5 MHz	650 m
	High Frequency (HF)	1.7 MHz	180 m
	Very High Frequency (VHF)	30 MHz – 300 MHz	10 m
Microwaves	Ultra High Frequency (UHF)	300 MHz – 3 GHz	10 cm – 1 m
	Super High Frequency (SHF)	3 GHz – 30 GHz	1 cm – 10 cm
	Extremely High Frequency (EHF)	30 GHz – 0.3 THz	1 mm – 1 cm
Infrared	Far Infrared	0.3 THz – 6 THz	50 μm – 1 mm
	Medium Infrared	6 THz – 100 THz	3 μm – 50 μm
	Near Infrared	100 THz – 385 THz	780 nm – 3 μm

Source: Author's own data.

6.1.3 Acoustic weapons

Acoustic weapons can use audible sound or they could use infrasound (20 Hz or below) or ultrasound (20 kHz or above). In here only infrasound and ultrasound weapons are relevant, since they have been claimed to be able to affect the human brain and CNS. Similar to microwave weapons decades of research into infrasound and ultrasound weapons have not resulted in the official deployment of any weapon of that kind. Experts such as physicist Jürgen Altmann have cast doubt about the general feasibility and practicality of acoustic weapons (Altmann, 1999). It is still worthwhile to consider acoustic weapons with respect to their ability to affect the human mind in more subtle ways than severe incapacitation.

The myth of the powerful infrasound weapon originated from the French acoustician Vladimir Gavreau, who made in 1968 the extraordinary claim that infrasound of 7 Hz would have strong physiological effects on the human mind and body and that it could be weaponized (Altmann, 1999: 15). A sound gun that he built based on the discovered principle, had supposedly near lethal effects, but further experiments undertaken elsewhere were not able to independently reproduce Gavreau's results (Volcler, 2013: 26–27).

Nevertheless, many references to infrasound weapons can be found in the NLW literature. For example, a US Air Force reference work on NLW suggests: 'Very low-frequency sound which can travel long distances and easily penetrate most buildings and vehicles. Transmission of long wavelength sound creates biophysical effects; nausea, loss of bowels, disorientation, vomiting, potential internal organ damage or death may occur' (Bunker, 1997: 2–3; similar: Lyell, 1997: 3; US DoD, 1972: 92–93; Koplow, 2006: 19).

The aforementioned declassified DoD summary on the bioeffects of certain NLW from 1998 also claims that human tests have indicated that infrasound can have disorientating effects on humans: 'Human subjects listened to very high levels of low-frequency noise and infrasound in the protected or unprotected modes. Two-minute duration as high as 140 to 155 dB produced a range of effects from mild discomfort to severe pressures sensations, nausea, gagging, and giddiness. Effects also included blurred vision and visual field distortions in some exposure conditions' (US DoD, 1998).

These statements can be contrasted with Altmann's conclusion that '[c]ontrary to several articles in the defense press, high-power infrasound has no profound effect on humans. The pain threshold is higher than in the audio range, and there is no hard evidence for the alleged effects on inner organs, on the vestibular system, for vomiting, or uncontrolled defecation up to levels of 170 dB or more' (Altmann, 1999:61).

This being said, there is some recent research that suggests that infrasound might be able to affect the moods and emotions of people in more subtle ways. Researchers from the University of Hertferdshire claim that exposure to infrasound was described by test subjects as ' "shivering on my wrist", "an odd feeling in my stomach", "increased heart rate", "feeling very anxious", and "a sudden memory of emotional loss"' (Amos, 2003). The researchers speculated that infrasound produced by organs may be responsible for strong religious experiences that some people have in church. Movie makers have already included inaudible infrasounds in horror movies to increase the emotional impact of the pictures (Stewart, 2013). Although Altmann has pointed out that infrasound is very unsuitable as a DEW, as it propagates in all directions and as the effective range would be too short (Altmann, 1999: 46–47), infrasound may be still useful for interrogations or in situations where targeted persons are in confined spaces, e.g. underground structures or inside buildings.

6.1.4 'Voice-of-God' weapons

The Air Force *Vistas* document projects for 2050 '[t]he ability to communicate directly to designated individuals, perhaps through bone conduction or through direct stimulation of the basilar membrane or the auditory cortex without the limitations of conventional communications equipment' (US Air Force, 1994: 109). It has been alleged that the capability of putting voices into people's heads could be misused for mind control (Binhi, 2009: 40). A 'voice-of-God' weapon would be a means of communication that can send speech directly into the heads of enemies for the purpose of impersonating a supranatural entity (God or a ghost) with the aim of confusing or commanding them. Interestingly, there are actually several potential methods of how one can project sound or speech in a manner that targets one person, but is inaudible to a bystander, which includes (1) bone conduction, (2) the microwave auditory effect and (3) ultrasound projection.

The oldest technology is called 'bone conduction' and it was discovered by the inventor Patrick Flanagan in 1958, which was popularized in a 1962 *Life* magazine article that celebrated the thirteen year old 'whiz kid' (Moeser, 1962: 69–72). The patent filed by Flanagan in 1968 describes the device he calls 'neurophone' in the following way: 'This invention relates to electromagnetic excitation of the nervous system of a mammal and pertains more particularly to a method and apparatus for exciting the nervous system of a person with electromagnetic waves that are capable of causing that person to become conscious of information conveyed by the electromagnetic waves' (US Patent 3393279 A). The neurophone transmits sound through skin contact and can even be heard by deaf persons (Moeser, 1962: 69). According to Flanegan's website *neurophone.ca*, the DIA temporarily classified the invention after the *Life* magazine article. It is a proven technology and the neurophone is commercially available from the inventor for $500.

The second method is called microwave auditory effect (MAE), which was discovered by radar operators during the 1940s, who often heard an inexplicable knocking or ticking noise. In 1962 the physicist Allan Frey was able to reproduce the effect using pulsed microwaves at 'extremely low average power densities', enabling 'the perception of sound' by 'normal and deaf humans' (Frey, 1962: 689). An article by Don Justesen from 1975 refers to an experiment by Joseph Sharp and Mark Grove at the Walter Reed Army Institute, which apparently showed that speech could be transmitted using the MAE. Sharp and Grove were able to understand nine out of ten words that they had recorded and then transmitted as 'voice modulated microwaves' (Justesen, 1975). This was also suggested in 1976 DIA report (US DoD, 1976: VIII), as well as in the aforementioned NLW report (US DoD, 1998). The US Army also referred to a MAE-based 'voice-to-skull' device on a webpage, claiming that the technology was already in use as an electronic scarecrow (Weinberger, 2008b). Sceptics have pointed out that the power levels required for transmitting sound by microwaves would be so high that brain damage due to microwave thermal effects would result (Heger, 2008) and that there would be no conclusive evidence for MAE at lower energy densities (Binhi, 2009: 39).

The third known technology is based on directed sound using ultrasound and was developed in the late 1990s. The company *Holosonics* founded by Joseph Pompei introduced the *Audio Spotlight* in 2004, which converts audible sound into ultrasound that can be projected in a narrow beam. Pompeii's invention is based on research undertaken by a Japanese research group in the 1980s (Lee, J., 2001). People standing within the ultrasound beam can clearly hear the sound, but people standing outside of it cannot. The technology was used in a New York advertising campaign for a TV series on paranormal phenomena. The *Audio Spotlight* loudspeakers were placed on a warehouse and aimed at passerby, whispering 'It's not your imagination' (Volcler, 2013: 121). DARPA announced in 2007 a program for developing a 'sonic projector', which would be based on the same technology and could be used as a 'voice-of-God' weapon (Volcler, 2013: 122).

6.1.5 Low energy EMF

While thermal bioeffects of microwaves are scientifically acknowledged, the issue of possible non-thermal effects remains a hotly debated one. Constant exposure to low energy electromagnetic fields (EMF) is suspected to affect the general health and well-being of people. There is even speculation that low energy electromagnetic waves could manipulate brain processes and could affect moods, perception and behaviour in some way. This section will discuss some of the evidence for non-thermal effects of microwaves and other sources of EMF.

A particular concern are *extremely low frequency* (ELF) waves and microwaves that are modulated in the ELF range, as they could interfere with biological cycles – brain waves are in the ELF range with their average between 0 and 30 cycles (Becker, 1990: 212). Apart from the highly controversial mind control aspect, a further complication arises from the fact that the $2.2 trillion a year global telecommunications industry has a major stake in the use of low power microwaves (6.8 billion cell phones and over one billion wi-fi connections as of 2013). Any health hazard from low energy non-ionizing radiation (electromagnetic fields or EMF) has to be categorically denied in order to discourage lawsuits, which still occur since '[s]ome experts and scientists in this field believe that these EMFs can cause such adverse health effects' (Anibogu, 1998: 11).

Regardless, there is a growing body of scientific evidence that EMF or electropollution through long-term EMF exposure is indeed harmful and is affecting many people, who suffer from electromagnetic hypersensitivity (some countries such as Sweden recognize it as a proven health condition or functional impairment). Recently, 195 scientists from 39 countries sent a letter to the World Health Organization that stated: '[b]ased on peer-reviewed, published research, we have serious concerns regarding the ubiquitous and increasing exposure to EMF generated by electric and wireless devices... The various agencies setting safety standards have failed to impose sufficient guidelines to protect the general public, particularly children who are more vulnerable to the effects of EMF' (Harkinson, 2015).

Many studies on microwaves from the 1960s and 1970s have already indicated many of the potential issues: long-term exposure to microwaves can cause genetic mutation, brain tumours and other cancers, behavioural abnormalities (suicide), alterations in biological cycles, weakening of the immune system and alterations in learning ability (Becker, 1990: 214–215; also Brodeur, 1977). A 1970 RAND metastudy on neurological effects of microwaves suggested: 'it seems likely that neural function, and therefore behavior, are indeed disturbed by low intensity microwaves...the studies consistently and repeatedly report that human beings do exhibit behavioral disturbances when exposed to low intensity microwaves' (MacGregor, 1970: 8).

These conclusions are not too different from some more recent studies that also indicate brain and behavior changes resulting from long-term EMF exposure. For example, a study from 2000 claims: 'Natural and man-made electromagnetic fields influence the mood and behavior of healthy and sick people. Considerable evidence suggests that electromagnetic fields affect sleep' (Sher, 2000). Another study from 2006 has linked ambient electromagnetic fields to human moods and behaviour, finding a significantly increased suicide rate during periods of geomagnetic storm (Berk et al., 2006). Continuous electromagnetic field exposure has been linked in scientific studies to 'cancer, heart disease, sleep disturbance, depression, suicide, anger, rage, violence, homicide, neurological disease and mortality' (Cherry, 2002). A leaked study on the British police radio system TETRA prepared for the Police Federation of England and Wales contends that microwaves pulsed in certain frequencies can induce paranoia, depression, suicide, manic behaviour and blindness (Trower, 2001: 30).

Russian scientists have claimed long ago that there are nonthermal bioeffects of EMFs possible at lower energy densities. A Russian military article by Major General Belous claims:

> Studies into the impact of electromagnetic radiation on the human organism show that even exposure to low intensity EM radiation results in an array of functional changes and disturbances. In particular, EM radiation can disrupt the heart rhythm and even, according to some scientists, can cause cardiac arrest. There are two types of impact: thermal and non-thermal. The thermal impact causes an overheating of tissue and organs, and with sufficiently long radiation can cause irreparable pathological changes. The non-thermal impact mainly leads to functional disorders in various organs of the human organism, especially in the cardio-vascular and nervous systems.
>
> (Belous, 2009: 72)

Similarly, national security expert Chuck De Caro wrote that 'a highly tuned RF device could act as a sort of negative pacemaker, inducing heart attacks by overriding the heart's normal "P" wave' (De Caro, 1985: B3). EMF, and in particular ELF and modulated microwaves, could be therefore weaponized. Of course, such weapons would often fail to produce any instant effects. However, constant and long-term exposure could be used for influencing moods and general mental capacity.

An article by Paul Tyler on 'The Electromagnetic Spectrum in Low-Intensity Conflict' discusses the potential uses of electromagnetic radiation, especially ELF radiation, in terms of their bioeffects in the context of 'dealing with terrorist groups, crowd control, controlling breeches of security at military installations, and anti-personnel techniques in tactical warfare' (Tyler, 1986: 251). After giving an overview of Soviet research, Tyler argues that scientists in the West have erroneously assumed that a 'microwave is a microwave', not taking into account that even the slightest variation in frequency can make a huge difference in bioeffects (Tyler, 1986: 256).

The general idea is that specific frequencies could have fairly specific physiological and possibly behavioural effects. For example, according to the medical doctor Robert C. Beck, 'ELF fields of 6.67 Hz and lower can produce symptoms of confusion, anxiety, depression, tension, fear, mild nausea and headaches, cholinergia, arthritis-like aches, insomnia, extended reaction times, hemispheric EEC desynchronization, and many other vegetative disturbances' (Beck, 1986: 47).

Chinese military analysts have suggested that electromagnetic waves could be used for producing specific genetic effects. They stated: 'high-intensity ultraviolet waves and electromagnetic waves can induce genetic locus cell mutation. If we determine the relationship between the specific frequency, wavelength, or power of the ray or wave and the specific gene or locus, we can cause injury by remote, radiation-induced, genetic function changes' (Guo and Yang, 2005: 77). It may also be possible to manipulate emotion with specific electromagnetic waves. Binhi suggests that weaponized EMF could 'cause pain, fear, or similar reactions in personnel, to influence the perceptions and attitudes of individuals and groups, which finally could disrupt military operations' (Binhi, 2009: 6–7).

A particularly frightening possibility in this respect has been discussed by neuroscientist Michael Persinger, who has extensively published on EMF influence on the brain. Persinger suggests that the brain uses frequencies for 'informational transactions' between brain structures: 'Consciousness would be associated with an electromagnetic pattern generated by a neural aggregate with invariant statistical features which are independent of the cells contributing to each feature' (Persinger, 1995: 793). According to this theory, 'specific neuropatterns can be evoked by extremely weak magnetic fields whose intensities are within the range of normal geomagnetic variations', as electrical processes in the brain are so weak (Persinger, 1995: 795). Persinger claims that he can produce altered states of consciousness, 'sense of presence' and spiritual experiences by generating weak EMF near the heads of test subjects (Persinger, 2003). Persinger argues:

> Within the last two decades…a potential has emerged which was improbable but which is now marginally feasible. This potential is the technical capability to influence directly the major portion of the approximately six billion brains of the human species without mediation through classical sensory modalities by generating neural information within a physical medium within which all members of the species are immersed.
>
> (Persinger, 1995: 797)

What he is suggesting is that through the precise electromagnetic manipulation of the earth's atmosphere it could be possible to affect the minds and well-being of all humans in the world. It is clear that Persinger's theories are very controversial in the scientific community. A group of Swedish researchers was unable to reproduce any of the effects claimed by Persinger (Granqvist et al., 2004). Many potential weapons applications of weak EMF therefore remain highly speculative apart from the realization that constant exposure to EMF is unlikely to be healthy.

6.1.6 Electromagnetic mind control?

Currently, there is no strong evidence that electromagnetic mind control is feasible in the way it has been suggested by conspiracy writers. At the same time, it is clearly within the bounds of existing technology to use high-powered microwaves (and also lasers), exploiting well-understood thermal effects, to cause discomfort, disorientation, depression and even death. Such weapons would violate international law since they would most likely cause 'unnecessary suffering and superfluous injury' (Rosenberg, 1994: 45), but terrorists and some state actors might develop and employ such weapons. At lower energy densities, constant exposure can result in genetic damage, cancer, sleeplessness and mental disorders in a time frame of years, if not decades.

It has been suggested that governments might focus their efforts with respect to electromagnetic mind control in the ELF and the GHz frequency ranges (Binhi, 2009: X-XI). Clearly, more research is needed before one can completely dismiss the theoretical feasibility of electromagnetic mind control. Binhi, who has conducted the most exhaustive scientific review on this subject, concludes

> there are no physical reasons to assert that EM mind control is not possible. Under the influence of a remote source of EMF, the brain is exposed to the EMF as a whole. Therefore one could expect that the EM impact would be similar to that of chemicals, but can induce particular states of mind...there are no fundamental constraints to realization of remote EM [electromagnetic] control over the human brain with the goal to bring the mind in obedience with the will or another vulnerable state.
>
> (Binhi, 2009: 93)

This possibility may be still far off, but lesser effects such as influencing moods, emotions, sleep, cognition and mental capacity could already be feasible with today's technology.

6.2 Bytes

Not all neuroweapons that influence human thoughts, emotions and behaviour have to be material in nature – some of them might be simply information or software that can 'hack' the human brain and/or target the cognitive system. Much of this relates to the ancient practices of psychological warfare and to modern military deception and propaganda, which can all be made much more effective with the insights provided by neuroscience research. Some newer technologies and approaches discussed below include subliminals (including silent sounds), narrative networks and the manipulation of perception, using holograms and invisibility cloaks.

6.2.1 Subliminals

Since Freud, it is known that much of our psyche remains subconscious. The subconscious fears and desires, nevertheless, have great influence on the human mind and human behaviour in ways that are beyond the understanding of the conscious mind. So the idea is if one could manipulate the subconscious, one could influence conscious behaviour without the person being aware of any external influence. Hence the concept of the 'subliminal' stimulus was born, which is not consciously perceptible, but is still registered by the mind. The marketing industry has been interested in subliminal advertising for a very long time. The practice of subliminal advertising has been first revealed by Vance Packard in 1957 in his book *The Hidden Persuaders*.

It has been reported in the news media that subliminals have been used successfully by department stores to reduce shoplifting (Time, 1979). Often advertising subliminals contain hidden references to 'the taboos of society – sex, death, incest, homosexuality, and at times, pagan icons' (Chen, 1990: 2). The communications scientist Wilson Bryan Key claimed that the advertising industry would spend upwards of $50,000 for checking every single detail in an advert. Key testified before Congress 'that there are some 500 published articles on the effects of subliminal suggestion in the psychology literature. While inconclusive, the research seems to indicate that subliminal messages "affect some people under some circumstances, some of the time"' (Chen, 1990: 2).

According to subliminal communication expert Eldon Taylor, a subliminal stimuli can be subdivided into four groups: (1) 'Below the level of registration', (2) 'Above the level of registration but below the level of detection', (3) 'Above the level of detection and discrimination, but below the level of identification' and (4) 'Below the level of identification only because of a defensive action' (E. Taylor, 2007: 30). These categories can be used for the analysis of subliminals and the development of research designs.

In the 1980s, there was a growing concern across Western societies about satanic subliminal messages that may be embedded in heavy metal music, which would encourage suicidal, homicidal, or other antisocial behaviour. These concerns were sparked by the case of two young adults, who shot themselves after listening to a *Judas Priest* record that had a suicide theme and a subliminal message that said 'do it'. This resulted in a lawsuit and a trial in 1990. The case was eventually dismissed on grounds that subliminals could not have led to the suicides. Some have taken the verdict to mean that subliminals just do not work and are therefore no threat to the freedom of will (Streatfield, 2008: 171–209; E. Taylor, 2007: 30–31).

Taylor claims '[c]ontrary to popular opinion, the literature and evidence supporting subliminal information theory is robust' (E. Taylor, 2007: 29). Although it is true that subliminals are very unlikely to override the most basic instincts of nature and principles of morality, some lesser effects are clearly possible. The ten billion dollar self-help industry and the much bigger advertising industry rely on the effectiveness of subliminals, which would suggest that subliminals can and do influence people under certain conditions in terms of their 'conscious perception, dreams, drives, emotions, memory, defense mechanisms of perception, value system, and verbal behavior' (E. Taylor, 1990: 11). This has been validated by several recent neuroscience studies that used EEGs or fMRIs to detect brain responses to subliminal stimuli or that observed behaviour of test subjects after they were given imperceptible stimuli.

The earliest research in this field was undertaken by neuroscientist Benjamin Libet in the 1960s. Using an EEG, Libet discovered that a brain can respond to a subliminal stimulus without conscious awareness of the stimulus: '1. Cerebral cortical activities, in response to a somatosensory stimulus, must proceed for about 500 ms in order to elicit the conscious sensation. 2. Activations of shorter durations at the same intensities can produce unconscious detection of that input' (Libet, 2002: 291). In other words, if neural activities in response to a stimulus remain below the 500 ms threshold, then the stimulus will not become conscious, although it was clearly registered by the brain.

A study carried out at University College London (UCL) also found that the brain responds to subliminal stimuli, if it is not too busy or distracted, which may suggest that 'subliminal advertising might affect our decisions about buying things' (Jha, 2007). A follow-up study at UCL discovered that subliminals work better with negative messages than with positive ones. Words displayed on a computer screen for just 0.2 seconds had a greater impact on test subjects if they were emotional and negative than positive terms, which might suggest that evolution has made brains responding to subconscious warning signs rather than subconscious signs that everything is OK (Bahrami et al., 2007: 509–513).

A Duke University study suggests that company logos displayed only a fraction of a second affect behaviour in terms of the traits associated with the brands. For example, the Apple brand encouraged people subconsciously to come up with unusual solutions to a problem, while the Disney logo made people behave more honestly (Science Daily, 2008). In another study at Kent University from 2014 it was shown that an athlete's performance could be affected by showing positive or negative subliminal visual cues before an exercise (Blanchfield et al., 2014). Obviously, subliminal stimuli could be used to enhance the effectiveness of PSYOPS messages.

6.2.2 Silent sounds

It is quite extraordinary that one can find no discussion of the exploitation of subliminals in PSYOPS manuals or literature, which either means that nobody in the PSYOPS field ever thought about the use of subliminals, or that such information remains highly classified. According to a story by the British channel *ITV*, subliminals hidden in 'silent sounds' may have been operationally used during the 1991 Gulf War. The *ITV* article stated: 'The clandestine station programming consisted of patriotic and religious music and intentionally vague, confusing and contradictory military orders and information to Iraqi soldiers in the Kuwaiti Theater of Command (KTO). The size and power of enemy forces was always intentionally exaggerated. Surrender was encouraged' (ITV, 1991). Most interesting is *ITV's* claim that the radio messages had embedded subliminals and inaudible sounds:

> According to statements made by captured and deserting Iraqi soldiers, however, the most devastating and demoralizing programming was the first known military use of the new, high-tech, type of subliminal messages referred to as ultra-high-frequency 'silent sounds' or 'silent subliminals'. Although completely silent to the human ear, the negative voice messages placed on the tapes alongside the audible programming by PSY-OPS psychologists were clearly perceived by the subconscious minds of the Iraqi soldiers and the silent messages completely demoralized them and instilled a perpetual feeling of fear and hopelessness in their minds.
>
> (ITV, 1991)

Many more details regarding the use of 'silent sounds' during the Gulf War can be found in an article by bioelectromagnetics expert Judy Wall in the *Nexus* magazine, which explains Oliver Lowery's invention 'Silent Sound Spread Spectrum' (S-Quad), which was allegedly used 'with great success in the Gulf War' (Wall, 1998). While the Patent does exist (#5,159,703) and Lowery confirmed the story to journalist Jon Ronson, there is no further independent confirmation that can be found anywhere (Ronson, 2004: 189–193)

Regardless whether this happened or not, it is notable that US PSYOPS were tremendously effective during the Gulf War: within 24 hours of the ground war already 10,000 Iraqi soldiers surrendered to coalition forces. The Iraqis were so desperate to surrender that they even surrendered to Western journalists and drones. Overall, 87,000 Iraqi soldiers surrendered during the conflict and 17,000 deserted, which has to be considered an enormous and often overlooked PSYOPS success (Wojtysiak, 2001).

6.2.3 The 25th frame technique

From Russia originated the theory or myth of the '25th frame effect' and it has remained quite persistent over the years. According to this theory, humans perceive 24 images per second. By adding a 25th frame shown on a screen a subliminal message can be included to subconsciously manipulate the viewers, who will not consciously register the existence of the added frame.

Timothy Thomas mentions in his *Parameters* article that the Russians claimed to have developed a computer virus that manipulates the computer display to hack a user's mind:

> According to Solntsev, one computer virus capable of affecting a person's psyche is Russian Virus 666. It manifests itself in every 25th frame of a visual display, where it produces a combination of colors that alleg-edly put computer operators into a trance. The subconscious perception of the new pattern eventually results in arrhythmia of the heart. Other Russian computer specialists, not just Solntsev, talk openly about this '25th frame effect' and its ability to subtly manage a computer user's per-ceptions. The purpose of this technique is to inject a thought into the viewer's subconscious.
>
> (T. Thomas, 1998)

It was reported by the *LA Times* that the Russian TV channel ATV based in Yekaterinburg has used this technique to manipulate its viewers into watch-ing it more. The use of subliminals on Russian TV had become so perva-sive and aggressive that the Russian Ministry for Press, Broadcasting and Communications decided to take action by developing a scanning software that can detect methods like the 25th frame and by banning violators from the airwaves (Dixon, 2002).

The BBC recently reported that the Russians have deployed the 25th frame technique in the Ukraine to influence Ukrainian audiences. The Ukrainian intelligence service SBU has found subliminal messages embedded in TV pro-gramming from the news channel *Rossiya 24*. They allegedly found hidden messages in the TV signal such as 'torched by the Right Sector', 'people killed by Banderites [followers of Ukrainian wartime nationalist leader Stepan Bandera]' and 'National Guard murderers' (BBC, 2014). Unfortunately, there is seemingly no scientific study in the West that has specifically looked into the 25th frame effect, which means that it remains speculative.

6.2.4 DARPA's Narrative Networks

In 2011, DARPA announced its Narrative Networks (N2) program, which is apparently aimed at 'master[ing] the science of propaganda' (Lim, 2011). According to DARPA program manager William Casebeer, the goal of narrative networks is to 'understand how narratives influence human thoughts and behaviour, then apply those findings to a security context in order to address security challenges such as radicalization, violent social mobilization, insurgency and terrorism, and conflict prevention and resolution' (Weinberger, 2014b). The original program announcement stated:

> DARPA is soliciting innovative research proposals in the areas of (1) quan-titative analysis of narratives, (2) understanding the effects narratives have

on human psychology and its affiliated neurobiology, and (3) modeling, simulating, and sensing – especially in stand-off modalities – these narrative influences. Proposers to this effort will be expected to revolutionize the study of narratives and narrative influence by advancing narrative analysis and neuroscience so as to create new narrative influence sensors, doubling status quo capacity to forecast narrative influence.

(DARPA, 2011)

The solicitation listed three sub-goals: (1) 'Develop new, and extend existing narrative theories', (2) '[i]dentify and understand the role of narrative in security', and (3) '[s]urvey and extend the state of the art in narrative analysis and decomposition' (DARPA 2011). The DARPA project website summarises the aims of the project in the following way: 'to understand how narratives influence human cognition and behavior, and apply those findings in international security contexts. The program aims to address the factors that contribute to radicalization, violent social mobilization, insurgency, and terrorism among foreign populations, and to support conflict prevention and resolution, effective communication and innovative PTSD treatments' (DARPA website).

The proposed research methods include neuroscience approaches such as '[a]nalyz[ing] the neurobiological impact of narratives on hormones and neurotransmitters, reward processing, and emotion-cognition interaction' (DARPA website). For example, a N2 grant worth $200,000 was awarded to researchers from Arizona State University in 2012, who have already produced a preliminary project report. The published report indicated the development of a 'persuasion protocol' and the use of EEGs for human tests (Corman, 2011). Other N2 researchers at Georgia Institute of Technology monitor the brain processes of test subjects with MRIs while they watch dramatic scenes from popular movies to determine their emotional responses to them. One of the findings is that in situations of great suspense the test subjects decreased their awareness of the world around them (Cha, 2015).

It is widely understood that 'narratives in the form of stories, rumors, biographies and pictures drive our behaviors and shape our convictions' (Petit, 2012: 26–27). The apparent idea behind N2 is to better understand how narratives motivate enemies and to eventually come up with more persuasive narratives that counter those other narratives and thereby reduce the enemy's motivation to engage in conflict in the first place.

The two legal scholars Cass Sunstein and Adrian Vermeule from Harvard University have argued even before N2 that there is a need to scientifically fight harmful narratives such as 'conspiracy theories' taking advantage of neuroscience. They warned that 'inevitably action will ensue' as a result of the spread of conspiracy theories on the Internet: even without directly causing violence 'such theories can still have pernicious effects from the government's point of view, either by inducing unjustifiably widespread public skepticism about the government's assertions, or by dampening public mobilization and participation in government-led efforts, or both' (Sunstein and Vermeule, 2009: 220).

In order to counter this threat the authors suggest to 'cognitively infiltrate' these groups by 'government agents' covertly participating in the debates and debunking them or 'to induce some cognitive diversity' for confusing, deceiving and discrediting people (Sunstein and Vermeule, 2009: 224–226).

That this idea has gone far beyond philosophical theorizing can be shown by the fact that the British intelligence service GCHQ commissioned in 2011 a study by behavioral scientist, Mandeep Dhami from then Cambridge University, to improve the art of Internet trolling (Dhami, 2011). In the study the following methods are suggested 'to discredit, promote distrust, dissuade, deter, delay or disrupt': upload Youtube videos containing 'persuasive communications', '[s]etting up Facebook groups, forums, blogs and Twitter accounts', '[e]stablishing online aliases/ personalities', '[p]roviding spoof online resources', '[s]ending spoof e-mails and text messages from a fake person or mimicking a real person', etc. (Dhami, 2011: 9). A whole arsenal of dirty tricks is revealed for influencing and manipulating people in the online world, which are undoubtedly already employed by many intelligence and security services around the globe.

The DARPA scientists envision N2 also as an early-warning device that can detect 'who has fallen prey to dangerous ideas' (Lim, 2011). For example, the use of certain language can function as an indicator for violence (Weinberger, 2014b). In this aspect, narrative networks sounds a bit like the DHS FAST program or other pre-crime programs, which all suffer from the same problem: human behaviour is complex and machines will end up generating lots of false positives, thereby putting innocent people at risk. It seems also likely that there is a serious risk of 'blowback' from introducing new narratives: people may interpret invented narratives in unforeseen ways, leading to surprising behaviour (Sunstein and Vermeule, 2009: 225). Furthermore, there is an ethics issue with respect to deliberately deceiving people, even if it is a 'noble lie' that helps reduce the motivation for violence, as 'deceit can hamper the exercise of rational choice at every step' and thereby hurt people (Bok, 1989: 26). Sunstein and Vermeule have, however, been widely criticized for advocating the wilful deception of the public as a means of neutralizing dissenters.

6.2.5 Holograms and invisibility cloaks

A slightly different approach of influencing an enemy's behaviour is to manipulate the enemy's sensory perception using advanced technology that creates holographic 3D images that are realistic and to alternatively render real objects invisible to an enemy. Some of these ideas will probably not be achievable or particularly useful in operational situations, but there are no reasons why they could not work in principle.

A hologram is a three-dimensional image of an object created by a laser, which under optimal light conditions may be mistaken for a real object. Holography or the science of creating holograms has been invented by Dennis Gabor in 1947. Potential applications for holograms range from entertainment to medical training and human machine interface technology. There are also obvious military applications with regard to military deception and PSYOPS. The USAF had temporarily a webpage that featured an 'airborne holographic projector'. The picture on the webpage showed an aircraft projecting its own image ahead of itself and the description stated: 'The holographic projector displays a 3-dimensional visual image in a desired location, removed from the display generator. The projector can be used for psychological operations and strategic perception management. It is also useful for optical deception and cloaking, providing a momentary distraction when engaging an unsophisticated adversary'. Claimed or desired capabilities of the holographic projector included '[p]recision projection of 3-D visual images into a selected area; [s]upports PSYOP and strategic deception management; [p]rovides deception and cloaking against optical sensors' (US Air Force, 2003). In 2010 Pentagon scientists proposed a 'face of Allah'-weapon that would project a realistic holographic image of a deity into the sky 'to incite fear in soldiers on a battlefield' (Axe, 2010). Holography is one of those military technologies where it is very challenging to find good information on the current state of the art. So it remains very hard to judge what is or might be possible in this respect.

Neuroscience could provide greater insights into the workings of human vision, resulting in new and revolutionary camouflage technology up to the point of making objects or soldiers completely invisible to the human eye, as well as optical and other kinds of sensors. Camouflage patterns could be tailored perfectly to the operating environment using software that 'crunch[es] meteorological data on typical lighting and visibility conditions, combined with information about the colours and predominance of shapes visible in cities, fields and wilderness areas' (Economist, 2008). Battledress uniforms have now light-absorbing features and patterns that are so efficient at concealing soldiers that observers need to be 40 per cent closer to them in order to see them than was the case back in 2000 (Economist, 2008).

Another approach to relative invisibility is called 'adaptive camouflage' and is based on adjusting the colour and texture of an object to the background, making it very hard to perceive, especially from distance. For example, an aircraft could have colour changing or electrochromatic panels on its fuselage and wings that adjust to the exact colour and brightness of the sky in the background, making the aircraft virtually invisible from below. Aviation enthusiasts have speculated about secret US 'daylight stealth' aircraft since the late 1990s. A more technologically advanced approach is to bend light waves around an object so that a viewer cannot see it from a particular position. However, once the viewer moves to the side, the object may become partially visible because of different light refraction.

The Canadian company *Hyperstealth* claims to have developed a nanotechnology-based material it calls 'Quantum Stealth', which supposedly makes an object invisible across the spectrum of visible light and infrared (Anthony, 2012). It seems that scientists all around the world are discovering new ways of making things invisible. For example, researchers from the University of Toronto have built an electromagnetic invisibility cloak. According to scientist George Eleftheriades, 'It's very simple: instead of surrounding what you're trying to cloak with a thick metamaterial shell, we surround it with one layer of tiny antennas, and this layer radiates back a field that cancels the reflections from the object' (Parnell, 2013). Such invisibility technology could be used defensively, depriving the enemy of targets, or offensively, sneaking up on the enemy in order to make an attack more likely to succeed.

6.2.6 'Mind viruses'

The term 'mind virus' appears in the previously cited Timothy Thomas article, which refers to the 25th frame effect (T. Thomas, 1998). However, a different concept of a mind virus was proposed by neuroscientist Rajesh Rao: 'Malicious entities could send a "virus" as part of a communication from a machine, resulting in cognitive impairment or cognitive manipulation' through a BCI (Rao, 2013: 274). In other words, a BCI could be hacked and be used against the wearer of the BCI, for example by stimulating specific brain regions in a malicious fashion. Writing in *Wired Magazine* analysts for the Australian army, Chloe Diggins and Clint Arizmendi, claim that 'the marriage between neuroscience and military systems will fundamentally alter the future of conflict' since BCIs create new vulnerabilities that can be exploited:

> We need to prevent BCIs from being disrupted or manipulated, and safeguard against the ability of the enemy to hack an individual's brain. The possibilities for damage, destruction, and chaos are very real. This could include manipulating a soldier's BCI during conflict so that s/he were forced to pull the gun trigger on friendlies, install malicious code in his own secure computer system, call in inaccurate coordinates for an air strike, or divulge state secrets to the enemy seemingly voluntarily.
>
> (Diggins and Arizmendi, 2012)

Although it seems like a long way before soldiers are outfitted with brain monitoring and brain stimulation devices or even brain chips that could be remotely hacked, the day may not be as far off as one might think. The US Army expects this to happen in the 2020 to 2040 time bracket (Brady, 2015: 6–7).

6.3 Conclusion

This chapter has reviewed possible electromagnetic/ acoustic and 'informational'/ cyber neuroweapons that could be developed. Stun weapons that hack the CNS through direct contact or by dazzling a person with an incoherent light source have been proven to work with several prototypes in existence. It was shown that RF/ microwave type weapons can impact mental capacity and otherwise incapacitate due to thermal effects occurring at high energy densities. These weapons would be limited mostly by range and atmospheric conditions, but they are clearly possible to develop, as are so-called 'voice-of-God' weapons. There is far less proof that acoustic weapons and RF/ microwave weapons with low energy density exploiting assumed non-thermal effects are feasible or practical. Infrasound seems to have some influence on human emotions and well-being. There is also a growing amount of studies that show that long-term exposure to EMF can have negative health effects (cancer) and may impact on behaviour (suicide). It is hard to say, whether more precise effects on human behaviour could be possible with low energy RF/ microwaves – physicist Binhi has argued that there is at least no physical principle that would make this impossible in the future. Equally controversial are other forms of 'mind hacking' such as the use of subliminals. Several experts such as Wilson Bryan Key and Eldon Taylor have argued that subliminals are effective in influencing human behaviour under certain conditions. The use of subliminals by the advertising industry has been publicly known since 1957 and only recently neuroscience has been able to confirm that there is such a thing as a subconscious or subliminal stimulus and that it can influence human behaviour. It seems obvious enough that this can be exploited in PSYOPS campaigns. The only unclassified program in this field is DARPA's N2, which includes the use of neuroscience for researching the influence of narratives on people. N2 is meant as a 'defensive weapon' that can neutralize enemy narratives and as an early-warning system that can detect radicalization. Finally, the chapter discussed the possibility of hacking the brain using BCIs that could enable others to take control over a person. This possibility is real, although it will take some time before such a threat can materialize – again there is no scientific reason why it would not be possible to 'roboticise' a human: it has been done to animals with great success (see chapter 1).

7 The strategic context

This chapter discusses the strategic context in which neuroweapons may be employed in the future by looking at the current spectrum of conflict. During the Cold War, the US military defined the spectrum of conflict in a straightforward way according to risk and cost involved for engaging in conflict from very low intensity conflict (terrorism) to very high intensity conflict (strategic NBC war), with guerrilla war, civil war, conventional war and theatre NBC war in the middle to high intensity range (US DoD, 1986: 4). The idea of a spectrum of conflict has undergone notable changes after the end of the Cold War. A paper by the Heritage Foundation divides the new spectrum of conflict as follows from 'Gray Zone/ Ambiguous' to 'Irregular/ Terrorism' to 'Hybrid' to 'Limited Conventional' to 'Theater Conventional' at the high end (Hoffman, 2015: 29).

The analysis presented here divides the spectrum of conflict into (1) *state-on-state warfare* (nuclear and conventional warfare at the high end), to (2) *state-on-nonstate warfare* of which hybrid wars are an aspect of, to (3) *non-obvious/ ambiguous warfare*, which are hostile acts that are difficult to attribute and may be disguised as natural events, technical glitches or accidents, to (4) *controlling populations/ security*, which is at the very low end spectrum of conflict and which is meant to protect societies against a range of man-made and natural threats.

It is argued that interstate wars are becoming increasingly unlikely, while indirect and ambiguous forms of aggression and destabilisation become more prevalent, necessitating an all-encompassing approach to protecting societies. In the wars of the past, the main objectives were the occupation of territory and the destruction of enemy forces. In the wars of the future, the main objective and target of military action will be to influence and control large populations, which reduces the need for physical destruction and other violence and which puts a premium on socio-political and psychological manipulation. In other words, it will be shown that military neuroscience has enormous relevance for the kinds of conflict that will likely define warfare in the first half of the twenty-first century.

7.1 State-on-state warfare

As long as there are sovereign nation states there will always be geostrategic rivalries. Therefore the possibility of war amongst states will endure. The only way of completely eliminating the threat of interstate conflict would be to create a world government that unifies all nations on earth. Some theorists such as Alexander Wendt have even argued that 'a world state is inevitable', as globalisation would reach its logical conclusion (Wendt, 2003: 491–542). Interestingly, this liberal dream of world government seems to have become much more palatable in recent years, even within major defence establishments. For example, the British MoD argued in a strategy paper that the world in 2040 will 'face the reality of a changing climate, rapid population growth, resource scarcity, resurgence in ideology, and shifts in global power from West to East' and that '[n]o state, group or individual can meet these challenges in isolation, only collective responses will be sufficient'. This would make it necessary 'to establish an effective system of global governance, capable of responding to these challenges', which 'will be a central theme of the era' (UK MoD, 2010: 10). The report anticipates that there will be resistance to US hegemony by rising powers, which is expected to endure (UK MoD, 2010: 46).

The new US 'National Military Strategy' has identified only four states that threaten US national security interests: Russia, Iran, North Korea and China (US DoD, 2015: 2). Although the strategy paper argues that '[n]one of these nations are believed to be seeking direct military conflict with the United States or our allies', it also suggests that 'the probability of US involvement in interstate war with a major power is assessed to be low but growing. Should one occur, however, the consequences would be immense' (US DoD, 2015: 2, 4).

This section will briefly discuss four general arguments why major interstate war is fading from world history: (1) the nuclear war dilemma, (2) the overall military superiority of the US, (3) the delegitimisation of the use of force and (4) growing global interdependence.

7.1.1 The nuclear war dilemma

Military historian Martin van Creveld has argued in his groundbreaking 1991 book *The Transformation of War* that nuclear weapons have made warfare so destructive that a total war between the superpowers could not even be imagined, far less rationally contemplated. He wrote: 'nobody has yet figured out how to wage a nuclear war without risk of global suicide' (van Creveld, 1991: 10). Van Creveld argued that nuclear weapons could not be conceivably or plausibly be used for anything other than deterrence and even they cannot deter lesser aggressions by nuclear and nonnuclear states (van Creveld, 1991: 3–4). The Russian General Makhmut Gareev agreed with van Creveld's analysis that nuclear weapons are a bluff (Gareev, 1998: 73). In addition, they would also make conventional attacks between Russia and the

West impossible since 'the opposing sides will always be compelled to take into account the danger of engendering nuclear war' (Gareev, 1998: 90).

In the 1980s, the combined nuclear arsenals of the superpowers had grown to 50,000 operational nuclear weapons with a destructive of 20 billion tons of TNT or over a million times the destructive power of the Hiroshima bomb (Cirincione, 2007: 85). Averaged out, this amounted to roughly four tons of TNT for every human being on earth at that time. A widely cited study commissioned by Congress in 1979 painted an extremely bleak picture of the consequences of nuclear war, estimating that a 25 Mt airburst over a city like Detroit would result in over 1.8 million deaths and over 3 million casualties (US Congress, 1979: 37). In addition to the blast and heat effects that are responsible for most of the immediate deaths and destruction that is caused by nuclear detonation, there is also the issue of radiation spreading far through fallout, making some regions permanently uninhabitable (US Congress, 1979: 5).

Although after the Cold War there have been political efforts to rid the world of nuclear weapons, it is highly likely that the possibility of global nuclear war will endure for many more decades, if not centuries. Even in today's world very substantial nuclear arsenals remain. The US has over 2,000 deployed nuclear weapons and Russia has about 1,780 deployed nuclear weapons with each power having an additional 2,000 to 3,000 warheads stockpiled (Kristensen and Norris, 2014: 96). As pointed out by the *Bulletin of Atomic Scientists*, there are plans to not only modernize existing nuclear weapons, but to also develop entirely new ones. The US anticipates spending $355 billion dollars on the maintenance and modernization of its nuclear arsenal, including the development of a new B61-12 gravity bomb and a long-range nuclear cruise missile (Kristensen and Norris, 2014: 102). Russia has recently deployed 40 new ICBMs and has leaked the existence of a long-range nuclear torpedo designed to circumvent US missile defences with the capability to 'destroy important economic installations of the enemy in coastal areas and cause guaranteed devastating damage to the country's territory by creating wide areas of radioactive contamination, rendering them unusable for military, economic or other activity for a long time' (BBC, 2015).

In the light of a perceived growing threat by Russia and the growing spread of nuclear weapons some think tanks and analysts have suggested that a future nuclear war could remain limited and that nuclear weapons could be used with restraint and discrimination (Larsen and Kartchner, 2014). This is, apparently, a very dangerous idea, as 'limited nuclear strikes...can provoke nuclear war' (Gareev, 1998: 89).

Furthermore, the aforementioned Congressional study claimed '[t]he impact of even a "small" or "limited" nuclear attack would be enormous' (US Congress, 1979: 4). More recent research has found that the consequences of nuclear war had been underestimated during the Cold War. Climate scientist Alan Robock has created environmental and climate models of nuclear explosions and found that even a small nuclear war with 100 nuclear weapons that are detonated 'could cause fires releasing 5 million tonnes of black carbon, with long-term temperature effects', resulting in a decade of cooling (Robock, 2011: 276).

The problem, already discovered by American scientists like Carl Sagan in the 1980s, based on advanced computer models, is called 'nuclear winter': the sunlight would be blocked for years by the massive amount of radioactive dust that gets thrown high up into the atmosphere in an above ground nuclear explosion, which would alter the climate, destroy crops and doom survivors to starvation (J.D. Hamblin, 2013: 237–240).

A regional nuclear war between India and Pakistan could have dramatic consequences for the global climate, due to 'massive ozone depletion, allowing more ultraviolet radiation to reach Earth's surface', which could threaten the survival of humanity (Robock and Toon, 2006: 74). As nuclear weapons spread to more countries, it will subsequently get more difficult for them to engage in any conflict without risking nuclear war and jeopardizing humanity's existence.

7.1.2 The military superiority of the US

The US still remains the only military superpower in the world that can globally project substantial force all by its own. Only the US has a fleet of 10 carrier groups on constant deployment rotation across the world. The only strategic 'near-peer competitors' that could potentially challenge US hegemony in the long-run are Russia and China, according to a new study by the *Center for Strategic and International Studies* (Murdock et al., 2014). But neither the Russian nor the Chinese navy could take on the US Navy directly, making any attempt of an invasion of the US mainland futile from the start. According to SIPRI, the US military budget for 2014 was $610 billion compared to China's of $216 billion and Russia's of $84 billion, which seems to give the US a clear military superiority over its strategic competitors.

However, as pointed out by the US 'National Military Strategy' paper, the American advantages in terms of 'technology, energy, alliances and partnerships, and demographics…are being challenged' (US DoD, 2015: 3). Russia's main military advantage remains its strategic nuclear forces, which give them almost parity to US nuclear forces, but its conventional military is deficient in terms of training, equipment and size. Although Russia is re-emerging as a military threat to the West, it seems very unlikely that Russia would risk any direct attack on NATO territory, especially as NATO continues to rely on tactical nuclear weapons as a 'core component of NATO's overall capabilities for deterrence and defence' of Western Europe (NATO, 2012).

Even more complex is the situation with respect to Chinese military ambitions, as American and Chinese economies are closely linked. Undoubtedly, Chinese military spending has grown dramatically over last ten years from less than $50 billion a year to over $200 billion today, but a direct military conflict with the US seems very improbable for many reasons. RAND analyst James

Dobbins has claimed that although there are potentials for conflict between the US and China over 'changes in status of North Korea or Taiwan, Sino-American confrontation in cyberspace, and disputes arising from China's uneasy relationships with Japan and India...a China-US military conflict is not probable in any of these cases' (Dobbins, 2012: 8–9). Both Russia and China are primarily investing in anti-access/ area denial capabilities that are meant to deter any outside intervention into their respective regions, which means that they maintain a defensive rather than an offensive posture.

It has been frequently pointed out that no state would dare to launch a direct military attack on the US and NATO, since they could not hope to win in such a confrontation. For example, the UK MoD's 'Global Trends' report claims '[w]hile it is unlikely that any state would directly challenge the US militarily, asymmetric attacks, from ideologically opposed state and non-state actors are likely' (UK MoD, 2010: 46). Similarly, the US National Military Strategy also anticipates that nonstate and hybrid conflicts will be more likely than state conflict (US DoD, 2015: 4).

7.1.3 The delegitimization of the use of force

Since the end of the First World War, there has been a process to delegitimize the use of force as an instrument of politics, which initially resulted in the League of Nations and later the United Nations. According to the UN Charter to which almost every state on earth is a party to, the use of force is only allowed either on the basis that it has been authorized by the UN Security Council (UN SC) to deal with a threat to the peace, or in the case that it is an act of self-defence according to Article 51. No state can launch a war of aggression against another state without risking military action by the international community to stop the aggression. Of course, UN peace enforcement has only happened twice in the UN's history (Korea 1950–53 and the Gulf War of 1991), but the prohibition of aggression remains the most important and most consistently followed norm in international relations.

The Bush administration proposed in 2002 a new doctrine of 'pre-emptive self-defence' in order to circumvent the requirement of a UN SC Resolution against Iraq to justify its attack on that country. The doctrine suggested that force could be used pre-emptively against terrorists and 'rogue states' even if there is no indication of an imminent or specific attack to which the act of 'self-defence' would respond. The administration argued: 'Iraq's demonstrated capability and willingness to use weapons of mass destruction... against the United States or its Armed Forces...and the extreme magnitude of harm that would result to the United States and its citizens from such an attack, combine to justify action by the United States to defend itself' (US Congress, 2002). However, if this was the standard for any nation to go to war with another nation, then it would do little to promote peace, as the mere suspicion of a threat could justify war. Many international law scholars have thus concluded that this doctrine does not conform to international law and even those governments that accept it as lawful, apart from the US, do so only in the context of nonstate threats (Reisman and Armstrong, 2006: 547).

Since the Arab Spring of 2011, NATO has tried to use a different legal rationale for using force against states that have not actually threatened them and this is the new UN doctrine of the 'responsibility to protect' (R2P), which proposes that the international community can intervene in states to prevent genocide. R2P was initially formulated in the World Summit Outcome document of 2005 and has since been included in a series of UN documents such as the *Implementing Responsibility to Protect* report of 2009 (Averre and Davies, 2015: 816). NATO invoked R2P in March 2011 and got a mandate from the UN SC to protect civilians during the Libyan civil war. Since NATO violated this very limited mandate contained in UN SCR 1973 by attacking Gaddafi forces, regardless whether they threatened civilians, Russia and China no longer agreed to a proposed 'no-fly-zone' over Syria. In summer 2013, the US tried to mount an attack on the Assad government in Syria based on the claim that Assad had used CW on civilians and that this would have crossed a 'red line'. Putin made it clear that Russia considered an attack on the Assad government a violation of the UN Charter (Averre and Davies, 2015: 824–825). In the end, the US military action against the state of Syria did not take place for the simple reason that Russia made a veiled threat to intervene on behalf of Assad (Roberts et al., 2013). In other words, engaging in open war against another state has become legally difficult even for the world's military hegemon.

7.1.4 Growing global interdependence

The liberal argument that growing global interdependence through trade and other exchanges makes interstate war less likely since it increases the relative cost of war, while diminishing any advantages that can be gained from war, seems to be accurate. In a global economy, states gain substantially more from cooperating with each other than from waging open war with each other. This is particularly true today since even conventional war has become so destructive that wars would destroy whatever states sought to protect or to economically gain from war (Mandelbaum, 1998: 21–22). Globalization has resulted in a situation in which hardly any state can afford to remain aloof and in which any disruption in global trade and finance can have system-wide impact. For almost every consumer and industrial good, there are global supply chains that, if disrupted, make these goods unavailable on the global market with serious implications for the world economy.

This interdependence and the resulting sensitivities and vulnerabilities to disruptions in supply even extends to the defence industries of the major powers. In 2011, the Senate Armed Services Committee discovered 1,800 counterfeit electronics parts in US military equipment of which 70 per cent originated from China (US Congress, 2011: 3). The committee found in one out of many examples that key electronics parts such as the FLIR system for the

Navy's SH-60 helicopter were procured from a Californian company owned by a British company that sourced the components from a Chinese company in Shenzhen, revealing the global connections (US Congress, 2011: 4). The investigators also found Chinese parts used in F-15s, V-22s and nuclear submarines (Reuters, 2012). The US military is so much dependent on electronics from China that it raises the serious question how the US could fight and sustain a war against this country, especially since supply chains tend to be 'on demand' and since so many pieces of US military equipment require Chinese components to function.

It would, theoretically, still be possible to fight major conventional wars against states that are less integrated into the global economy such as Iran and North Korea. At the same time, even such limited wars could potentially trigger a global recession if the attacked states and their major allies, most importantly Russia and China, would respond with economic sanctions and other economic warfare such as closing the Straits of Hormuz, which would cause the price of oil to skyrocket (Krepenevich, 2009: 22–23). It can be expected that further globalization and economic interconnectedness will increasingly constrain direct large-scale military conflict.

7.1.5 *Military neuroscience and state-on-state conflict*

Military neuroscience will be relevant in future conventional wars in the unlikely case that they will still occur. However, as the amount of automation in war grows, the value of military neuroscience in interstate conflict would be subsequently diminished since most decisions would be made and implemented by machines rather than humans. Considering the enormous amount of firepower at the disposal of modern armed forces, it would be far too risky to rely on nonlethal modes of attack that may be less effective than lethal modes of attack. It can be expected that future conventional conflicts would be fairly short in duration since they cannot be sustained for long and since enormous destructive force can be unleashed almost instantly. They would be localized and high-intensity, taking place across all military domains, including space and cyberspace. A likely strategy in such wars would be to quickly overwhelm the enemy in a 'shock-and-awe' style condensed campaign designed to psychologically manipulate the enemy into quickly accepting defeat. This would be also critical for controlling escalation in a localized war against another nuclear weapons state, for example in a US-China conflict scenario. Here are some areas to which neuroscience could contribute to future conventional wars:

- Psychological preparation of the population and armed forces for war: neuroscience can help improving war propaganda and strengthening morale that is critical in sustaining the political will for war in democracies. It was after all a major lesson of Vietnam that the war was lost as much on the home front as it was lost on the battlefield. Foreign audiences also need to be influenced into accepting the use of force as legitimate as it could affect alliances, partnerships and soft power more generally.

- The development of strong or human-level AI for battle management that can predict enemy behaviour and generate complex battle plans: it seems likely that AI modelled after the human brain will eventually lead to computer systems that can make better decisions in complex situations than humans and do so much faster. This creates strong incentives to transfer much decision-making to intelligent machines, which will increasingly minimize the human element in conventional or nuclear wars (Adams, 2001).
- Neurally-controlled unmanned systems: the US and other advanced militaries already operate a large number of unmanned vehicles (UVs) for reconnaissance and combat missions with the prospect of future 'robot armies'. Since the UVs still lack human-level AI, which limits their ability to discriminate and overall effectiveness, it will remain preferable for the time being to have a human in the loop. By using neural interfaces for controlling UVs there could be gains in terms of speed and precision.
- Enhanced warfighters and intelligence personnel: SOF and human agents can infiltrate enemy territory and could sabotage enemy defences ahead of an attack. For example, they could sabotage radars or command and control facilities, blinding the enemy, or disable nuclear missiles and other WMD. Neuroscientific enhancement could improve their chances of successfully completing their missions, as they could operate for many days without rest, communicate silently using 'synthetic telepathy' and be better able to withstand stress and pain.
- Degrading the enemy's capacity for sound decision-making: if the enemy is not capable of making good decisions it is a lot easier to defeat them. This could be achieved in numerous ways such as the use of light and sound weapons on the battlefield that put enemy soldiers into a state of trance. For example, laser dazzlers could be used to disrupt the brains of pilots, causing them to lose control over their aircraft. It has been also suggested that '[l]ow-frequency weapons and application of bioelectrics may severely reduce the effectiveness and alertness of enemy forces, commanders, and political leaders' (Bellamy, 1990: 243).
- Battlefield PYSOPS: neuroscience could improve PSYOPS to the point that enemy forces can be mentally coerced to lay down their weapons and surrender. This could include subliminals, acoustic or RF weapons that induce fear or panic, holograms, or more advanced methods of 'mind hacking'. Most likely such attacks would only be useful on a tactical level and may not be scalable.
- Nonlethal biological attack against enemy forces and populations: in order to prepare the battlefield an enemy might release 'stealth viruses' that can be triggered at a later point to affect enemy logistics and their ability to resist more lethal attacks. Using nonlethal rather than lethal agents appears to the world public to be more humane, while also

demonstrating capability for a lethal attack, which could persuade the enemy to give in to demands.

- 'Cyborged' and transgenic animals as sensors and fighters: coopted insects and birds could be used for reconnaissance and intelligence missions and new transgenic species could be created to be used as expendable fighters. Insects could be genetically engineered to produce toxins and incapacitating biochemicals and could be released over enemy territory to attack enemy forces and populations.

7.2 State-on-nonstate warfare

The type of conflicts that characterize warfare in the early twenty-first century differ from the major wars of the twentieth century in numerous aspects, most importantly they are wars in which state actors fight nonstate actors of various kinds such as terrorists, insurgents, mercenaries, criminal syndicates or warlords. The new reality of contemporary warfare has been described with terms like '4th generation warfare' (Lind et al., 1989) 'low-intensity wars' (van Creveld, 1991), 'gray area phenomena' (Manwaring, 1993), 'asymmetric threats', 'unconventional warfare', 'neomedievalism' (Cerny, 1998), 'new wars' (Kaldor, 1999) and most recently 'hybrid wars' (Freier, 2009), all of which convey the idea that states now have to fight irregular forces in which regular armies are often at a disadvantage. The new enemies can disguise themselves as civilians and attack only when they have the element of surprise on their side. They may hide in remote and difficult-to-access parts of the world or in the middle of major Western cities. They can move across borders and do not present many obvious targets against which the massive firepower of modern armed forces can be leveraged. Traditional distinctions of external defence and internal security, military and police, intelligence and law enforcement have all lost relevance as all of it gets blended into a very broad notion of 'security'. It is therefore misleading to assume that the task of defence would be limited to the armed forces or even that security would be limited to security forces and agencies (UK MoD, 2010: 76).

It will be argued that state-on-nonstate wars will remain the dominant form of armed conflict for many decades to come, which has many potential causes: the general decline of the nation state and the emergence of powerful nonstate actors since the second half of the twentieth century, the growing impossibility of great power war, the proliferation of technology among lesser powers, the growth of ungoverned spaces in the world and the new tendency of state actors to use nonstate actors as proxies as part of 'hybrid warfare' strategies. In short, the Westphalian international order is being deliberately undermined by various actors from below and above and this is fundamentally transforming the nature of war and conflict, creating a 'new security dilemma' caused by the weakness of states due to fragmentation from below (Cerny, 1998: 38–40).

7.2.1 'New wars' and 'weak states'

Van Creveld started in the early 1990s the debate about the new dominance of 'low-intensity conflicts'. He pointed out that only one quarter of the about 160 armed conflicts from 1945 to 1990 were conventional interstate wars, while the rest falls into the broad category of 'low-intensity conflicts' (van Creveld, 1991: 20). None of the currently ca. 16 ongoing armed conflicts in the world is an interstate war. In the 1990s, there was much debate about the security challenges caused by weak and failing states since they could not maintain control over the entirety of their territories, thus allowing illegitimate nonstate actors to gain a foothold and safe haven. Mary Kaldor theorized about 'new wars' that she distinguished from the earlier 'Clausewitzian' interstate wars. She argued: 'I use the term 'war' to emphasize the political nature of this new type of violence…the new wars involve a blurring of the distinctions between war…organized crime…and large-scale violations of human rights' (Kaldor, 1999: 2).

In her analysis, she identified four main characteristics: (1) actors: the 'new wars' are fought by networks of state and nonstate actors (compared to states only in 'old wars'), (2) goals: they are often motivated or justified by identity politics (instead of geopolitical objectives in 'old wars'), (3) methods: the focus is on politically capturing a population (instead of capturing territory in 'old wars') (4) forms of finance: they are financed through private predatory activities such as looting, smuggling, drug trafficking, etc. (instead of state taxation in 'old wars') (Kaldor, 2013: 2–3).

Other theorists such as Herfried Münkler have emphasized the privatisation aspect in these conflicts, suggesting that the dynamics and long duration of these conflicts result from the greed of 'war entrepreneurs', who have an incentive of keeping the conflicts going, causing the permanent disintegration of states (Münkler, 2004). Although the academic debate shifted significantly after 9/11 towards terrorism and its roots, many of the key arguments of the 'new wars' proponents seem to remain remarkably relevant to the present day, if one just looks at the biggest ongoing conflicts such as Afghanistan (NATO vs. Taliban), Syria (Syrian government vs. insurgent groups and international coalition vs. ISIS), the Ukraine (Ukrainian government vs. separatists), Iraq (Iraqi government vs. ISIS), Libya (Libyan government vs. Islamic groups), Turkey (Turkish government vs. Kurds) and Nigeria (Nigerian government vs. Boko Haram/ ISIS). The Pentagon seems to prefer the term 'violent extremist organizations' (VEO) to characterize the new opposition forces. The US 'National Military Strategy' of 2015, claims that VEO

> led by al Qaida and the self-proclaimed Islamic State of Iraq and the Levant (ISIL) – are working to undermine transregional security, especially in the Middle East and North Africa. Such groups are dedicated to radicalizing populations, spreading violence, and leveraging terror to impose their visions of societal organization. They are strongest where

governments are weakest, exploiting people trapped in fragile or failed states. In many locations, VEOs coexist with transnational criminal organizations, where they conduct illicit trade and spread corruption, further undermining security and stability.

(US DoD, 2015: 3)

Currently, there is no victory in sight against VEOs such as al Qaeda or ISIS. The British MoD's 2010 *Global Trends* report argues that radicalization and terrorism will remain a major challenge in the long-term: 'out to 2040, international terrorism *will* persist. Terrorist organisations, such as al-Qaeda, are *likely* to evolve, while maintaining their overall strategic aims. Al-Qaeda's pursuit of global objectives *will* rely upon radicalisation to provide support and to generate the conditions in which they can operate' (UK MoD, 2010: 32). The Obama administration expects the War on Terror to continue for at least another ten to 20 years (Greenwald, 2013).

7.2.2 *'Hybrid wars'*

The 'Global Trends' report argues that 'inter-state rivalries are *likely* to be expressed through proxies that have linked or complementary objectives' (UK MoD, 2010: 85). In other words, states will tend to sponsor proxy forces instead of seeking any direct military conflict with an adversary. Furthermore, '[m]any of these proxy forces are *likely* to employ irregular tactics including terrorism, while concealing and refuting links to state sponsors in order to preserve their freedom of action and maintaining a degree of deniability for the state. Proxies are *unlikely* to follow predictable paths and are *likely* to prove difficult to control over time' (UK MoD, 2010: 85).

What is discussed here has been described as 'hybrid warfare' elsewhere, which is currently a new military buzzword. The term was coined by Bill Nemeth, who defined it as 'the contemporary form of guerrilla warfare' that 'employs both modern technology and modern mobilization methods' (quoted from Freier, 2009). Hybrid wars are largely unconventional, although belligerents may use major conventional weapons systems and conduct some more conventional attacks. Unlike in wars against nonstate actors not supported by a major state sponsor one cannot know what to expect in a hybrid war: opponents may fight irregularly, but may also field tanks, air defence systems and engage in electronic warfare activities (Tilghman and Pawlyk, 2015).

The primary example of hybrid warfare is Russia's role in the Ukraine, supplying advanced weapons to rebels in the Eastern Ukraine, supported by clandestinely operating SOF. In addition, '[h]ybrid warfare involves a state or state-like actor's use of all available diplomatic, informational, military, and economic means to destabilize an adversary. Whole-of-government by nature, Hybrid Warfare as seen in the Russian and Iranian cases places a particular premium on unconventional warfare (UW)' (US Army, 2014: 3). UW is defined as '[a]ctivities conducted to enable a resistance movement or insurgency to coerce, disrupt, or overthrow a government or occupying power by operating through or with an underground, auxiliary, and guerrilla force in a denied area' (US Army, 2014: 41). A hybrid war strategy would thus blend state and nonstate capabilities, conventional and irregular forces to create ambiguity and uncertainty, which makes it very difficult to counter.

To some extent hybrid warfare overlaps with covert action as it has been conducted by the CIA during the Cold War, most notably the arming and training of the Mujahedeen in Afghanistan during the 1980s. In recent years the US has been pursuing a 'hybrid war' strategy in Syria to overthrow Assad by providing covert 'nonlethal' assistance to insurgents since at least August 2012 (Reuters, 2012), training 5,000 fighters and resupplying them via Turkey (Entous, 2015: 12). Although these indirect US interventions in conflicts are publicly known, they are still carried in a manner that provides a fig leaf of deniability, which avoids daunting legal issues, public accountability and possibly direct retaliation by the attacked party.

Mainstream media and major US think tanks suggest that this kind of covert warfare should be intensified. For example, Richard L. Russell from the National Defense University argued in *The National Journal* that the shift towards covert warfare is under current circumstances advisable. He wrote: 'To make American ends and means match, we will have to be smarter at using instruments other than military power. These instruments include diplomacy, intelligence and covert action' (Russell, 2015). He furthermore advocates understanding covert action in a broad sense to include a great variety of operations aimed to 'influence global situations to the advantage of American national interests that require the hiding of the American role to minimize risks and investments, while maximizing the chances for success' (Russell, 2015).

States that are threatened by superior conventional forces may also emphasize unconventional warfare as a key aspect of their deterrence and defence strategy. Hussein's Iraq created the paramilitary resistance fighters *Fedayeen Saddam* in the lead-up to the Iraq War, as part of its defence strategy, which showed its true effectiveness only after the defeat of Iraq's conventional forces during the years of American occupation. Iran also has paramilitary organisations that are trained to asymmetrically resist invading armies and it uses Hezbollah as its deterrent force. Unconventional/ hybrid warfare strategies are particularly attractive to weaker opponents and they have proven to defeat powerful conventional forces in Algeria, Vietnam and Afghanistan. It can be thus expected that most future wars will fall into this UW/ hybrid warfare/ covert action paradigm, as great powers will seek to avoid direct military confrontation that could easily end in nuclear war and as lesser power try to exploit asymmetry. Using proxy forces for achieving military strategic ends provides a fig leaf of deniability, as state sponsors will claim that they cannot control their proxies and it also puts clear limits to possible escalation, as support to proxies cannot be endlessly expanded without completely undermining the idea of deniability.

7.2.3 Military neuroscience and state-on-nonstate warfare

From the review of military literature and military strategy papers it is clear that most military analysts believe that the future of war is going to be unconventional and 'hybrid', where state forces are pitted against nonstate forces that may or may not be supported by state actors. The object of war will be the political control of populations rather than the physical control of territory. In such conflicts, military neuroscience could make a huge difference since it can be employed for political and military mobilization. In their work on 4th generation warfare, Lind et al. already anticipated a greatly expanded role of psychological and information operations in warfare: 'Psychological operations may become the dominant operational and strategic weapon in the form of media/ information intervention...Television news may become a more powerful operational weapon than armored divisions' (Lind et al., 1989: 24). There are many ways in which neuroscience could help in manipulating perceptions, emotions and the political will of the enemy, enabling victory with little or no bloodshed.

- Super soldiers: hybrid wars typically involve SOF as they can operate clandestinely in small units behind enemy lines and as their main function is training and leading indigenous forces. SOF may be enhanced in numerous ways, most importantly they have to be able to remain in the field for an extended of time, coping with sleep deprivation, high amounts of stress and high physical demands. It is already the case that Western SOF encounter 'enhanced' nonstate fighters on the battlefield. Mercenaries for the Mexican drug cartels are known to use cocaine, while Syrian fighters take an amphetamine cocktail known as 'captagon', which 'quickly produces a euphoric intensity in users, allowing Syria's fighters to stay up for days, killing with a numb, reckless abandon' (Holley, 2015).
- Cultural awareness and predicting enemy behaviour: key to success in unconventional conflicts is having cultural awareness and the ability to understand what the enemy really wants. Arguably, the US lost the Vietnam War because it lacked any understanding of Vietnamese culture and the motivations of the enemy they encountered. Cultural neuroscience can contribute to cultural awareness and help determine what the enemy holds sacred and will not compromise on and what can be negotiated. Of course, cultural awareness is key to successful PSYOPS, as an enemy will not react positively to messages that are culturally inappropriate or that are culturally insensitive.
- Political mobilisation: in hybrid wars it is often necessary to mobilize parts of the public or to shape public perceptions in a way conducive to the goals of the sponsors of hybrid wars. A political objective can be to either overthrow an existing government by fomenting discontent and civil unrest or it could be the opposite to garner support or sympathy for the actions of a government. As pointed out by military analyst Brian

Petit, with 'the weapon of the narrative' it becomes possible to over-throw a regime of 30 years in a matter of 18 days (referring to Egypt). 'Narratives in the form of stories, rumors, biographies and pictures drive our behaviors and shape our convictions' (Petit, 2012: 26–27). Thus the need for initiatives such as N2.

- Influencing external perceptions of a conflict: in addition to manipulating perceptions and narratives within a conflict zone, it is also important to influence domestic and other foreign audiences since this determines the amount of support that can be given openly by any state. For example, Russia has been reasonably successful in influencing certain Western audiences in terms of their perception of the Ukraine conflict, eroding public support for the sanctions against Russia that have been pushed by the US. At the same time, Russia's propaganda offensive has made it difficult for Western governments to take a more active role in the Ukraine conflict.

7.3 'Non-obvious' / ambiguous warfare

The US military conceptualises hybrid warfare as a form of unconventional warfare or psychological warfare (US DoD, 2014: 3), which implicitly assumes that there is an armed conflict in the first place in which outside parties indirectly intervene. However, states may engage in other hostile activities against each other that fall outside the unconventional warfare paradigm and that occur during times of relative peace. The strategist Martin Libicki has therefore introduced the concept of 'non-obvious warfare' in 2012. He suggested: 'Although non-obvious warfare can be epitomized by cyber warfare, states can attack one another in many ways without the victim being certain exactly who did it or even what was done' (Libicki, 2012: 88). He lists numerous possibilities for how a state could attack another state with the possibility that the hostile action might not be attributable or may not even be registered as a deliberate attack, such as 'cyber warfare, space warfare, electronic warfare, drone warfare, sabotage, special operations, assassins and mines, proxy attacks, weapons of mass destruction and intelligence support to combat operations' (Libicki, 2012: 88–89).

This list is not even exhaustive, as one could imagine the use of environmental modification methods for controlling the weather in another country (e.g. causing droughts), or seismic weapons that trigger earth quakes or volcanoes, or hidden financial manipulations to engineer a currency or stock market collapse. A covert attack may be also long-term or may have a delayed action mechanism, as suggested by Russian General Vladimir Belous. He discussed the idea that established WMD 'could be replaced by "slow action" systems and combat assets producing a latent damaging effect on the human organism, gradually destroying it, eroding the vital organs, impairing protection against meteorological and infectious impacts, and thus gradually killing it or causing its long-term incapacitation' (Belous, 2009: 66).

What is critical to non-obvious warfare is ambiguity: the target should have no certainty about who is responsible or whether an action is indeed hostile. This ambiguity has advantages and disadvantages, as pointed out by Libicki. The main advantage is obviously that if a hostile act cannot be attributed it reduces the chances of retaliation since the target would not know against whom to retaliate. This also creates limitations for escalation: in order to be non-obvious, hostile acts need to be single events (if they become a sustained campaign, they become obvious) or they need to be undetectable (for example, by disguising them as 'natural' events or 'glitches'). If such incidents happen too often they also become obvious: 'many accidents will start to smell like attack' (Libicki, 2012: 101).

Another downside is that non-obvious warfare is not very suitable for coercion: if the target cannot attribute the attack and thus does not understand whether there is an attack or why an attack happened, then it is less likely that it will change the behaviour of the target in a desired manner. Ambiguity could result from using proxies, such as nonstate actors like terrorist groups, mercenaries, or independent hacker groups, who act on behalf of a state, but do not have any visible links to that state. The need for secrecy also imposes organizational limitations. According to Libicki, '[n]on-obvious warfare, almost by definition, has to be the work of small teams that must isolate themselves from the larger community, much like intelligence operatives, lest word of their adventures leak out' (Libicki, 2012: 100).

Based on the discussion by Libicki, non-obvious warfare attacks would have the following characteristics:

- They are ambiguous in terms of attribution and intention.
- They occur in times of relative peace.
- They are carried out by proxies or small teams within a national security establishment.
- They may use non-military means.
- They may be long-term or may have a delayed action mechanism.
- They are designed to affect either the capability and or the behaviour of the target in a specific way.

The following section will discuss some 'non-obvious warfare' approaches that are likely to become more prominent in the twenty-first century, namely cyber warfare, other sabotage and 'grey terror' and environmental modification.

7.3.1 Cyber warfare

According Richard Clarke and Robert Knake, cyber war refers to 'actions by a nation-state to penetrate another nation's computers or networks for the purposes of causing damage or disruption' (Clarke and Knake, 2010: 6). Cyber war is uniquely suited for non-obvious warfare since correct attribution is generally very difficult and since a cyber attack may be indistinguishable from technical glitches. Libicki explains:

> The ones and zeroes that constitute the attack do not bear the physical residues of their operators (especially if these ones and zeroes are copied from others' tools). Successfully attacked systems, almost by definition, cannot distinguish an attack from completely benign inputs at the time (with a distributed denial-of-service attack, it is volume, not content, that matters; the attacking bytes generally come from 'innocent' machines that have been tricked into spamming the victim). Forensic methods such as tracing the attack back to its sources can be easily frustrated by bouncing the attack through enough portals, using the services of an innocent machine, or jumping on a third-party Wi-Fi connection.
>
> (Libicki, 2012: 92)

For example, distributed denial of service (DDOS) attacks, which are an established tool in the cyber war arsenal, typically rely on 'botnots' comprised of hijacked computers that send nonsensical requests to a server to shut down an Internet service (Clarke and Knake, 2010: 14–15). Estonia suffered a massive DDOS attack involving a million zombie computers in April 2007 directed against online banking services, online newspapers and government websites, which disrupted public life for several days (Clarke and Knake, 2010: 12–16). The Estonian government traced back the attacks to Russia, but apparently '[n]either experts from the Atlantic Alliance nor from the European Commission were able to identify Russian fingerprints in the operations. Russian officials called accusations of involvement "unfounded"' (Rid, 2011: 12). The problem is that in some countries Internet security is weak as most of their computers use non-licensed versions of Windows that do not receive security updates and are thus particularly vulnerable to getting infected with malware that can turn these machines into 'zombies'. It is therefore not really surprising that most cyber attacks originate from Russia and China, which provides these governments with some plausible deniability for the attacks that they do perpetrate themselves.

The Stuxnet worm is usually considered to have been the first real incident of cyber war, since it was apparently the first time that a malware could cause physical damage to machinery. The worm targeted the Iranian nuclear program in a non-attributable way. It has been estimated that the Stuxnet worm destroyed 1,000 centrifuges or about 20 per cent of the total number of centrifuges between 2009 and 2010, setting Iranian uranium production back by up to two years. Unfortunately, the Stuxnet worm unintentionally ended up on the Internet, infecting 100,000 computers worldwide, 60 per cent of them located in Iran (Rid, 2013: 44). The highly unusual sophistication of the malware (four 'zero-day exploits' and two stolen certificates) pointed at small number of state actors that had both the motive and the capabilities for conducting this high-tech sabotage (Libicki, 2012: 92).

Currently, the main cyber warfare threat posed by states like China relates to cyber espionage, but there are concerns that cyber attacks on critical infrastructure could cause massive societal disruption if it was possible to disable the electric grid, steer nuclear power plants out of control, or shut down the Internet. Clarke and Knake have argued that the issues that cyber espionage and cyber sabotage overlap: hackers can 'map' an enemy's computer network for later attacks or they could leave behind 'logic bombs' that can be triggered at some later point. They define 'logic bomb' as a 'software application or series of instructions that cause a system or network to shut down and/or erase all data or software on the network' (Clarke and Knake, 2010: 287). In other words, 'logic bombs' are 'delayed action weapons' that could remain undetected on a network until they are activated. Even if they are discovered in time, there is little chance of proving that an act of aggression has been perpetrated by a particular government.

7.3.2 Other sabotage and 'grey terror'

Apart from cyber warfare, state actors could use 'grey terrorism' as a method of non-obvious warfare, aimed at weakening and destabilizing another state. Terrorism is most commonly defined as violence that is deliberately directed against noncombatant targets for psychological impact to achieve some political objective. Terrorism expert Bruce Hoffmann argued: 'Terrorism is specifically designed to have far-reaching psychological effects beyond the immediate victim(s) or object of the attack' (B. Hoffmann, 1998:43–44). Terrorism is typically a strategy of the weak against the strong, but sometimes terrorism is state sponsored, meaning that states provide '*intentional assistance* to a terrorist group to help it use violence, bolster its political activities, or sustain the organization' (Byman, 2005: 10).

The idea that terrorism is *mostly* state sponsored as a 'type of covert or surrogate warfare' was widely accepted in terrorism studies during the 1980s, but since the 1990s there has been the perception that terrorism has become self-sustaining through criminal activity such as narcotics trafficking (B. Hoffman, 1998: 27–28). Although al Qaeda has been considered to have acted without direct state-sponsorship when they conducted the 9/11 attacks, it would be wrong to completely disregard the possibility of state-sponsorship of terrorism, either in relation to 9/11 or any other major international terrorist incident.

Many defectors have confirmed that the Soviet Union had been supporting a very diverse set of terrorist groups such as the Irish IRA, the Basque ETA, the German RAF and the Arab PLO. It is uncontroversial that the KGB, through its proxy the East German Stasi, logistically supported the West German terrorist group Red Army Faction (RAF), which carried out a few spectacular attacks against German politicians and industrialists. This fact has been now validated by documentary evidence from the Stasi archives proving these links, although there is disagreement to the extent of Stasi support (Vielhaber, 2013: 534).

Romanian General Ion Pacepa, who defected to the West to the US in 1978, claimed: 'Today's international terrorism was conceived at the Lubyanka, the headquarters of the KGB, in the aftermath of the1967 Six-Day War in the Middle East. I witnessed its birth in my other life, as a Communist general'. He quoted the head of the KGB's First Directorate (foreign intelligence) General Alexander Sakharovsky: 'In today's world, when nuclear arms have made military force obsolete, terrorism should become our main weapon' (Pacepa, 2006). This is the same KGB general, who claimed in 1971 '[a]irplane hijacking is my own invention' in reference to the Palestinian hijackings of the late 1960s (Lockwood, 2011).

Terrorism studies experts have tended to argue that state-sponsored terrorism is actually less dangerous than terrorism without state-sponsorship, as state sponsors have to be concerned about escalation and the possibility that they themselves become targets of retaliation, if a terrorist attack is excessive and can be traced back to them. For this reason state sponsors would be unlikely to supply WMD to a terrorist group that they may not be able to fully control and that has ostensible links to them (Byman, 2005: 52). However, a state may attempt to conduct a WMD terrorist attack in a manner that implicates an innocent third party, which is sometimes called 'false-flag terrorism'.

According to Soviet defector Victor Suvorov, the Soviets intended to conduct false-flag terrorist attacks in the West in an effort to demoralise, distract and prepare a larger military attack. He argued: 'The overture is a series of large and small operations the purpose of which is, before actual military operations begin, to weaken the enemy's morale, create an atmosphere of general suspicion, fear and uncertainty, and divert the attention of the enemy's armies and police forces to a huge number of difficult targets, each of which may be the object of the next attack' (Suvorov, 1988: 176). Furthermore,

> The overture is carried out by agents of the secret services of the Soviet satellite countries and by mercenaries recruited by intermediaries. The principal method employed at this stage is 'grey terror', that is a kind of terror which is not conducted in the name of the Soviet Union. The Soviet secret services do not at this stage leave their visiting cards, or leave other people's cards. The terror is carried out in the name of already existing extremist groups not connected in any way with the Soviet Union, or in the name of fictitious organizations.
>
> (Suvorov, 1988: 176)

The biggest concern would be that a state might carry out a 'grey' terrorist attack against another state with WMDs, including bioterrorism or a nuclear terrorist attack. The psychological impact of a nuclear terrorist attack would be so severe that it could cause societal chaos, as argued by researchers Charles Ferguson and William Potter in their book:

Consequences stemming from a terrorist-detonated nuclear weapon in an American city would emanate beyond the immediate tens or hundreds of thousands of fatalities and the massive property and financial damage. Americans who were not killed or injured by the explosion would live in fear that they could die from future nuclear terrorist attacks. Such fear would erode public confidence in the government and could spark the downfall of the administration in power. The tightly interconnected economies of the United States and the rest of the world could sink into a depression as a result of a crude nuclear weapon destroying a city.

(Ferguson and Potter, 2005: 3)

The only thing that can prevent the scenario of a small nuclear bomb going off in a ship container in a major port would be if there was sufficient certainty that the real perpetrators can be accurately identified after the fact. This does not seem to be the case. According to military analyst Andrew Krepinevich, nuclear forensics 'is not foolproof': 'it is far more capable of telling us where the materials in a nuclear weapon did not come from than where they did' (Krepinevich, 2009: 71). Even if a nuclear terrorist attack never happens, 'grey terrorism' and systematic sabotage can be a 'strategy of a thousand cuts' that continuously weakens and demoralises a society.

7.3.3 Environmental modification

Although nuclear weapons clearly dominated strategic thought of the Cold War period, there were also substantial resources and thinking devoted to the development and use of other types of WMD, namely environmental modification (ENMOD) weapons. Historian Jacob Darwin Hamblin has chronicled in his book *Arming Mother Nature* how the US and the Soviet Union have toyed and indeed experimented with ENMOD since the Second World War, which included 'the development of weapons of massive death, the contamination of large areas of land, the poisoning of crops, the harnessing of geotectonic forces, [and] the steering of storms' (Hamblin, J.D. 2013: 9). The superpowers explored the ideas of triggering earthquakes and tsunamis with powerful nuclear explosions and of contaminating large territories through biological and radiological warfare. General Belous even includes more exotic concepts in his list of new WMD such as 'ozone weapons', 'asteroid de-orbiting' and 'psychotronic weapons' (Belous, 2009).

Since the 1960s, there was a particular interest in controlling the weather and the climate, as there were legitimate peacetime applications and as it promised to interfere with the enemy's food production and other logistics in war. A US Air Force publication on weather warfare mentions a project undertaken during the Vietnam War:

A pilot program known as Project Popeye conducted in 1966 attempted to extend the monsoon season in order to increase the amount of mud on the Ho Chi Minh trail thereby reducing enemy movements. A silver iodide nuclei agent was dispersed from WC-130, F4 and A-1E aircraft into the clouds over portions of the trail winding from North Vietnam through Laos and Cambodia into South Vietnam. Positive results during this initial program led to continued operations from 1967 to 1972. While the effects of this program remain disputed, some scientists believe it resulted in a significant reduction in the enemy's ability to bring supplies into South Vietnam along the trail.

(House et al., 1996: 28)

By 1977, both superpowers were sufficiently concerned about the reality of ENMOD warfare that they negotiated the ENMOD treaty, which comprehensively outlaws any hostile environmental modification. The convention specifically prohibited ENMOD techniques such as 'earthquakes, tsunamis; an upset in the ecological balance of a region; changes in weather patterns (clouds, precipitation, cyclones of various types and tornadic storms); changes in climate patterns; changes in ocean currents; changes in the state of the ozone layer; and changes in the state of the ionosphere' (UN, 1978).

The convention was certainly a step in the right direction, but like many other arms control agreements it lacks verification mechanisms and it does contain some major loopholes that may allow states to develop ENMOD techniques, which could be used covertly against other states. The biggest loophole is that ENMOD is not entirely prohibited, only 'military' or 'hostile' ENMOD. However, it is unclear how peaceful ENMOD could be distinguished from 'hostile' ENMOD as ENMOD can easily affect other states negatively and as it is hard to determine whether this is accidental or whether there was hostile intent. Hamblin quotes an arms control specialist, who

admitted to not understanding the difference between weather modification and environmental warfare, except that the latter seemed to include a variety of 'less probable technologies – making holes in the ozone layer, melting ice caps; manufacturing earthquakes…and…fiddling with lightning in such a way as to enhance low frequency electrical oscillations in the atmosphere and thereby interfere with the alpha activity of the brain!'

(J.D. Hamblin, 2013: 206)

Even after the Cold War government concerns about hostile ENMOD have not subsided. The UK MoD 2010 'Global Trends' report states: 'Weather modification will continue to be explored. The aims are to obtain more water, reduce hail damage, eliminate fog, or other similar practical result in response to a recognized need. Manipulation of the weather may affect changes in operating conditions, limit aviation flight envelopes, generate poor visibility while providing concealment and disrupt lines of communications. Weather modification may also affect morale' (UK MoD, 2010: 156). Recently, the climate scientist Alan Robock was asked by the CIA '[i]f another country were trying to control our climate, would we be able to detect it?', which indicates at the very least that the US national security establishment takes the threat of covert ENMOD seriously (Robock, 2015).

7.3.4 Neuroscience and non-obvious warfare

There are many good reasons to believe that apart from hybrid wars non-obvious warfare will become the prevalent mode of conflict between great powers, which continue to compete fiercely over global influence and diminishing natural resources: (1) this form of conflict avoids a direct military confrontation, which could escalate to global nuclear war, (2) the ambiguity inherent in non-obvious attacks makes retaliation against the guilty party less likely, (3) non-obvious attacks can affect both the capabilities of another state and its behaviour, and (4) it might be difficult for the attacked party, especially if it was a democratic state, to come up with any effective or open response to a major threat that may not even be perceptible as such by the public at large. Neuroscience can enhance capabilities for non-obvious warfare in numerous areas:

- Covert attacks on leaders: attacks may be directed against enemy leaders to degrade their ability for making sound decisions or for psychologically manipulating them, if not recruiting them surreptitiously. Claims have been made that the Soviets successfully recruited foreign leaders with the help of drugs to turn their countries into Soviet puppet states (Douglass, 2001: 130–133). Although this seems exaggerated, such kind of covert mind control of foreign leaders might become possible in the future.
- Mind hacking: in the future, some cyber attacks may be directed against users rather than networks and hardware. Sophisticated psychological operations could be conducted on the Internet with the advantage that it can reach a very large number of people, who could be influenced in a desired way. It has been frequently noted that the Internet has become an ideal radicalization and brainwashing tool by VEOs. People could be manipulated into conducting terrorist attacks with little or no central support or coordination. Several Islamic terrorist attacks of recent years seem to fit this pattern. Neuroscience could assist nefarious actors with identifying and radicalising vulnerable individuals.
- Prediction of social dynamics: computer modelling may enable the accurate prediction of how a population would react to non-obvious warfare attacks, which can be used for fine-tuning for the desired psychological effects, such as the overthrow of another government.
- Large-scale mood manipulation and 'psychopolitical warfare': some ENMOD techniques such as weather and climate modification or the manipulation of the ionosphere seem to be particularly promising in terms of the possibility of affecting the moods and mental capacity of

entire populations. Possible methods that have been discussed by experts include biological warfare, the introduction of psychoactive chemicals in the food supply, the spraying of metallic nanoparticles that are ingested and debilitate brain and CNS, the use of ionising and non-ionising radiation, or the deliberate creation of weather conditions that affect mental capacity, moods and well-being in a specific way (Bushnell, 2001). Climate change has been linked to psychological disorders and propensity to violence (Doherty and Clayton, 2011: 265). For example, research indicates that warm and sunny weather reduces depression; that heat and extreme rain amplify aggression; and that serious depression and suicide are influenced by seasonal changes.

7.4 Controlling populations/ security

The biggest security threat to Western societies will probably not be related to any external military aggression by other state actors, but rather to the prospect of collapse due to insurmountable future challenges and poor governance. Robert Kaplan wrote an influential article with the title 'The Coming Anarchy' in 1994, in which he claimed that the whole world would be slowly sliding into chaos. He wrote:

> West Africa is becoming the symbol of worldwide demographic, environmental, and societal stress, in which criminal anarchy emerges as the real 'strategic' danger. Disease, overpopulation, unprovoked crime, scarcity of resources, refugee migrations, the increasing erosion of nation-states and international borders, and the empowerment of private armies, security firms, and international drug cartels are now most tellingly demonstrated through a West African prism.
>
> (Kaplan, 1999: 7)

Twenty years later Kaplan's predictions remain relevant and his ideas are referenced in Western strategy papers. World population is now at 7.2 billion, which is already two billion people more than in 1994. By 2040, no fewer than 8.8 billion are likely to live on earth (UK MoD, 2010: 94). It is predictable that there will be huge problems in terms of providing an adequate living standard to them, as resources are finite and are diminishing at ever faster rates.

There are huge concerns with respect to the possible depletion of fresh water reserves due to the impact of climate change (UK MoD, 2015: 21). People need water, food, clothes, shelter, energy, transportation, health services, education and many other things, or civilisation collapses. Mass impoverishment, as seen in the developing world, leads to rapid and uncontrolled urbanisation (the growth of slums), rising crime rates, environmental degradation, spreading diseases and political extremism.

It is expected that by 2030, two thirds of the world's population will live in cities with a population of more than a million and that by 2025 there will be at least 27 megacities with ten million or more living in them (from currently 17) (Liotta and Miskel, 2012: 4). Governments in the South are already struggling to control their populations to prevent the depletion of critical resources such as water and to reduce rampant crime. Some parts of the world are effectively ungoverned spaces that are controlled or occupied by nonstate actors such as drug cartels, militias and jihadists. In 2003, DCI George Tenet identified 50 'ungoverned spaces' around the world and the problem seems to get worse. More recently a State Department official stated: 'Quite frankly, the problem of un- and under-governed spaces is one of the toughest ones this and future administrations will face' (Patrick, 2010).

7.4.1 The crisis of the West

Many of the problems of the south will eventually reach the West through immigration, global economic crises and out-of-control debts of many Western governments that will make current welfare systems unsustainable. The US government has an official national debt of $19 trillion (exceeding the $17 trillion GDP) and it has grown in average by a trillion dollars a year over the last nine years. It is clear that the debt can never be repaid and that at some point the US currency system may collapse (Rickards, 2014). Many European countries are similarly burdened with debt while their economies are stagnating, making the debt unsustainable, which leaves them only with two terrible options: default or hyperinflation. The National Intelligence Council warned in its 2012 'Global Trends' forecast:

> A return to pre-2008 growth rates and previous patterns of rapid globalization looks increasingly unlikely, at least for the next decade. Across G-7 countries, total nonfinancial debt has doubled since 1980 to 300 percent of GDP, accumulating over a generation. Historical studies indicate that recessions involving financial crises tend to be deeper and require recoveries that take twice as long. Major Western economies – with some exceptions such as the US, Australia, and South Korea – have only just begun deleveraging (reducing their debts); previous episodes have taken close to a decade.
>
> (NIC, 2012: VI)

Mass impoverishment in the West also becomes likely because of growing automation and rationalisation, which will threaten jobs not only in the manufacturing sector, but also in the services sector. Driverless cars will make taxi drivers and truck drivers redundant while AI software will take over many functions from telemarketing to giving health or financial advice. A study by researchers from Oxford University suggests that 47 per cent of occupations are at a high risk of being automated out of existence in the US in the future (Frey and Osborne, 2013). In the Western world hundreds of millions of jobs may be lost due to automation.

Even if the AI and robotics revolution does not happen, the current state of the world economy makes it unlikely that there will ever be a return to low unemployment rates. There are 94 million working-age Americans out of job compared to 144.3 million employed Americans, which brings the real unemployment rate closer to 38 per cent rather than the about 5 per cent claimed by the official government statistics (WSJ, 2013). Wealth is getting more and more concentrated while the gap between rich and poor is widening across the G7 nations. An article in the *Scientific American* claims that wealth inequality is 'far worse than you think': 'The top 20% of US households own more than 84% of the wealth, and the bottom 40% combine for a paltry 0.3%. The Walton family, for example, has more wealth than 42% of American families combined' (Fitz, 2015).

At a certain level of unemployment combined with severe cuts in the welfare system, as currently experienced in Greece and Spain, the situation may become dangerous, as food riots, political extremism and general lawlessness ensue. The main security challenge for governments around the world in the next decades could be to control their populations and to maintain their political stability in the face of generally declining living standards and impoverishment, especially in formerly affluent societies.

7.4.2 *Mass surveillance*

Western governments have greatly intensified the mass surveillance of their populations often in violation constitutionally guaranteed civil rights, supposedly with the intention of fighting terrorism. Documents leaked by NSA whistleblower Edward Snowden have provided solid proof of the existence of NSA domestic surveillance that collects communications metadata of 'US persons' in bulk and that can be queried by NSA analysts to find terrorism connections.

The extent of the NSA surveillance is truly mindboggling. For example, it is known that the NSA built for this purpose its own version of Google that can query a communications database containing '850 billion records about phone calls, emails, cellphone locations, and internet chats...' (Gallagher, 2014). The NSA monitored 125 billion phone calls in a 30-day period, including 3 billion calls that originated in the US (4 billion per day). It records at least one billion mobile phone calls per day. Former NSA technical director William Binney estimated in 2012 that the NSA collected 15 to 20 trillion 'transactions' in eleven years since 2001 (Binney, 2012). NSA expert James Bamford claimed that the $2 billion dollar NSA facility in Bluffdale, Utah has supercomputers so powerful that they can handle yottabytes of data – this is a thousand times bigger than the size of the current Internet (Bamford, 2012).

However, it is not just the NSA spying on everybody, but rather many governments participating in this effort of global surveillance. Internet security expert Bruce Schneier recently wrote in *The Atlantic* that governments are united by their desire to conduct mass surveillance globally, which would create strong incentives

> to join the most extensive spying network around. And that's the United States. This is what's happening right now. US intelligence agencies partner with many countries as part of an extremely close relationship of wealthy, English-speaking nations called the Five Eyes: the US, UK, Canada, Australia, and New Zealand. Other partnerships include the Nine Eyes, which adds Denmark, France, the Netherlands, and Norway; and the Fourteen Eyes, which adds Germany, Belgium, Italy, Spain, and Sweden.
>
> (Schneier, 2015)

It has since emerged that NSA global mass surveillance has not prevented a single terrorist attack (Peterson, 2013). This is not really surprising as untargeted mass surveillance is unlikely to discover any terrorists, who are obviously aware that their communications could be monitored and who are quite capable of evading electronic surveillance by rather simple measures such as using couriers or using inconspicuous coded language that will not be picked up by an automated surveillance system. So what is the real purpose of mass surveillance? At the moment it is likely little more than a PSYOP designed to produce a chilling effect to curb criticism of governments and to keep citizens in line.

7.4.3 Preparing for civil unrest

In the US and in Europe governments already prepare for some kind of major future disruption and mass civil unrest (Ahmed, 2014). The US Army has conducted war games and exercises such as Unified Quest 2011 in preparation of '"large scale economic breakdown" in the US impacting on food supplies and other essential services, as well as how to maintain "domestic order amid civil unrest"' (Ahmed, 2013). Traditional constitutional barriers to the domestic use of the military are eroded as laws are changed and plans for extensive domestic security operations are drawn up. Previously, the Posse Comitatus law created in the aftermath of the American Civil War prevented military operations inside the country. However, since 9/11 Northern Command was established, which the President can use in national emergencies, such as an attack with WMD or any other undefined major disaster. There are currently 20,000 troops attached to NORTHCOM available for domestic contingencies. Army Chief Staff General Ray Odierno pointed in a Foreign Affairs article towards a greater domestic role of the US Army:

> Army forces will continue to be ready to contribute to broader national efforts to counter those challenges at home, if needed. Our reserve component soldiers remain the bedrock of the army's domestic response

capability, but where appropriate we will dedicate active-duty forces... to provide civilian officials with a robust set of reliable and rapid response options.

(Odierno, 2012)

The US military has since drawn up a still classified contingency plan for domestic civil unrest, codenamed CONPLAN 3502 (Hudson, 2011). There are widespread concerns that the US military would be deployed if there was domestic civil unrest. These speculations are fuelled by military publications like Kevin Benson's and Jennifer Weber's article 'Full Spectrum Operations in the Homeland' published in the *Small Wars Journal*, where they develop the hypothetical scenario of a TEA Party insurrection in 2016, caused by rising unemployment, political discontent and growing racism (Benson and Weber, 2012). That the US is ripe for mass civil unrest is indicated by opinion polls, which reveal that many Americans are so fed up with their government that 29 per cent would even support a military coup (Brait, 2015). The very idea of a military coup in a major Western democracy was previously considered unimaginable.

While neither a TEA Party insurrection, nor a military coup, are very likely, the militarisation of the police is a much more concrete and worrying tendency. The police become increasingly militarized in terms of their training, equipment and culture. Tremendous amounts of military grade equipment have been transferred to police departments under the '1033 program' that began in 1997 (Balko, 2013: 209). From 2006 to 2014, police departments received 600 MRAP 18-ton tanks, 79,288 assault rifles, 205 grenade launchers, 11,959 bayonets, 3,972 combat knives, $124 million worth of night-vision equipment, including night-vision sniper scopes, 479 bomb detonator robots, 50 airplanes, including 27 cargo transport airplanes, 422 helicopters and $3.6 million worth of camouflage gear (NPR, 2014).

Investigative journalist Radley Balko also noted in his book that law enforcement agencies are adopting military tactics and acquire a similar confrontational outlook as military forces. This is particularly apparent with respect to the proliferation of so-called 'Special Weapons and Tactics' (SWAT) teams, who are trained like the military and who specialize in unannounced 'no-knock'-raids. According to Balko, the percentage of SWAT teams for towns between 25,000 to 50,000 people grew from 25.6 per cent in 1984 to 80 per cent in 2005 while the number of drug raids exploded from 30,000 in 1995 to 50,000–60,000 in 2005 (Balko, 2013: 308).

The Department of Homeland Security (DHS) also seems to be expecting widespread domestic unrest, as it made some otherwise inexplicable massive orders for military-grade equipment and ammunition. DHS plans to purchase 14 more *Predator* drones for domestic surveillance bringing its current fleet to the size of 24 *Predators*. It also put in an order for 1.6 billion rounds of ammunition, which is somewhat out of proportion for normal law enforcement purposes (Bell, 2013). Equally puzzling is the order for 2,700 MRAP vehicles that are more suitable for urban warfare in Baghdad than policing tranquil American suburbs (Bell, 2013).

It becomes more and more evident that Western governments do not fear so much the few Islamic radicals that are living in their midst and who can be relatively easily monitored, but rather the spreading of other extremist anti-government sentiments in response to rising unemployment and declining living standards. For example, right-wing extremism seems to be on the rise across Europe as there are fears that the masses of migrants from the Middle East could further burden and ultimately collapse their existing welfare and healthcare systems. Regardless of whether these fears are justified, the reality is that extremism is growing and is already destabilising Western societies.

7.4.4 Military neuroscience for controlling populations

There are strong indicators that controlling large and impoverished populations will become a key challenge in the twenty-first century, as there is a combination of factors that will over time exacerbate already existing problems of overpopulation, resource scarcity, unemployment and failing systems of governance around the world. These problems will be made worse by covert aggression and deliberate destabilization by a variety of state and nonstate actors. The impacts will not be limited to the developing world, although they will be felt most severely there. Western governments seem to be already preparing for mass civil unrest, if not the threat of civil war. For this reason, any technologies and methods that can reduce or otherwise combat political extremism and generally calm populations will be extremely important for the future. There can be little doubt that neuro S/T will play an important role in the homeland security domain, which could make the use of direct force unnecessary.

- New NLW for riot control: military neuroscience could help develop new technologies for crowd control such as NLW that can influence the moods and behaviour of populations. This can include biochemicals and DEW that do not only temporarily capacitate, but that may permanently alter the behaviour of violent extremists. Populations could be calmed and psychologically discouraged to riot, or could be influenced to go along with government policies that are socially divisive.
- Combating violent extremism: military neuroscience can assist with fighting extremism by better understanding the motivations of extremists and by undermining their narratives in a more sophisticated and effective way.
- Using PSYOPS on own populations: as conditions in western democracies may be worsening governments may feel compelled to misdirect the public with contrived events and narratives that distract from real problems and issues. Diversionary politics aimed at directing public anger towards an imagined external threat may become more common. For example, the Russian government has been very successful in blaming

Table 7.1 The new spectrum of conflict

Conflict at societal level	Conflict at state level (Clausewitzian Paradigm)			
Controlling Populations (very low intensity conflict)	State-on-Nonstate (low intensity conflict)	State-on-State (medium intensity conflict)	State-on-State (high intensity conflict)	State-on-State (very high intensity conflict)
• Political instability • Economic instability • Crime • Terrorism • Environmental degradation	• Insurgency • 'Hybrid wars'	• Conventional war • Coercive diplomacy	• Conventional war against a peer competitor	• Conflict involving WMD
• Homeland security • Counterterrorism • 'Non-obvious war': Netwar, cyber war, space war, etc. • Social engineering	• COIN • Covert action (pro-insurgency ops)	• Conventional operations	• Conventional operations • Anti-Access/ Area Denial	• Nuclear war

Source: Author's own data.

the West for its currently difficult economic situation, which has to some extent galvanized the Russian population to support Putin.

- New mass surveillance technologies: the development of strong AI will be critical to perfecting the mass surveillance of millions, if not billions of people. Many experts, such as William Binney, believe that the NSA still lacks the necessary analytical capabilities for making any sense of the vast amounts of data that it collects. Without these capabilities the massive 'big data' collection is pointless. By training computers to imitate the human mind and human reasoning effective mass surveillance becomes much more feasible.

- Predicting violent behaviour: using brain scans, genetic sequencing (DNA analysis) or the analysis of social media and other surveillance data dangerous individuals could be identified before they turn violent and put on watchlists. Using psychological conditioning (brainwashing), drugs or genetic modification individuals with a recognized propensity for extremism could be made to conform to social standards.

7.5 Conclusion

This chapter has argued the following: (1) major conventional war, especially amongst the remaining Great Powers, is becoming increasingly unlikely because of the nuclear dilemma, the continuing military superiority of the US, the delegitimisation of the use of force and growing global interdependence; (2) the currently dominant form of armed conflict is state-on-nonstate warfare, which are generally internationalized internal conflicts and which are sometimes 'hybrid wars', where state actors support nonstate proxy forces in a variety of ways, including material and training support, economic support, diplomatic support and propaganda support; (3) a third type of conflict, which relies generally on ambiguous non-military means (e.g. cyber warfare, grey terrorism and ENMOD) and which occurs in times of relative peace, is called 'non-obvious warfare' and it seems to become more important in the future; and finally (4) the main security threat to states around the world is their destabilisation and fragmentation through a variety of negative trends and societal challenges, including economic decline, demographic dynamics, climate change, mass impoverishment, declining resources, etc., which has already resulted in Western governments trying to find new approaches to controlling their populations and suppressing mass civil unrest even before it occurs.

It seems obvious enough that conventional military capabilities and technologies, such as tanks, aircraft carriers and fighter jets, will not be particularly important to the kinds of conflicts and security challenges that lie ahead. In 1991 van Creveld claimed: 'The shift from conventional war to low-intensity conflict will cause many of today's weapons systems, including specifically those that are most powerful and most advanced, to be assigned to the scrapheap. Very likely it also will put an end to large-scale military-technological research and development as we understand it today' (van Creveld, 1991: 205). Van Creveld's prediction about the uselessness of not only nuclear, but also major conventional weapons, seems accurate. Superiority in major weapons systems apparently does not contribute much to resolving 'local wars' or 'new wars': western militaries have hardly achieved victory and long-term stability in any of the conflicts they intervened in since 1990, which includes Somalia, the occupations of Afghanistan and Iraq and more recently Libya and the war on ISIS.

A very prescient analysis by Steven Metz and James Kievitz from the US Army's Strategic Studies Institute dating back to 1994 already identified 'psychological technology' as key to success in the most likely future 'conflicts short of war'.

> Advances in sensors and other elements of information technology may bring great benefits to conventional, combined-arms warfare, but will have less impact in conflict short of war, which is most often won or lost through the manipulation of images, beliefs, attitudes, and perceptions. These things rather than troop concentrations, command and control nodes, and transportation infrastructure are the key military targets in conflict short of war. This makes psychological technology much more important than strike technology. Ways must be found to use emerging technology, including advanced artificial intelligence and information dissemination systems, to help military strategists develop, implement, and continually improve methods of influencing opinion, mobilizing public support, and sometimes demobilizing it. There is also the potential for defensive pyschotechnology such as 'strategic personality simulations' to aid national security decision makers. To date, most analysts feel that the RMA has not generated adequate advances in such 'soft war' capabilities or even the promise of such gains. But ultimate success in applying the RMA to conflict short of war hinges on the development of psychotechnology. As this emerges, it could be tested for political acceptability by using it first in non-lethal operations other than war like humanitarian relief and nation assistance.
>
> (Metz and Kievit, 1994: 17)

The next chapter will thus look at the various aspects of how 'psychological technologies' could be applied in what might be called *neurowarfare*.

8　Neurowarfare

This chapter aims to discuss some of the developing theory behind what could be called 'information warfare', 'neocortical warfare', 'psycho-informational warfare', or simply 'neurowarfare'. The main argument is that future wars are unlikely to be won by traditional weapons systems that dominated twentieth century warfare, namely tanks, bombers and aircraft carriers. The entire mode of conducting hostility is shifting from the physical realm to the informational and psychological realm, marking also a shift from 'hard power' and geopolitics (the control of territory) to 'soft power' and biopolitics (the control of life itself). It is argued that the objective of war is no longer destroying enemy forces and seizing territory, but shaping perceptions and beliefs and this way gaining political and social control over populations, including the use of populations as new WMD. In these new types of conflicts that are already beginning to unfold, traditional military strength is meaningless since attacks on the enemy's mind skips the battlefield or rather make's the enemy's society the main battlefield by turning the own citizens against their government or by turning the government against their people. Neuroweapons and neurowarfare may even make in some more distant future physical violence unnecessary altogether. The chapter will rely largely on Soviet/ Russian ideas of political warfare and information warfare in order to develop some key concepts and possible approaches to future neurowarfare since Western strategic thought has devoted little attention to the whole concept of politically and culturally subverting societies as a method of war.

8.1　What is neurowarfare?

The word neurowarfare is naturally a composite of 'neuro' as in neuroscience and neurotechnology and of warfare. In order to clarify what neurowarfare could amount to it is important to look at the meaning of the concept of warfare. The term warfare is defined in the *Oxford's Advanced Learner's Dictionary* as '(1) the activity of fighting a war, especially using particular weapons or methods' and '(2) the activity of competing in an aggressive way with another group, company, etc.' In the first sense neurowarfare is warfare that would use 'neuroweapons', which have been defined in the introduction as 'weapons that specifically target the brain or the central nervous system in order to affect the targeted person's mental state, mental capacity and ultimately the person's behaviour in a specific and predictable way'. In the second sense neurowarfare would mean to compete aggressively with others in the neuroscience domain.

However, as pointed out by social scientist Cioffi-Revilla, there is a richer meaning to term 'warfare' than given in most dictionary definitions. He argues that war and warfare are temporally different: wars are events and warfare is a long-lasting process. He states:

> Warfare…is an aggregate, relatively more complex process that generally lasts at least decades and is far less systematized than wars ($10^3 - 10^4$ days). Social scientists and historians have focused relatively less attention on this larger scale belligerence – 'warfare' – although arguably this is the scale that has greater impact on the long-range development of societies and civilizations. If war is what 'forges' a state, warfare is what forges an entire civilization.
>
> (Cioffi-Revilla, 2000: 257)

Cioffi-Revilla gives a list of examples to illustrate the point. The Spanish-Aztec War, the French and Indian War and the Six Day War would be examples for wars as events while the Spanish conquest of the New World, the Napoleonic Wars and the Cold War would be examples of warfare as social processes lasting decades. He suggests that there are four main types of warfare (1) protracted warfare 'is defined as a sequence of recurring wars between the same belligerents fighting for similar objectives'; (2) integrative warfare, in which is 'a violent conflict process that begins with several belligerents and eventually ends with one victor'; (3) disintegrative warfare, which is 'the reverse process of integrative warfare, beginning with one belligerent and ending with many'; and (4) sporadic warfare 'is a series of unrelated wars fought among different belligerents for a period of time, typically over different objectives' (Cioffi-Revilla, 2000: 260–262).

Neuroscience and neurowarfare is likely to reshape human civilizations in very fundamental ways. Neurowarfare can be understood as a battle over the minds of men that extends over many decades, which may turn out to be a protracted, integrative, disintegrative or sporadic type of conflict, which can result in a world that is either politically unified or politically much more fragmented than it is today. Nations and civilizations may use neurowarfare for reshaping the world order according to their own designs. Unlike other forms of warfare, the domination of the mind domain promises dominance across *all* military and social domains. It could become the ultimate 'high ground' that when seized by a belligerent party guarantees victory overall since it can reduce, or potentially eliminate, the adversary's will to resist.

Military analyst Robert Chandler has argued in his book *Shadow World* that there is currently a race for geopolitical world dominance by three main

Table 8.1 The domains of warfare

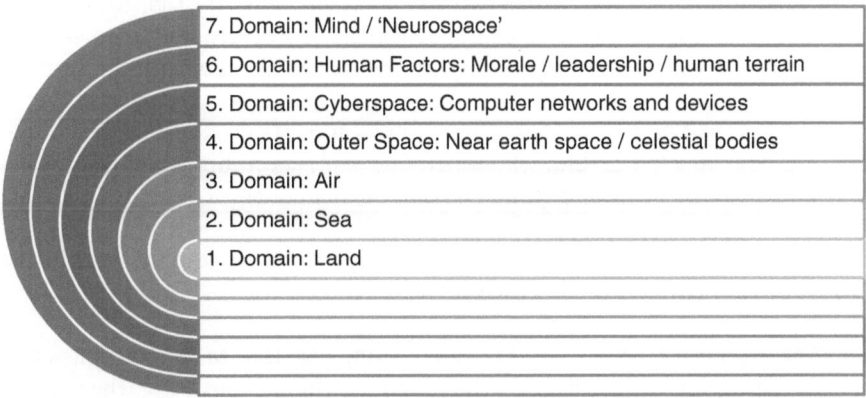

| 7. Domain: Mind / 'Neurospace' |
| 6. Domain: Human Factors: Morale / leadership / human terrain |
| 5. Domain: Cyberspace: Computer networks and devices |
| 4. Domain: Outer Space: Near earth space / celestial bodies |
| 3. Domain: Air |
| 2. Domain: Sea |
| 1. Domain: Land |

Source: Author's own data.

contenders: (1) the capitalist West led by the United States; (2) a progressive-socialist-Marxist group led by Russia joined by a geostrategic quadrangle of China, Iran, Cuba-Brazil-Venezuela; and (3) radical Islam centred in Saudi Arabia with its influence expanding into the West through immigration (R. Chandler, 2008: XIV-XV). According to Chandler, the 'three major political contenders, each with sufficient resources to establish a new world order, are locked in a long-term, winner-take all competition. Their weapons are words and culture, economic strength and diplomatic skill, political action in the open and from the world's shadows, as well as guns and bombs. The ultimate victor will win the right to exercise its power and political authority over all of the Earth's six billion people' (R. Chandler, 2008: XIII). The question then becomes: should these activities described above be characterized as war or warfare?

8.1.1 What is war?

A discussion of neurowarfare has to come to terms with the most basic question, namely, what is war? The military thinker who defined Western military thinking for nearly 200 years is the Prussian general and military theorist Carl von Clausewitz, whose posthumously published book *On War* is still taught at all military academies in the Western world and beyond. In his famous first chapter Clausewitz proposed the following definition: 'War therefore is an act of violence to compel our opponent to fulfil our will'. What the will is specifically would be dictated by politics, which can use war if other political means such as diplomacy fail.

Prominent in the definition is the instrumental use of violence as an act of coercion. Clausewitz considers war to be 'a duel on an extensive scale' and an 'act of mutual destruction'. Furthermore, it is defined as a competition in the use of force that is ultimately a contest of will: the party that can inflict most destruction or that perseveres in the fight will win. In order for one side to concede defeat so that war can end, war has to be cruel and potentially unlimited in destructiveness. Only when one side is faced with total destruction a total victory is possible.

The Clausewitzian view of war has been commented on by military theorists like Martin van Creveld, who noted that war is becoming unconventional and thus extends warfare beyond the narrow definition of regular interstate war. Although van Creveld's *Transformation of War* was written as a giant critique of Clausewitz, van Creveld still stays in many respects within the Clausewitzian paradigm. He claims:

> The essence of war is fighting. Everything else that takes place in war – the gathering of intelligence, the planning, the maneuvering, the supplying – either acts as prelude to fighting or exploits it. To use Clausewitz' own metaphor, fighting and bloodshed are to war what cash-payment is to business. However rarely they may take place in practice, they alone give meaning to the rest.
>
> (van Creveld, 1991: 161)

For van Creveld, violence and in particular the risk of death is essential to war: 'In fact, war does not begin when some people kill others; instead, it starts at the point where they themselves risk being killed in return' (van Creveld, 1991: 159). The implication is that any hostile social activity pursued by a political entity that does not involve killing, or rather the risk of getting killed, would not qualify for being called either war or warfare. As a result, nonlethal warfare (if it was possible at all) that could subdue an enemy without resort to violence or even the threat of violence would be beyond the scope and understanding of the Clausewitzian paradigm.

A new paradigm of war and warfare that allows incorporating nonlethal warfare (informational-, psychological warfare) has been proposed by the French postmodern thinker Michel Foucault, who is famous mostly for his history of the prison (*Discipline and Punish: The Birth of the Prison*). Foucault inversed Clausewitz by stating that 'politics is a continuation of war by other means' (Foucault, 1997: 15). War becomes the model for understanding social relations, which are all governed in Foucault's view by power. Power is defined as a 'strategic situation' in a society, 'more or less taken for granted and consolidated by means of a long-term confrontation between adversaries' (Foucault, 1997: 15).

Power emerges in a continuous battle over power with the structural feature of domination of 'a group, a caste, a class':

> together with the resistance and revolts which that domination comes up against, a central phenomenon in the history of societies is that they manifest in a massive and universalizing form, at the level of the whole

social body, the locking together of power relations with relations of strategy and the results proceeding from their interaction.

(Foucault, 1983: 225)

What comes out of domination is a process that Foucault called 'subjection' or the creation of subjects through various means of discipline, surveillance and discourse. In Foucault's view, power is violent and abusive in the way that it produces 'subjects' and negates individualism and freedom. What Foucault means by subjection is the formation and control of consciousness through external influences that mould the individual according to the needs of the power structure at the time. This war for control over consciousness is continuous and never ends. The main insight that can be gained from Foucault's war paradigm is that everything counts as war or warfare that is designed to subjugate consciousness (that 'subjectifies').

8.1.2 What is consciousness?

Foucault shed some more light on his conceptions of consciousness and freedom in his lecture 'Technologies of the Self' from 1982. For Foucault, there were four major types of technologies:

(I) technologies of production, which permit us to produce, transform, or manipulate things; (2) technologies of sign systems, which permit us to use signs, meanings, symbols, or signification; (3) technologies of power, which determine the conduct of individuals and submit them to certain ends or domination, an objectivizing of the subject; (4) technologies of the self, which permit individuals to effect by their own means or with the help of others a certain number of operations on their own bodies and souls, thoughts, conduct, and way of being, so as to transform themselves in order to attain a certain state of happiness, purity, wisdom, perfection, or immortality.

(Foucault, 1988: 17)

The self is consciousness and can be changed from the outside through application of 'technologies of power' or internally through 'technologies of the self': the conscious effort of self-improvement, most importantly through introspection and writing diaries that explain and steer the self. Here Foucault emphasises the great importance of moral rules and morality as means of self-limitation and self-control, as well of knowing oneself as a foundation for taking care of oneself. What Foucault seems to advocate is the idea that true freedom paradoxically arises from critical moral self-observation and self-limitation.

The term consciousness will become very important for the discussion below, which necessitates to somewhat define it better, which is not easy. The problem is that philosophers and psychologists have struggled with the concept of consciousness for centuries. Still, it remains so underdefined that most neuroscientists tend to avoid it altogether (Crick and Koch, 1998: 97). Neuroscientist Adam Zeman has investigated the meanings and interpretations of consciousness and how they relate to the findings of neuroscience. In the summary of his excellent article on 'Consciousness', he wrote:

> Consciousness is an ambiguous term. It can refer to (i) the waking state; (ii) experience; and (iii) the possession of any mental state. Self-consciousness is equally ambiguous, with senses including (i) proneness to embarrassment in social settings; (ii) the ability to detect our own sensations and recall our recent actions; (iii) self-recognition; (iv) the awareness of awareness; and (v) self-knowledge in the broadest sense. The understanding of states of consciousness has been transformed by the delineation of their electrical correlates, of structures in brainstem and diencephalon which regulate the sleep–wake cycle, and of these structures' cellular physiology and regional pharmacology…Whether scientific observation and theory will yield a complete account of consciousness remains a live issue. Physicalism, functionalism, property dualism and dual aspect theories attempt to do justice to three central, but controversial, intuitions about experience: that it is a robust phenomenon which calls for explanation, that it is intimately related to the activity of the brain and that it has an important influence on behaviour.
>
> (Zeman, 2001: 1264)

Neuroscience research suggests (amongst many other things) the following about human consciousness: (1) consciousness relates to information processing within the brain (e.g. 'Global Workspace Theory'); (2) brain states correspond to mental states; (3) behaviour ultimately results from mental states (will or intentionality); and (4) brain plasticity means that the brain can be rewired and mental processes altered, for example through neurotropic drugs, through brain stimulation, by providing different external stimuli and by changing cognitive habits. These general ideas are largely in line with what Foucault and other philosophers have written about consciousness, namely that consciousness relates to self-awareness, memory, experience, perception, values, desires and behavioural control, which can be externally manipulated and which can change the self or identity. This is still extremely sketchy, but it may suffice for the purposes of this chapter.

8.1.3 'Neocortical warfare'

While Clausewitz emphasized the need for violence and, in fact, cruelty as a requirement for victory, the Chinese military thinker Sun Tzu stressed the importance of intelligence ('Know your enemy and yourself; and a hundred battles you will never be in peril') and deception ('All warfare is based on deception'). Although Sun Tzu covers the use of force extensively in his work *The Art of War*, he also suggests that '[f]or to win one hundred victories in one hundred battles is not the acme of skill. To subdue the enemy without fighting is the acme of skill'. Sun Tzu proposes to attack the enemy's strategy before it can be carried out – in the strong belief of certain defeat the enemy would surrender before any combat takes place. In the end, this boils down to what is nowadays called psychological warfare: the manipulation of the enemy's perception of reality and the enemy's morale (willingness to fight and determination) to the point that even an attempt of resistance appears futile to the enemy. It is widely known that Sun Tzu's book has been required reading for intelligence officers around the world, including the CIA, the KGB and Mossad.

Military analyst Richard Szafranski has suggested the term 'neocortical warfare' for hostile actions that are primarily designed to attack the enemy's will. He argued:

> The object of war is, quite simply, to force or encourage the enemy to make what you assert is a better choice, or to choose what *you* desire the enemy to choose. Said another way, the object of war is to subdue the hostile will of the enemy. We cannot meet the immediate objective of war until or unless we subdue hostile will.
>
> (Szafranski, 1997: 397)

The previous paradigm of war suggested '[d]estroy enough brains, or the correct brains, our studies seem to encourage us, and "will" necessarily dies along with the organism' (Szafranski, 1997: 398). A better approach is obviously to attack the will of the enemy directly instead of the indirect way of destruction. Western military thinking and practice is too much focused on 'kinetic' effects or, in other words, on '[d]estroying brains' to the neglect of the real objective of war: the subjugation of will. Szafranski raised the question: 'What if we viewed war not as the application of physical *force*, but as the quest for metaphysical *control*?' (Szafranski, 1997: 399).

The difficulty would be to determine what exactly 'will' is and how one could go about altering the will of others in a desired way. He proposes that 'will' is essentially brain-centred. Therefore the enemies' brains need to be the focus of attack. 'Neocortical warfare is warfare that strives to *control* or *shape* the behavior of enemy organisms, but without destroying the organisms. It does this by *influencing*, even to the point of regulating, the consciousness, perceptions and will of the adversary's leadership: the enemy's neocortical system' (Szafranski, 1997: 404). In order to do so, it would be necessary to understand the adversary's culture and world view up to the point of understanding 'how the adversary receives, processes and organises auditory, visual and kinesthetic perceptions', so that 'the adversary chooses the nonfighting alternative voluntarily' (Szafranski, 1997: 405). Since 'will' is fluid and changeable, neocortical warfare becomes an endless conflict with 'endless engagements' (Szafranski, 1997: 407). War would, therefore, be no longer an aberration but a constant competition of wills extending into all domains of human life with the aim of making the enemy choose not to fight, not to resist and not to contest an aggressor's designs. Similarly wrote PSYOPS specialists Paul E. Vallely and Michael Aquino 'that wars are fought and won or lost not on battlefields but in the minds of men', which means that in the end result, it is always up to the enemy to concede defeat, or not (Vallely and Aquino, 1980: 6).

The Swedish military analyst Henrik Friman has proposed the concept of 'perception warfare': 'Perception warfare is the concept of how to create occurrences that give illusions of all as winners in their own way. It is a combat of the commanders' minds' (Friman, 1999: 1). In other words, Friman suggests manipulating the perceptions of enemy commanders in such a way that they believe they have won so that hostilities can be terminated – there is no reason to fight if victory is already achieved in the commander's mind. He observes '[p]erception warfare is not the same as information warfare, but there are many similarities' (Friman, 1999: 2). However, the main difference would be that IW activities would be aimed at controlling information, while perception warfare declares the mind of the commander to be the real target. Friman quotes Mao, who said: 'In order to win victory we must try our best to seal the eyes and the ears of the enemy, making him blind and deaf, and to create confusion in the minds of enemy commanders, driving them insane' (Friman, 1999: 3–4). In the end, perception warfare is not so much about making enemies insane, but rather about distorting their perceptions to such an extent that they can no longer come to any rational conclusions about what is good for them and their own side.

The difficulty with concepts like neocortical warfare, perception warfare, or what is suggested here, neurowarfare, is that to an observer it may not appear to be warfare at all since no open hostilities would ever take place. No tanks would be racing over the plains of Eastern Europe and no ICBMs would be flying towards their targets. Indeed, it might be most prudent for a belligerent using a neurowarfare approach to hide the attempt of mental or psychological manipulation and coercion. Ideally, the victims of brain interference should never be aware of covert attacks on their perception, decision-making and behaviour. This is easily helped by the all-too-human flaw of vanity: as the saying goes, it is easier to fool others than to convince them that they have been fooled.

While 'neocortical warfare' was only more recently developed as a concept in Western strategic literature and is still little appreciated by the Western military culture, Eastern political and military strategists have long understood that the ability to shape consciousness is more important than the ability to cause physical destruction. The Bolsheviks, in particular Lenin and Trotsky, have theorized about the art of subversion and its use against the Soviet Union's much more powerful enemies. These ideas are still reflected in Russian theories of political warfare and information warfare (IW), which will be discussed below since they are leading up to a future of neurowarfare that will determine the fate of humanity.

8.2 Soviet/Russian concepts of political warfare and subversion

The Bolsheviks were always interested in methods and technologies of mind control for bringing about their vision of a communist utopia that they hoped to spread across the world through a socialist world revolution. It has been mentioned in an earlier chapter that Lenin brought the famous pioneer of behaviouralism Ivan Pavlov to the Kremlin in order to develop a psychological warfare technique that the Russians would call 'reflexive control' (Hunter, 1956: 40). Reflexive control is a method for shaping the thinking of others by manipulating the cultural context and perceptions in a way that the people reach the conclusions and make the decisions that their leaders want them to make (Chotikul, 1986: 47). Russian IW theorist Aleksandr Dugin defined reflexive control as 'a means of conveying to a partner or an opponent specially prepared information to incline him to voluntarily make the predetermined decision desired by the initiator of the action' (quoted from Perry, 2015: 7). In other words, the Bolsheviks wanted to mind control others into accepting their world view and getting them to internalize Marxist values to control their everyday thinking and behaviour.

Of course, the communists did not stop at 'brainwashing' their own populations, but were early on committed to exporting socialist revolutions. From a Leninist point of view, a state of peace cannot in principle exist between socialist and capitalist societies – peace can only result from a communist revolution that eliminates all class contradictions (Kintner and Kornfeder, 1962: 85). For Leninists, peace is therefore just a continuation of warfare by other means with the ultimate goal of politically subverting the enemy so that actual warfare is unnecessary.

8.2.1 Political warfare

The term 'political warfare', which was used to describe these Soviet activities in the West, was coined by the American diplomat George Kennan, who defined it as the

> logical application of Clausewitz's doctrine in time of peace. In broadest definition, political warfare is the employment of all the means at a nation's command, short of war, to achieve its national objectives. Such operations are both overt and covert. They range from such overt actions as political alliances, economic measures (as ERP – the Marshall Plan), and 'white' propaganda to such covert operations as clandestine support of 'friendly' foreign elements, 'black' psychological warfare and even encouragement of underground resistance in hostile states.
>
> (quoted from Boot et al., 2013)

In older literature from the 1950s and 1960s, political warfare has been frequently acknowledged as a key aspect of Communist strategy. According to policy analysts William Kintner and Joseph Kornfeder,

> t]he non-military weapons system of communism may be described in various terms: political warfare, 'psywar', institutional conflict, psychosocial combat. By any name, its primary purpose is to disorient and disarm the opposition. At the grassroots level, political warfare seeks to induce the desire for surrender – in opposing peoples. At the strategic level, political warfare seeks to corrode the entire moral, political, and economic infrastructure of a nation, particularly by affecting governmental decisions...Political warfare aims to weaken, if not to destroy, the enemy by use of diplomatic proposals, economic sorties, propaganda and misinformation, provocation, intimidation, sabotage, terrorism, and by driving a wedge between the main enemy and his allies.
>
> (Kintner and Kornfelder, 1962: XIII)

In short, political warfare is a strategy of divide and conquer that relies on tactics of infiltration and subversion with the help methods such as the recruitment of 'agents of influence' (foreign journalists and politicians), the use of front organisations such as political parties, political movements, labour unions, think tanks and other political organisations and the peddling of propaganda and disinformation through official and unofficial channels to create and exploit societal and political divisions that are meant to cause political paralysis and eventually shift perceptions in a way conducive to the goals of the aggressor.

8.2.2 'Cultural Marxism' and demoralisation as method of psychosocial combat

Robert Chandler has claimed that the Soviet designs against the capitalist West were heavily influenced by Italian Marxist Antonio Gramsci, who produced in his prison notebooks of the 1930s a Marxist strategy for world domination. It shall be mentioned that Foucault was himself a Marxist, who got much inspiration from Gramsci's work on cultural hegemony and who obsessed about the subversion of power. Gramsci theorized that the bourgeoisie was using ideology for maintaining a cultural hegemony, which he suggested needed to be subverted first before the communist revolution could succeed. His solution was a 'cultural Marxism' that was aimed at destroying the influence of religion in society and of promoting instead 'secular humanism' by quietly infiltrating civil society institutions (R. Chandler, 2008: 29–30). The strategy of cultural Marxism is thus one of demoralisation: existing morals and values in a targeted society are systematically undermined, which paves the way for the introduction of a new socialist value system that then replaces the old value system of the bourgeoisie.

Cultural Marxism was not just a philosophical theory, but it became a blueprint for communist subversion during the Cold War. According to the Czech defector Major-General Jan Sejna, Brezhnev (General Secretary 1964–82) created a 'long-range strategic plan' for defeating the West through political warfare and demoralisation. The plan had four stages: (1) 'peaceful coexistence', (2) 'peaceful coexistence struggle', (3) 'period of dynamic social change' and (4) the 'era of global democratic peace', which was supposed to arrive in 1995 (Sejna, 1982: 106). The objective in phase two was to 'accelerate the social fragmentation of the Capitalist countries' (Sejna, 1982: 106).

The House of Un-American Activities Committee (HUAC) also suggested that there was indeed a Soviet 'master plan' to corrupt American society, in order to eventually merge it with the communist Soviet Union once Western societies have been sufficiently 'Marxized'. In 1963, HUAC presented 45 communist aims in the US, which included taking control of political parties, the press, the education system, the promotion of obscenity and homosexuality, the discrediting of the Bible and religion and the portrayal of the US Constitution as 'inadequate, old-fashioned, out of step with modern needs, a hindrance to cooperation between nations on a worldwide basis' (US Congress, 1963).

The publisher and British government advisor Christopher Story, who is well-known for his collaboration with Soviet defector Anatoly Golitsyn, resulting in the book *Perestroika Deception*, summarized the Soviet demoralisation strategy in the following way:

> The ambition to control the Western mind is a long-standing objective of Soviet policy, embracing the ideas of the Italian Communist Antonio Gramsci, who argued that mastery of the human consciousness should be paramount political objective...control of the Western mind is to be achieved not only by the dishonest use of language, but also to demoralize the West through corrosive attacks on society's institutions, the active promotion of drug abuse, and the spiral of agnosticism, nihilism, permissiveness and concerted attacks on the family in order to destabilize society. Religion and the traditional cultural and hegemony must first be destroyed, before the revolution can be successful.
>
> (quoted from R. Chandler, 2008: 71)

While HUAC and the McCarthy years, during which it prospered, are now seen as an episode of somewhat irrational American anti-Communist paranoia, it is at least interesting that many Soviet defectors have indicated that there had been indeed a massive Soviet attack on western culture and consciousness with the aim of thoroughly demoralizing western societies. For example, Yuri Bezmenov, a former KGB officer who officially worked for the Soviet news agency *RIA Novosti* until his defection in 1970, confirmed that there was a large-scale Soviet psychological warfare operation against the West (Bezmenov, 1985; Schuman, 1985). Bezmenov claimed that an

'APN-KGB subversion may be painless, but its long-term result is more devastating than a nuclear explosion'. A country becomes effectively disarmed and is no longer able to defend itself. Bezmenov explains: 'It effects an irreversible (at least within one generation) change in the public's perception of social, political and economic reality, to such an extent that the concept of destroying individual and collective property, safety, freedom and often life itself...no longer seems to be such a bad idea' (Schuman, 1985: 15).

8.2.3 The Western failure to recognize the Soviet grand design

The Soviets concentrated their efforts on promoting socialist ideas in the Western education and on converting teachers and professors to socialism. Although the process is necessarily very slow, since it takes decades before people, who passed through an education system, will get into positions of power and influence in society, it is also tremendously effective. Communications researcher Jacques Ellul made in his famous book on propaganda the central argument that propaganda works best when people can be imprinted with it at a time when they still lack critical faculties. Once people have gone through that process of school indoctrination at an early age they will not question basic 'truths' and are more susceptible to propaganda later in life (Ellul, 1965: 109–110).

In the interview with Edward Griffin in 1984, Bezmenov discussed how leftwing indoctrination through the education system can distort the perception of a population so much that they cannot even see the danger that results from ideological subversion. The irony is that the more advanced the demoralisation, the less there is any societal ability to understand what is happening. He stated pessimistically:

> The demoralization process in the United States is basically completed already for the last 25 years. Actually, it's over fulfilled because demoralization now reaches such areas where not even Comrade Andropov and all his experts would even dream of such tremendous success. Most of it is done by Americans to Americans thanks to lack of moral standards. As I mentioned before, exposure to true information does not matter anymore. A person who was demoralized is unable to assess true information. The facts tell nothing to him, even if I shower him with information, with authentic proof, with documents and pictures. ...he will refuse to believe it... That's the tragedy of the situation of demoralization.
>
> (Bezmenov, 1984)

What Bezmenov describes here is called 'cognitive dissonance', which is a well-known limitation of the way the human mind works and which can be exploited for improving the success of deception and propaganda. As noted in an earlier chapter (chapter 4), cognitive dissonance theory, which was first developed by psychologist Leon Festinger in 1957, postulates that people have a strong desire for consistency and will ignore or avoid information that is inconsistent to their fundamental beliefs (Festinger, 1957). Of course, people can comfortably hold two contradictory beliefs in their minds, but they find it distressful when made aware of the contradiction. Psychologist Kathleen Taylor explains: 'So if we notice that our beliefs – that we do after all take to represent how the world is – contain contradictions, then we can be justified assuming that something has gone wrong with our representation of reality... humans will go to some length to remove inconsistencies among their deeply held beliefs' (E. Taylor, 2004: 128). This can go so far as the 'willful suspension of disbelief', where people confronted with strong evidence contradicting their fundamental beliefs will simply choose not to accept the evidence as valid. Often they cannot simply quit believing because they have become emotionally and otherwise invested and quitting would mean to admit that they have been wrong all along, which is psychologically very painful. They will only wake up to reality when the consequences of their delusion can no longer be ignored. As Bezmenov said, an indoctrinated person will refuse to believe the dangers of socialism 'until he receives a kick in his fat bottom... but not before that' (Bezmenov, 1984).

8.2.4 *Illegal drugs and demoralisation*

In addition to the use of infiltration and propaganda, the Soviets may have even used much more nefarious means for corrupting Western minds. According to national security analyst Joseph Douglass, the Soviets deliberately flooded Western societies with illegal psychotropic drugs to demoralize and weaken the West. Douglass based his assertions on claims made by the high-ranking Czech defector Sejna. In his 1990 book *Red Cocaine*, Douglass made a case that the Soviet Union was a major player in the international illegal drug business (Douglass, 1999).

Sejna was the second-highest ranking person to defect from the Soviet Bloc (after Ion Pacepa) and he had unprecedented insights into KGB operations since the Czech intelligence service was used as a KGB proxy. The Soviets got the idea of using illegal drugs as a 'form of clandestine chemical warfare, in which the victim voluntarily exposes himself to chemical attack' from the communist Chinese, who started using illegal drugs as a weapon to corrode enemy morale in 1928 (Douglass, 1999: 11, 13). The Chinese may have gotten some cues from the Opium Wars of the nineteenth century, which were fought over the British desire to sell massive amounts of opium to the Chinese, which they knew corrupted Chinese society.

Khrushchev allegedly initiated the drug trafficking operation in 1964 under the codename 'friendship of nations'/ 'people's friendship' ('Druzbah Narodov'), which targeted the United States and Western Europe with heroine (Douglass, 1999: 47). Khrushchev hoped 'drugs and narcotics would lead to a decrease in the influence of religions and, he added, under certain conditions, could be used to create chaos' (Douglass, 1999: 37).

Douglass claimed that the Soviets deliberately spread illegal narcotics to US soldiers in Vietnam to massively impact their combat readiness to the point that 90 per cent of them were using some kind of drugs, most commonly marijuana (Douglass, 1999: 61). Furthermore, Douglass argued that illegal drugs were also a common approach of the KGB for recruiting and controlling 'agents of influence' in other countries. The drugs were used to drive people insane, blackmail them and brainwash them into accepting communist views. He stated:

> Mind control drugs were used to recruit almost the entire Indonesian cabinet. They were first used on Minister Subandrio's wife when she was in Prague for a medical operation. After she was recruited, with her help and 'the little pills,' Minister Subandrio was recruited. This process continued through most of the Indonesian cabinet. By the end of the 1950s, Indonesia was a Soviet puppet. India's Defense Minister, Khrisna Menon, was similarly recruited, as were a number of North Vietnamese officials in the 1960s, most notably Minister Pham van Dong.
>
> (Douglass, 2001: 132–133)

The information provided by Douglass and his source Sejna is impossible to verify, but it does align with earlier work such as A.H. Stanton Candlin's book on *Psycho-Chemical Warfare* of the Chinese communists from 1974 (Stanton Candlin, 1974) and, of course, with the CIA's drug experimentation of the 1950s and 1960s, which was based on the assumption that the communists were working on mind control drugs (Hunter, 1956: 232–237). Sejna's claims also dovetail with Alibek's information regarding a KGB drug mind control program codenamed Flute (Alibek and Handelman, 1999: 171–172). Finally, even if there was never any major communist involvement in the illegal drug business, it does not rule out the possibility that other governments could use illegal drugs as weapons for demoralizing and destabilising other countries.

8.2.5 'Maskirovka' or large-scale deception

An important aspect of Soviet political warfare that has not yet been discussed was large-scale deception that is designed to manipulate foreign societies and their political leaderships in a major way, even to the extent of creating entire false realities. The Soviets used the term 'maskirovka' in reference to military and political deception. As pointed out by military analyst Timothy Thomas, there was even a Russian 'maskirovka school' in the early 1900s, which was disbanded and 1929, although maskirovka theory seems to be alive today (T. Thomas, 1997). It can be used in wartime and peacetime and can be implemented on the strategic, operational and tactical levels of conflict (Heikerö, 2010: 21). It literally means masking something and in the context of military deception it can refer to a variety of things, including 'camouflage, concealment, deception, imitation, disinformation, secrecy, security, feints, diversions, and simulation' (Smith, 1988).

Maskirovka has been defined as 'a set of processes employed during the Soviet era designed to mislead, confuse, and interfere with anyone accurately assessing its plans, objectives, strengths, and weaknesses' (Shea, 2002: 63). When used in peacetime it is a political ruse that can be directed at domestic and foreign audiences, designed to alter perceptions about the Soviet Union and its allies in a desired way. A famous example of peacetime Soviet maskirovka was Operation Anadyr or the secret Soviet plan to deploy IRBMs MRBMs on Cuba in summer of 1962. Maskirovka is closely tied to the concept of 'reflexive control', which is about manipulating the enemy's perceptions in a way that the enemy will make decisions detrimental to the own interest – in this case not prevent the deployment of Soviet missiles (T. Thomas, 1997).

Maskirovka includes what Hitler called the 'big lie' – a lie so blatant and outrageous that ordinary people cannot believe that their trusted leaders would say something like that, if it was not true. This technique is helped by the fact that in the Soviet Union there was a culture of deceit, which still prevails in post-Soviet Bloc countries, especially in their militaries: 'Lying routinely occurs at the most senior uniformed levels, even when an argument is clearly untenable or contradicted by obvious facts' (Shea, 2002: 64). Some Soviet analysts and several high-level defectors have concluded that much of the open diplomacy and propaganda of the Soviet Union was little more than theatre for the purpose of deceiving the West about Soviet intentions and capabilities. The orchestrated disinformation worked because Western analysts could not fathom the scale of it, or imagine that hidden political agendas could be pursued over decades.

For example, Soviet defector Golitsyn claimed that the Sino-Soviet split in 1966/ 67 was a Soviet strategic deception meant to manipulate the West into making concessions detrimental to the West's interests (Golitsyn, 1984: 153–182). Golitsyn also predicted in 1984 that the Soviet Union would be faking liberalization by implementing ostensibly far-reaching political and economic reforms in the country and by creating a controlled opposition that allowed the communists to stay in power even after the official collapse of the Soviet Union (Golitsyn, 1984: 230; R. Chandler, 2008: 59–100). This was the so-called 'Andropov plan' named after Yuri Andropov, a former chairman of the KGB (1967–82) and General Secretary of the Soviet Union (1982–84). However, the chaos of the 1990s makes it doubtful that the Andropov plan was successfully implemented, as a group of 50 oligarchs took control of the country (R. Chandler, 2008: 82).

Another major, possibly unpredictable political change in Russia, resulted from the ascendance of Vladimir Putin to power in 2000, following a series of 'apartment bombings' across the country (that have been described by many as FSB 'false-flag operations') and a short political campaign that propelled Putin from being FSB chief to becoming Yeltsin's designated successor (van Herpen, 2015: 53). While the Andropov plan is a very controversial thesis, it is a widely accepted reality that many key components of the Soviet system survived in today's Russia, including the KGB, which was merely renamed. There is a continuation of authoritarian government, which extends to the tight state control of all the major media and state control of key components of the Russian economy (the energy and resources sector). Despite its democratic constitution, the existence of multiple parties and its periodic elections, many Western analysts have concluded long ago that 'Russia is not a democracy' (Rosefielde and Hlouskova, 2007).

Already in 2008, some authors like Edward Lucas claimed that there was a 'new Cold War' on the horizon. Western relations with Russia worsened in 2012, the year of Putin's 'return' to the position of President after serving four years in the officially subordinate role of Prime Minister. Relations really soured after Putin's move to annex the Crimea, following the 'Euromaidan revolution' in the Ukraine in February 2014.

Quickly afterwards Western militaries and prominent think tanks, such as the Royal Institute of International Affairs and the US Army's Strategic Studies Institute, acknowledged that the West is now subjected to a massive and potentially dangerous Russian IW attack (Sher, 2015: 23–32; US DoD, 2014: 4). NATO commander General Philip Breedlove stated in 2014 that Russia is waging 'the most amazing information warfare blitzkrieg we have ever seen in the history of information warfare' (Pomerantsev, 2014). Many people in the West still remain ignorant to the reality of Russian subversion and IW and many are deceived into accepting Russian narratives of world events.

8.3 Russian information warfare: concepts and ideas

Russian writers distinguish two types of IW: 'information-technical' IW, which includes cyber warfare and electronic warfare, and 'information-psychological' IW, which can be described as psychological warfare (T. Thomas, 2014: 101–102). Most relevant here are Russian ideas about 'information-psychological' combat. As noted by many analysts, there is nothing fundamentally new in contemporary Russian strategic thinking about 'information-psychological' IW or the current Russian propaganda offensive, since it is all firmly grounded in ideas that originate from Soviet times (Snegovaya, 2015: 7). The only thing that changed is that the ideological language of class struggle has been replaced by a narrative that portrays Russia as a victim of Western informational, psychological and cognitive attack (S. Blank, 2013: 33). In other words, Russia's new aggressive propaganda offensive is justified on the grounds that it is merely a defensive manoeuvre to protect Russia against Western subversion (van Herpen, 2015: 9). In essence, Soviet concepts of political warfare are 'mirror imaged' on the West and updated according to new technologies such as computers and the Internet.

8.3.1 IW as a new weapon of mass destruction

In Russian writing, IW is considered a weapon that can be used as a standalone effort during peacetime or in conjunction with military force, usually as preparation of the battlefield, e.g. for 'shaping a favorable response from the world community to the utilization of military force' (Russian Federation, 2010). When used in conjunction with military force, IW becomes a force multiplier that can make military actions more effective to the point that IW becomes a potential new WMD. According to General Victor Samsonov, a former Chief of the Russian General Staff,

> [t]he high effectiveness of information warfare systems, in combination with highly accurate weapons and non-military means of influence, make it possible to disorganize the system of state administration, hit strategically important installations and groupings of forces, and affect the mentality and moral spirit of the population. In other words, the effect of using these means is comparable with the damage resulting from the effect of weapons of mass destruction.
>
> (quoted from Fitzgerald, 1996)

These ideas are actually similar to Western military writing about a psychological effect of 'shock and awe' resulting from massive precision attacks that paralyze the enemy and rapidly convinces them that they have been defeated. Where Russian theorists present very different ideas, however, concerns the use of IW as a standalone weapon. Like the Soviet concept of perpetual ideological conflict, 'Russian doctrine sees information war as permanent "peaceful war", not necessarily related to military activity' (T. Thomas, 2015: 16). Furthermore, IW as a standalone weapon is directed against other populations with the aim of subverting them. Major-General Vladimir Belous has suggested that

> [i]n the future, the battlefield should be expected to shift increasingly into the intellectual realm, impacting on the consciousness and feelings of many millions of people. By deploying space-based relay stations in low earth orbits and using the great potentiality of television and the Internet, an aggressor country can work out and, under certain conditions, carry out a scenario for continuous, round-the-clock information warfare against a particular state, in an effort to blow it up from within. Provocative programming will be designed to affect not only the people's intelligence but primarily their senses, especially with the public's low political awareness, insufficient information and unpreparedness for such warfare.
>
> (Belous, 2009: 76)

In IW, the mass media are weaponized in the sense that they can be used to massively destabilize another country to the breaking point. As Belous pointed out, IW operations goals can include: 'to discredit a state's foreign and domestic policy and the socio-economic status of its population, to exacerbate ethnic contradictions, to distort its history, to foment religious strife among representatives of different confessions, to demoralize the population, to encourage antisocial behavior, and so on' (Belous, 2009: 75). The end result of such activities could be civil war or genocide.

8.3.2 Key concepts and tactics in Russian non-military IW

A major concept is what the Russians call 'informatization' in relation to using the Internet as a 'battlespace'. Military analyst Richard Zoller described it as follows:

> More than any other nation-state, Russia uses the cognitive domain of cyber as much as the technical domain. Where Western definitions of cyberspace focus on technical aspects of information technology, 'informatization' takes on a much broader definition. 'Informatization' can be broadly defined as, applying modern information technology into all fields of both social and economic development, including intensive exploitation and a broad use of information resources. What this means is that Russia uses cyberspace more to disrupt an adversary's information than to steal or destroy it.
>
> (quoted from S. Blank, 2013: 32)

Informatization is inspired by Western IW concepts, such as 'netwar'. The Russians have adopted this and other concepts and adjusted them according to the Russian situation and Russian needs. The idea of netwar, which is a contraction of 'network war', was developed by RAND analysts John Arquilla and David Ronfeldt in the 1990s, is now interpreted by the Russians as a strategic blueprint for the subversion of foreign societies with network-driven 'color revolutions' (Korybko, 2015: 38–41). Russian analysts have claimed that the Arab Spring, which got much of its initial traction through social media such as Facebook and Twitter, was a new American method for regime change (Korybko, 2015: 27). 'In Russian literature, the colour revolutions are regarded as "the most socially dangerous form of clashes between intelligence services"' (Darczewska, 2015: 32).

In May 2014, the Russian Ministry of Defence held its third high-level International Security conference in Moscow during which prominent Russian politicians and military leaders determined 'color revolutions' as one of the greatest threats that Russia would currently face (Cordesman, 2014; Korybko, 2015: 10). For example, General Gareev argued: 'The breakup of the Soviet Union and Yugoslavia, the parade of "color revolutions" in Georgia, the Ukraine…show how principal threats exist objectively, assuming not so much military forms as direct or indirect forms of political, diplomatic, economic, and informational pressure, subversive activities, and interference in internal affairs' (quoted from S. Blank, 2013: 34). Not surprisingly, the Russian 'Information Security Doctrine' indicates as a main domestic threat 'the illegal use of special measures to influence the individual, social, and group consciousness' (quoted from Darczewska, 2015: 13). Subsequently, the

Russian government has increased media control and Internet censorship, has instituted tougher laws against foreign NGOs in Russia and has vigorously persecuted political activists that are seen as foreign agents.

The Russian strategist Vladimir Karyakin has theorized that '[t]he mental sphere, a people's identity, and its national and cultural identity have already become battlegrounds. The first step in this direction is the discrediting of and then the destruction of a nation's traditional values. And in order for external aggression to be perceived painlessly to the mass consciousness, it must be perceived as movement along the path of progress' (quoted from S. Blank, 2013: 33). In other words, any attempt of democratization is considered by the Russian government as an attack on the country's 'national and cultural identity' and its 'traditional values'. It is hard to avoid the conclusion that the Russian government at the highest levels is so overprotective with respect to their national identity because they have waged war on Western consciousness for so long and because they understand the damage that can result from such activities.

Another concept that the Russians have adopted from the West and changed for their purposes is 'soft power', which was first formulated by political scientists Joseph Nye in the late 1980s to describe how the West can use the attractiveness of its values (democracy and human rights) for achieving its foreign policy objectives without the need of coercion. In the absence of a similar appeal in Russian culture, the Russian government has developed its own methodology for expanding its soft power in recognition of the fact that the value of 'hard power' is declining.

8.3.3 The Kremlin's new 'soft power' offensive

In recent years, the Russian government has rebuilt its strategic communications capabilities that had been in decline in the 1990s and has launched a 'soft power' offensive that is aimed at supporting Russian foreign policy. Most notably, Russia has created *Russia Today* (RT) in 2005, which is well-funded with $380 million dollars per year (2011) and which has foreign language programming and digital content, including in English, Spanish, Arabic, French and German (van Herpen, 2015: 71). RT and its sister organizations has a staff of 2,000, bigger than *Fox News* (van Herpen, 2015: 71). Its offices are located in London, Washington, DC, New York, Madrid, Paris, Delhi and Cairo, regularly reaching 15 million viewers worldwide. The Russian government also maintains, apart from RT, various other web platforms, such as *Sputnik News* (a English-language news website), *Ruptly* (an online news agency and video on demand service) and *Pravda* (English) with numerous other websites that seem independently run, but contain a lot of pro-Russia content with lots of links to RT and other known Russian propaganda outlets.

The Russian government's idea of 'soft power' includes a variety of methods that are rather unusual, such as the use of 'troll armies' and espionage, including dirty tricks (blackmail and bribery). Marcel van Herpen claims that Kremlin's arsenal of 'soft power' tools would consist of the following methods:

- 'Disseminating official Russian state propaganda *directly* abroad via foreign-language news channels, making use of TV and the Internet';
- 'Disseminating official state propaganda *indirectly* via Western media';
- 'Takeovers of Western papers';
- 'Gaining a hold over the new social networks and setting up Kremlin-friendly websites';
- 'An active presence in blogs and discussion forums, as well as the publication of organized postings by "Kremlin trolls" on the websites of Western papers';
- 'Financing Western politicians and/ or political parties';
- 'Reactivating spy rings, which had the task to penetrate influential political circles'; and
- 'Activating the Russian Orthodox Church as a soft-power tool' (van Herpen, 2015: 70).

The point behind these efforts is not just to promote a positive image of Russia, but rather to increase and exploit divisions in the West. Most controversial about Russian propaganda is its propensity for pushing various 'conspiracy theories' and other divisive narratives about the West (Pomerantsev, 2014). RT's motto is 'question more', which the network uses as an excuse for exploring the implausible and the absurd, as long as it achieves its goal of confusing its audience about what is true.

Many Russia analysts have pointed out that the Russian propaganda offensive often relies on half-truths and even outright lies for manipulating Western minds (Pomerantsev, 2014; van Herpen, 2015: 4). Pomerantsev goes as far as stating, '[t]he new Russia doesn't just deal in the petty disinformation, forgeries, lies, leaks, and cyber-sabotage usually associated with information warfare. It reinvents reality, creating mass hallucinations that then translate into political action' (Pomerantsev, 2014). Kadri Liik, a Russia expert from the European Council on Foreign Relations claimed 'Russian propaganda is sometimes so crazy, it says such impossible things, it doesn't have the effect of making people believe them but it breaks down people's defences…It's not just lies, in the way of Soviet propaganda. It's more sophisticated. A kind of violence against the mind' (McGreal, 2015). What makes Russian propaganda dangerous is not necessarily that many people in the West believe it, but rather that the propaganda confuses them so much that they can no longer tell difference between what is true and what is not, resulting in them embracing anti-government narratives by default.

8.3.4 Beyond propaganda: Russian visions of future war

Russian military theorists have asserted that information weapons 'make it possible to exert a specified effect on large social groups but also reshape the consciousness of entire peoples' (quoted from S. Blank, 2013: 32). Information weapons include mass media and cyber weapons, as well as 'psychotronic weapons' (Lopatin and Tsyngankov, 1999: 93). Western reviews of Russian IW literature often indicate Russian interest in psychotronic weapons that are also alluded to in official Russian military documents through the term 'information weapons'. For example, a report on Russian IW by the Swedish Defense Research Agency stated:

> The Russian IW toolbox also includes means such as radio-frequency weapons to disturb the human brain and nervous system, and electro-magnetic energy weapons to knock out electronics and components. Other areas of a specifically Russian character are microorganism-damaging electronic components and 'psychotropic' or 'psychical' weapons. The purpose of the latter is to affect human physiology and the brain by using neurolinguistic influences and audiovisual effects through computer programming.
>
> (Heikerö, 2010: 21)

Although no hard evidence for Russian psychotronic weapons has surfaced, there are quite a few former high-ranking Russian military officers, who have specifically discussed these weapons in their military writing. For example, in a review of future technological developments Russian General Makhmut Gareev claimed that

> [m]ore attention is being given to sciences connected with the study of personality, with the aim, on the one hand, to develop a means to promote the enhancement of inner moral-psychological stability to the stress of armed struggles of the future; and on the other hand to develop means to influence the central nervous system and mentality of people, with the aim of amplifying fear and paralyzing the will.
>
> (Gareev, 1998: 50)

Of particular interest here is an article by two top Russian military analysts, S.G. Chekinov and S.A. Bogdanov on what they envision future war to be. Apart from many ideas that are similarly formulated in Western military theory, such as the growing importance of precision-guided munitions, network-centric warfare and cyberwar, there are also important deviations from Western concepts. They claim that 'the early twenty-first century *is really the beginning of a new "military age" for humanity – an age of high-tech wars'*. They speculate about new types of weapons that will be instrumental in these future wars:

> Intensive fire strikes against seats of national and military power, and also military and industrial objectives by all arms of the service, and the employment of military space-based system, electronic warfare forces and

weapons, electromagnetic, information, infrasound, and psychotronic effects, corrosive chemical and biological formulations in new-generation wars will erode, to the greatest extent possible, the capabilities of adversary's troops and civilian population to resist. It is also expected that untraditional forms of armed struggle will be used to cause earthquakes, typhoons, and heavy rainfall lasting for a time long enough to damage the economy and aggravate the sociopsychological climate in warring countries.

(Chekinov and Bogdanov, 2013: 14)

They argue that a 'new generation' of information and psychological warfare 'will seek to achieve superiority in troops and weapons control and depress the opponent's armed forces personnel morally and psychologically. In the ongoing revolution in information technologies, information and psychological warfare will largely lay the groundwork for victory' (Chekinov and Bogdanov, 2013: 16). They also point out that a psychological attack would rely on nonmilitary components such as 'an effort to involve all public institutions in the country it intends to attack, primarily the mass media and religious organizations, cultural institutions, non-governmental organizations, public movements financed from abroad, and scholars engaged in research on foreign grants' (Chekinov and Bogdanov, 2013: 17). In other words, they claim that agents of civil society can be co-opted by foreign powers and weaponized in the sense of manipulating them into the deliberate promotion of chaos and disorder. Furthermore, '[d]epending on the obtaining situation, the *aggressor may use nonlethal new-generation genetically engineered biological weapons affecting human psyche, moods, and will to intensify the effect of mass-scale propaganda to drag the target country deeper into chaos and further out of control*' (Chekinov and Bogdanov, 2013: 19).

While these are the views of just two analysts, they are certainly no fringe views in Russian military circles. Russia expert Timothy Thomas, for example, quotes Anatoliy Tsyganok, the Director of the Center for Military Forecasting in Moscow, who suggested 'that in 20 years destructive, penetrating and fragmentation weapons will not be employed. In their place will be weapons that stop, calm or frighten people'. Tsyganok even claimed 'that the US used stink weapons against the Taliban a few years ago and that the US used a psychogenic weapon in Africa, employing a laser to write a message on a cloud or wall from God' (T. Thomas, 2014: 123–124).

Another interesting statement by retired Russian Major-General Vasiliy Burenok in Thomas' article is worth quoting in length:

According to Burenok, armed violence will assume a secondary position in the future as different 'forms and methods of adversely influencing a state, a society, or an individual' will appear. Such developments include changing the 'technogenic shell' of civilization, making a distinction between living and nonliving things uncertain. Here the real issue is cyber life, he notes. New gene combinations will be designed that do not currently exist, while nanobots will alter the characteristics of an organism. New conflicts will, in Burenok's opinion, not be so much 'wars between people as wars of artificial intellects and the equipment and virtual reality created by this kind of intellect'

(T. Thomas, 2014: 125)

In other words, Burenok argues that biotechnology and nanotechnology could be employed for remaking human life and reshaping human civilization as a method of war. The Russian strategist also seems to believe that military decision-making will be largely delegated to AI, turning warfare in a contest of competing military AI systems.

Some Russian authors refer to 'weapons based on new physical principles' (a term used by Soviet Marshal Nikolai Ogarkov in the 1980s) that could fundamentally alter warfare (Fitzgerald, 1996; Belous, 2009: 66–67). The Russian government officially announced the development of such new weapons in 2012. Defence Minister Anatoly Serdyukov said: 'The development of weaponry based on new physics principles – direct-energy weapons, geophysical weapons, wave-energy weapons, genetic weapons, psychotronic weapons, and so on – is part of the state arms procurement programme for 2011–2020' (Leake and Stewart, 2012). President Putin also commented '[s]uch high-tech weapons systems will be comparable in effect to nuclear weapons, but will be more acceptable in terms of political and military ideology' (Leake and Stewart, 2012).

The described weapons were already under development during the times of the Soviet Union and a number of Soviet defectors or former Soviet defence officials have ascertained that they already exist. Most notably, former KGB General Oleg Kalugin claimed on a broadcast of the US TV channel *ABC News* in August 1991 that psychotronic weapons may have been used against opposition leader Boris Yeltsin during the coup that brought down the Soviet Union (Ostrander and Schroeder, 1997: 331–332). Yeltsin suffered a heart attack at the time, which Kalugin attributed to the use of remote psychotronic weapons. In a *Pravda* article from 1991 the President of the League of Independent Scientists of the USSR Victor Sedlecki stated in relation to that event: '[a]s an expert and a legal entity I declare that mass production … of psychotronic biogenerators was launched in Kiev…the fact that they were used is obvious to me. What are psychotronic generators? It is an electronic equipment producing the effect of guided control in human organism. It especially affects the left and right hemisphere of the cortex' (quoted from Babacek, 2013). Russian media also reported about a strange antenna in Yeltsin's office that, according to experts, 'has been installed to provide psychological impact on the president' (quoted from Kernbach 2013: 9).

For some reason Russian military theorists seem to believe that psychotronic weapons are mostly a Western phenomenon. The Russian politician Vladimir Lopatin (a former member of the Russian State Duma) wrote a book with Russian scientist Vladimir Tsygankov with the title *Psychotronic*

Weapons and the Security of the Russian State in 1999, in which they claim that the US would be actually leading in this technology field (Lopatin and Tsygankov, 1999; T. Thomas, 2000). They suggest that psychotronic weapons can 'cause the blocking of the freedom of will of a human being on a sub-liminal level' or 'instillation into the consciousness or subconsciousness of a human being of information which will trigger a faulty or erroneous percep-tion of reality' (Lopatin and Tsygankov, 1999: 107).

A particular concern that Russian authors seem to have is that the Americans may use civilian systems like the High Frequency Active Auroral Research Program (HAARP) for nefarious military purposes such as weather control or even mind control. The Russian State Duma expressed concerns in 2002 that '[u]nder the High Frequency Active Auroral Research Program (HAARP), the US is creating new integral geophysical weapons that may influence the near-Earth medium with high-frequency radio waves...The significance of this qualitative leap could be compared to the transition from cold steel to fire arms, or from conventional weapons to nuclear weapons' (FAS, 2002). Stephen Blank comments on these Russian ideas: 'We may dismiss this kind of analysis as being both unoriginal and also literally fantastic, if not paranoid. But, despite Moscow's systematic self-disinformation and paranoia, the vis-tas presented here merit our close attention because they are utterly pervasive throughout the military-political leadership' (S. Blank, 2013: 37).

8.3.5 Psychotronic weapons as 'cognitive weapons'?

In Russian literature one can even find claims that the West would be employing 'cognitive weapons' against Russia, which are defined as 'the introduction into an enemy country's intellectual environment of false scientific theories, paradigms, concepts, and strategies that influence its state administration in the direction of weakening significant national defense potentials' (T. Thomas, 2015: 18). For example, Russian politicians (including Putin) view climate change theory as such a 'cognitive weapon' or a deliberate Western scientific fraud designed to weaken other countries by imposing detrimental restrictions on their economic growth (Kuzmin, 2015).

So one could speculate that Russian theorising about 'psychotronic weapons' is merely a 'cognitive weapon' meant to trick the West into wasting resources and efforts on technological concepts that will never work. However, there are a few reasons why this seems very unlikely. This book alone contains enough evidence that neuroweapons could work in principle, so it would be a mas-sive gamble for the Russians to indirectly encourage Western governments to develop them, as they may succeed. Reports of Soviet/ Russian psychotronic weapons span half a century and come from a great range of reliable sources. Furthermore, there is little that can be gained by the Russians through such a deception since the West has enough money for unorthodox research, which would therefore not impact other areas of defence. It may thus be better to take Russian ideas in this field more seriously.

8.3.6 *Some characteristics of neurowarfare*

Western strategists have a completely different understanding of war and warfare, which continues to be shaped by the Clausewitzian paradigm and which literally blinds them to the new realities of warfare as outlined in the Russian military literature. According to Clausewitz: (1) war is seen as a temporary aberration from the normality of peace; (2) war mainly consists of calculated acts of violence aimed at politically coercing an adversary; (3) war is primarily conducted openly by regular armed forces and is governed by strict rules; (4) war is mostly about seizing territory and destroying the military capabilities of an adversary; and (5) war ends when one side has achieved its political objective.

In contrast, the kind of warfare that has been practiced by totalitarian regimes in the twentieth century and that could be much refined by insights gained from neuroscience is following a completely different pattern for conducting hostilities:

- Neurowarfare is likely to extend over many decades with the distinction between peace and war becoming not just blurred, but meaningless;
- Neurowarfare primarily targets an adversary's minds and ultimately seeks to fundamentally alter the adversary's consciousness until the adversary perceives the world the same way as the sponsors of neurowarfare – the aim is to cognitively assimilate other societies;
- Hostilities are carried out by proxies in a mostly covert, indirect manner and will generally not even amount to violence – much of it will be merely IW or propaganda in combination with other techniques of covert subversion, including espionage, sabotage and the use of 'agents of influence';
- Populations and decision-makers are the main targets in neurowarfare and populations are also used as new WMD for destroying an enemy state through the calculated psychological instigation of internal chaos – no international rules exist or are observed in the new 'information-psychological' combat;
- Neurowarfare ultimately ends when all other societies are assimilated and a world state has been created – otherwise it could turn out to be a long disintegrative process of endless political destabilization, fragmentation and Balkanization.

The above spectrum of neurowarfare aims to delineate it from traditional psychological warfare (PSYOPS) or IW, which is limited to the use of communications, while neurowarfare would include more direct means for attacking the mental capacity or of manipulating the consciousness of targeted individuals, groups and populations. The objectives of such activities would be to make the adversary either incapable of effective resistance or to shape their consciousness in such a way as to manipulate them into not wanting

Table 8.2 The spectrum of neurowarfare

	Technique	Target	Objectives
Neurowarfare	Manipulation of the brain	Attack on consciousness	• Directly control human behaviour • Insert/ manipulate thoughts, perceptions, dreams • Manipulate emotions
		Attack on mental capacity	• Degrade cognitive functions • Interfere with perception • Alter mental states
	Predictive algorithms, simulations, brain imaging	Attack on brain functions/ consciousness	• Steer society at large • Influence individual or collective behaviour • Predict threats/ developments/ behaviours • Access memory • Detect deception
PSYOPS	Communication (mass media, Internet)	Attack on perception	• Influence behaviour • Influence beliefs • Influence perceptions • Influence emotions

Source: Author's own data.

Table 8.3 The evolution of political warfare

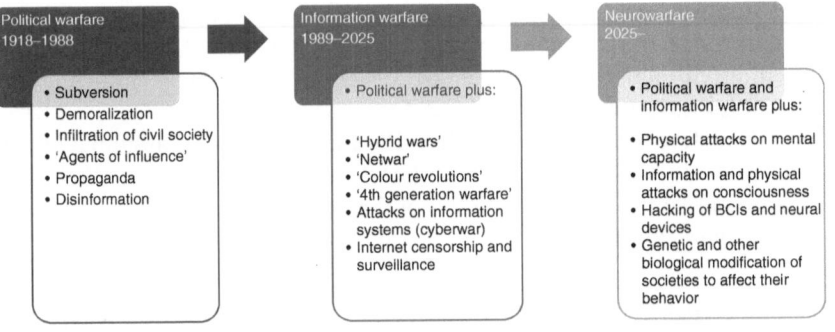

Source: Author's own data.

to resist at all. This may seem extreme or exaggerated, but is not too much a stretch considering how successful Communist political subversion was in the 1940s and 1950s with many countries seemingly voluntarily embracing communist ideology and accepting Moscow's leadership without traditional military conquest (Kintner and Kornfeder, 1962: XII–XIII). It is not hard to predict that as political warfare of the twentieth century has evolved into networked IW of today, it will eventually become neurowarfare proper – a scientific method of psychologically subverting minds by the way of sophisticated neuroweaponry that directly attacks brain function and consciousness to either cause social chaos or alternatively to suppress resistance and assimilate populations into larger political systems.

8.4 Conclusion

It is important to understand that neurowarfare will likely follow the approach of political warfare developed by the communists during the Cold War that aims at subverting the consciousness of enemy populations ('cultural Marxism') with the goal of either collapsing societies or of politically assimilating them, so that they can be incorporated into a larger political bloc, if not a world state. The Soviets have waged political warfare against the West and the threat clearly continues in the form of a resurgent Russia that is now employing sophisticated IW tactics against Western populations. The US may be leading in terms of military neuro S/T, but the US defence establishment is at least with respect to the published literature far behind Russia with respect to IW strategy and doctrine. The West excels in computer network attacks, but is clueless with respect to 'information-psychological' cyber combat, which is emphasized by the authoritarian regimes. Western governments, unlike the governments of Russia, China and Iran, do not seem to understand that information in conjunction with other overt and covert activities can be used as a powerful weapon that is designed to confuse other populations and to deliberately destabilize their societies to the breaking point. The development of future neuroweaponry can speed up the process of IW destabilization and may cause more severe and permanent damage if it reaches the point of civil war and genocide. In short, subverted populations can be turned into WMD more destructive than nuclear weapons. Russian strategic writing indicates that a lot of thinking has gone into methods that are now inaccurately described in the West as 'hybrid warfare'. It seems inevitable that neuroscience will contribute much to the refinement of methods of political subversion and destabilization, creating the possibility that the technologically advanced Western societies could be collapsed by much lesser powers with methods short of war, ranging from the infiltration of political organizations and the covert dissemination of disinformation to much more sophisticated methods such as attacks with brain-targeted genetic bioweapons or 'mind-viruses'. The final chapter will investigate various other dangers and also approaches of defending against IW and neurowarfare.

9 Dangers and solutions

The final chapter will discuss some of dangers that may arise from military neuroscience and neurowarfare and will propose some possible solutions. The dangers are extremely severe in the sense that the survival of humanity or at least the freedom of humanity is at stake. At the same time, the situation is not hopeless, as many risks can be mitigated. The advancement of neuroscience and related scientific areas will not only lead to advances in mind manipulation, but also in technologies that can protect individuals and society at large from nefarious interference with brain functions and the abusive moulding of their consciousness.

9.1 Human experimentation

The development of neuroweapons and other NLW inevitably requires some human experimentation. As Jonathan Moreno pointed out, '[b]ecause a new generation of weapons is being developed that are intended to incapacitate rather than kill an enemy, computer simulations and animal models can only go so far' (Moreno, 2000: 289). However, there are serious ethical and legal obstacles to conducting such experimentation.

9.1.1 The Nuremburg Code

The main ethical framework to which all researchers are obligated to adhere is the Nuremburg Code, which was established during the Nuremburg Nazi war crime trials to address the issue of Nazi medical experimentation on concentration camp inmates (Moreno, 2000: 79–80). The Nazis had subjected inmates to extremely brutal and often terminal experiments, mostly related to advancing military medicine and weapons development. But at the time there was no international law that prohibited governments from experimenting on their citizens or on prisoners (only PoWs were protected through the Geneva Conventions). The resulting consensus of the post-war Nuremburg Code was that 'voluntary consent of the human subject is absolutely essential' (Moreno, 2000: 80).

Further restrictions included that human experimentation should 'yield fruitful results for the good of society, unprocurable by other methods', should not result in death or disabling injury and 'the human subject should be at liberty to bring the experiment to an end'. The Nuremburg Code has been incorporated into US federal regulations and research ethics guidelines. This puts heavy, if not impossible, constraints on human experimentation involving neuroweapons.

For example, the more dangerous human experiments are typically conducted on ill people, who are agreeing to them in the hope that the experimentation will lead to a treatment or cure. However, in NLW development it would be more interesting to see how they affect healthy individuals rather than unhealthy ones. NLW that target the brain could easily result in permanent brain damage and would thus be unethical. Furthermore, sometimes one would want to see how neuroweapons work on people who are not aware of being test subjects (a rationale for some of the CIAs more unorthodox human experiments during the MK ULTRA years). It is obvious that such experiments could never pass internal ethics reviews required in a normal academic setting. This leaves national security establishments with only two options: either they decide to not develop neuroweapons and certain types of NLW at all, or they have to develop them in complete secrecy without the knowledge and consent of many test subjects.

If neuroweapons research was abandoned, then there is the risk that other nations and nonstate actors that are more unscrupulous will have an advantage: 'lax rules about human experimentation in other countries could mean that advances abroad might "parallel or even outstrip the ... work being done in the West"' (Foreign Policy, 2008: 29). Without own neuroweapons research, not even the level and nature of the threat could be adequately understood. In the light of recent neuroscience breakthroughs it is almost ludicrous to say that Western mind control research could be comfortably abandoned since mind control would be theoretically impossible. Furthermore, considering that involuntary and secret human experimentation in the name of national security has happened in the past it is not a stretch to suggest that it may happen again in the future (Welsh, 2012).

9.1.2 Legal and oversight loopholes

In order to address the issue of the controversial chemical and biological warfare tests that occurred during the Cold War Congress changed US Code 50, §1520 in 1997, which makes it illegal for the DoD to conduct such tests, except for peaceful, defensive and law enforcement purposes with informed consent of test subjects and prior notification of Congress. While this seems to comprehensively prohibit even defensive chemical and bioweapons tests, as they were carried out in earlier decades, there are somewhat different rules for classified human subject research. The final report of the Advisory Committee on Human Radiation Experiments from 1995 stated:

> The Common Rule [for all federally funded research] grants IRBs the authority to approve modifications in, or to waive entirely, informed consent requirements, but only for research involving no more than minimal risk. A separate provision grants an agency head the authority to waive any requirement of the Common Rule for any kind of human subject research as long as advance notice is given to OPRR [Office for Protection from Research Risks] and the action is announced in the *Federal Register*.
>
> (ACHRE, 1994: chapter 14)

According to Welsh, many of the ACHRE recommendations were never implemented, leaving test subjects inadequately protected. Writing in *The Bulletin of Atomic Scientists* Welsh argued:

> So while it seems crazy, it's true: Today, in a country that for the last eight years has been defined by questionable intelligence-gathering techniques and interrogation methods, the US government is no more restricted in carrying out nonconsensual, classified research on human subjects than it was after World War II. Thus, it's time for the Obama administration to reexamine the guidelines for classified experimentation on human subjects and close any loopholes that would allow a person to be unknowingly subject to experimentation. In fact, a prohibition on waiving informed consent in classified human-subject experiments should be the centerpiece of any new legislation – the waiver of this right being perhaps the most offensive aspect of the Cold War research programs.
>
> (Welsh, 2009)

If individuals that are subjected to classified and potentially illegal experimentation without their consent and try to sue the government, they have usually no chance in succeeding. Information related to secret experimentation is naturally classified and the government can invoke the State Secrets Privilege to keep it out of the courts.

For example, during the court case *CIA vs. Sims*, which was aimed at forcing the CIA to disclose its MK ULTRA research collaborators, the judge accepted the agency's argument that this information related to 'sources and methods' and would therefore be exempted from disclosure through the National Security Act of 1947 (Condon, 2014: 1146). The CIA similarly blocked the release of information on experimentation with brain implants it conducted in the 1960s (Welsh, 2012: 9). Without the classified government documents it is hardly possible to prove that the government is responsible for any health damage that victims may have received. Human rights activist Cheryl Welsh argued:

Because of the huge legal and governmental obstacles, the US court system provided access to legal liability or compensation to only a very few of the hundreds of thousands of Cold War victims. Legal experts explained that most Cold War victims were unable to overcome government secrecy and rules that consistently favoured national security.

(Welsh, 2012: 8)

Even when the government admits that experiments on human subjects have taken place, it is very difficult to prove that any existing health conditions resulted from the experimentation and had not other unrelated causes, especially if they materialize substantially after the experiments have taken place (GAO, 1994: 7).

9.1.3 The Presidential bioethics commission

Interestingly, there are several thousand individuals worldwide, who claim to be victims of involuntary human experimentation involving mind control. Defence journalist Sharon Weinberger published a story on illegal mind control experimentation in *The Washington Post* in 2007 for which she interviewed Harlan Girard, who claims to have been a victim of mind control experimentation since 1983 (Weinberger, 2007a). While remaining generally sceptical about the phenomenon, Weinberger also argued 'given the history of America's clandestine research, it's reasonable to assume that if the defense establishment could develop mind-control or long-distance ray weapons, it almost certainly would. And, once developed, the possibility that they might be tested on innocent civilians could not be categorically dismissed' (Weinberger, 2007a).

The *Presidential Commission for the Study of Bioethical Issues*, chaired by Amy Gutmann, heard numerous testimonies of so-called 'targeted individuals' (TIs), many of which were fairly similar (Gutman, 2011a). People complained about getting attacked with DEW in their homes, which was severely damaging their health. Some also mentioned voices and involuntary implantation. A medical doctor from Texas, John Hall, testified:

As a physician, relative to some of what you are hearing today, in the community we are seeing an alarming rate of complaints of use of electromagnetic weapons. Microwave auditory effects, silent sound spectrum, EEG cloning, which has taken the lab out of the laboratory and into the home. Most of these from the research we have reviewed can be done remotely. It seems to be more weapons research than medical research. I personally corresponded with upwards of 1500 victims all complaining of identical complaints from every state in the nation of being exposed to electromagnetic radiation, non-ionizing radiation for the use of cognitive control of behavior.

(Gutmann, 2011a)

John Hall has written a book that details his own experiences and that suggests that secret government technology may be abused by criminals, who had gained access to it (Hall, 2009: 98). No reference was made of any of the extensive TI testimonies in the final report of the commission published in December 2011. The commission concluded 'that current regulations generally appear to protect people from avoidable harm or unethical treatment, insofar as is feasible given limited resources' (Gutmann, 2011b: 5). Unfortunately, it is likely that involuntary mind control experimentation and 'psychoterrorism' will be affecting more and more people around the world as the technology proliferates and becomes more available to various governments, terrorists, criminals and even individuals with a grudge.

9.2 'Psychoterrorism' and other future crime

Criminals, terrorists and hostile intelligence agencies might use neuroweapons for torturing, coercing, manipulating, or otherwise harming innocent individuals. This means that the future of torture, crime, terrorism and political repression could look very different from the past. These activities will be more hidden and very difficult to detect and oppose.

9.2.1 'No-touch' torture

Torture usually serves three main purposes: interrogation, behavioural modification ('brainwashing') and punishment. This section will explore how neuroscience can alter torture practices.

Interrogational torture: There is already concern by neuroscientists that brain research can be potentially abused for torture. Indeed the George W. Bush administration authorized so-called 'enhanced interrogation techniques' for obtaining information from captured terrorism suspects (McCoy, 2012: 4). Many of these techniques involved 'no-touch' or psychological torture, including isolation, sensory deprivation, sleep deprivation, stress positions, psychotropic drugs and psychological humiliation. Most well-known is the use of waterboarding on terrorists like Khalid Sheik Mohammed, who was waterboarded no fewer than 183 times in a single month (McCoy, 2012: 42). When torture was carried out on behalf of Western intelligence services in third countries, such as Egypt, Morocco, Syria or Pakistan even more brutal methods were employed, including severe beatings, cutting, electro-shocks and ripping fingernails (Pyle, 2009: 66–75). Neuroscientists have already made a convincing argument that torture does not work since heightened release of stress hormones through intense fear and pain in conjunction with sleep deprivation can damage brain tissue, can result in compromized neurobiological function and in long-term memory loss (O'Mara, 2009). However, apart from the medieval methods used for getting information from uncooperative prisoners, the CIA also experimented with more sophisticated psychological or 'no-touch' torture. Historian Alfred McCoy wrote about the CIA methodology of torture: 'Refined over the next forty years, the CIA's methods came to rely on a mix of sensory overload and sensory deprivation via the manipulation of seemingly banal factors – heat and cold, light and dark, noise and silence, feast and famine – all meant to attack the sensory pathways into the human mind' (McCoy, 2012: 22). According to journalist Jon Ronson, the CIA used some very unusual methods for interrogating some prisoners in Guantanamo Bay such as the use of music that likely contained subliminal messages or infrasound for softening up the interviewees (Ronson, 2004: 177–180).

Behavioural modification: Torture is also instrumental in behaviour modification. Timothy Melley pointed out: ' "Brainwashing" is accomplished through torture. This conclusion was independently confirmed by three civilian researchers – Albert Biderman, Edgar Schein, and Robert Jay Lifton – who showed that Maoist "brainwashing" was not a new method, just a brutal combination of isolation, physical deprivation, and the nearly interminable revision of a personal confession' (Melley, 2011: 29). Communist dictatorships, most importantly China, have relentlessly tortured millions of dissidents in 're-education camps' to mentally coerce them into accepting communist ideology since the 1950s (Hunter, 1956). This continues to the present day. In 2013, the Chinese government announced that it would close these camps. However, according to *Amnesty International*, effectively little has changed since the old detention system was merely replaced by new detention systems called 'enforced drug rehabilitation centers' and 'legal education classes', which still empowers the police to detain individuals for up to four years without trial, but without the forced labour component (Amnesty International, 2013). The 'legal education classes' or 'law education centers' have been described as 'brainwashing centers', since the 'education' consists of forcing inmates to constantly watch propaganda videos, while sometimes also being subjecting to medical torture, such as the injection of drugs that affect the nervous system (Minghui, 2014).

Punitive torture: Torture is also an ancient method of punishment. Neuroscience can lead to new discoveries that can be abused for punishing an individual through mental torture, which could become a substitute for imprisonment. In an article in the *Journal of Neuroscience* researchers found that levels of the inhibitory neurotransmitter gamma-Aminobutyric acid (GABA) in the visual cortex predicts time perception (Terhune et al., 2014). Based on these findings, University of Oxford Philosopher Rebecca Roache has argued that the manipulation of time perception through this neurotransmitter could be a cost-effective method of punishment. Instead of locking convicts up for years at taxpayers' expense, they could be given the drug and spend a few days or months in prison, which they could perceive as a 1,000 years of incarceration (Love, 2014). On her practical ethics blog Roache explores the idea of 'virtual prisons':

Similarly, uploading the mind of a convicted criminal [into a computer] and running it a million times faster than nomal would enable the uploaded criminal to serve a 1,000 year sentence in eight-and-a-half hours. This would, obviously, be cheaper for the taxpayer than extending criminals' lifespans to enable them to serve 1,000 years in real time. Further, the eight-and-a-half hour 1,000 year sentence could be followed by a few hours (or, from the point of view of the criminal, several hundred years) of treatment and rehabilitation. Between sunrise and sunset, the vilest criminals could serve a millennium of hard labour and return fully rehabilitated either to the real world (if technology facilitates transferring them back to a biological substrate), or perhaps to exile in a simulated computer world.

(Roache, 2014)

For whatever reason torture may be used, neuroscience might make it possible for governments to use torture more widely because it becomes easier to hide (no incriminating physical evidence) and because they may pretend that psychological torture is more humane.

9.2.2 Criminal mind hacking

Humans become more and more intimately connected to the technology that surrounds them. Many people constantly keep their smart phones close to them or remain otherwise connected to the Internet. Many devices that are used every day from TVs, dishwashers and cameras have become 'smart', networked and can be accessed remotely via Bluetooth or from the Internet. This means that increasingly personal and revealing information is readily available in digital form and can be used for new forms of cybercrime. More people have pacemakers, advanced prostheses and other implants that can be hacked. Technology analyst Marc Goodman describes in his book *Future Crimes* many possibilities how criminals can hack technology users. In one example, he describes how he was able to use an iPhone app to remotely control the robotic hand of an amputee (M. Goodman, 2015: 262).

More frightening is the possibility that hackers could interfere with pacemakers. Since the medical device has to be surgically implanted into the body it would require a surgical operation to update software, unless it has some wireless access. Indeed, for convenience medical device manufacturers equip pacemakers with remote wireless access that still do not have any encryption or other defensive features in place (Wadhwa, 2012). Former Vice President Richard Cheney, who has a pacemaker, was apparently afraid that somebody would assassinate him by exploiting a security vulnerability of the implant (Peterson, 2013).

DBS implants or tDCS neural devices may be also vulnerable to hacking and may enable a criminal hacker to alter mental states, i.e. incapacitate a user. EEGs and other BCIs that could be used for controlling machinery such as switching TV channels, moving an avatar on a screen in computer gaming, biofeedback-controlled relaxation programs, all of which could be used for surreptitiously spying on the minds of users. Marketers may wish to know how potential customers react to various stimuli they receive on the screen in the form of advertising and may want to use this information to make advertising more effective or manipulative by designing stimuli that are more likely to produce a positive response from a particular individual. According to Zach Lynch, there are two ways the information can be exploited: 1) to build a brand's identity and 'wire that brand into the brain's happiness circuit'; and 2) to produce feedback about the effectiveness of marketing campaigns (Lynch, 2009: 67).

This approach is called 'neuromarketing' and it is gaining some traction. What is disconcerting is that corporations have already been caught collecting sensitive biometric information from consumers surreptitiously. For example, Microsoft's Xbox listens to conversations, has a camera and can potentially measure a user's heartbeat (Hollister, 2013). Voice data can also be analyzed for emotional responses. Even if Microsoft itself does not exploit data produced through interactive gaming interfaces, criminals might use brain monitoring and other biometric data to steal sensitive information from users or to surreptitiously manipulate users.

Marc Goodman explained that a future mind hacker could steal a PIN code for a debit card or guess a birth date with cheap EEG headsets. Researchers from Oxford University, UC Berkeley and the University of Geneva used the P300 response to guess PINs and birth dates: 'The results were powerful: by reading the brain waves emanating from these $300 headsets, researchers were able to figure out a subject's PIN number with 30 percent accuracy and her month of birth with 60 percent accuracy' (M. Goodman, 2015: 331). Greater dangers could arise from implanted BCIs that may allow criminals to take complete control over the thinking and behaviour of a user, for example by stimulating the motor cortex with a TMS to perform an unintended action (see chapter 3).

9.2.3 Manchurian candidates

Richard Condon told in his 1959 novel *Manchurian Candidate* the story of an American soldier Raymond Shaw, who was captured in Korea and then hypnotically programmed to assassinate the presidential candidate as part of a Communist coup in the US. Shaw becomes a sleeper agent, whose programming could be triggered by presenting him with a specific visual cue (the playing card Queen of Diamonds). The CIA's own documents on hypnosis from the MK ULTRA period indicate that it was, in principle, possible to override the basic morality and instincts of individuals and program them to carry out actions they would never do consciously. Although it does remain questionable that any of the exotic techniques tried by the CIA would have worked well in the field, there is good reason to believe that Manchurian candidates can be created using a more conventional brainwashing technique widely known as 'trauma-based mind control'. According to former Nebraska

State Senator and CIA intelligence operative John DeCamp, the CIA had investigated 'trauma-based mind control' in the 'Monarch Project', which aimed to create multiple personalities through methods of 'satanic ritual abuse' often involving children, which had been confirmed by none other than DeCamp's superior in Vietnam, former DCI William Colby (DeCamp, 1996: 323–331).

Terrorist groups have successfully managed to brainwash ordinary people into conducting suicide attacks and there is no reason to believe that anybody would be immune to brainwashing (systematic efforts of changing the perceptions and beliefs of a person). Former British Home Secretary Charles Clarke claimed that Muslim youths in Britain would not be motivated by the free spirit of revolutionaries, but would rather be brainwashed into joining extremist organisations. He suggested that 'anti-brainwashing techniques' should be used to 'deprogramme' the terrorists (quoted from Melley, 2011: 19). 'Deprogramming' was a buzzword in the 1980s that described controversial methods of reversing cult brainwashing used by 'deprogrammers'. Future cult-like terrorist groups might use neurotechnologies to more effectively brainwash their members so that they can be radicalized to take up arms against outsiders or non-believers.

Psychological destabilisation can be easily achieved using psychotropic drugs that can cause paranoia and psychotic behaviour. In fact, psychotropic drugs seem to be a common denominator for many cases of spree killings that occurred in the US in recent years, including the 'Batman shooter' James Holmes and the Navy Yard shooter Aaron Alexis, both of whom had been prescribed anti-depressants (Holmes was on Zoloft and Alexis on Trazodone). It is also noteworthy that none of the two had a prior criminal record: Holmes was a graduate student enrolled in a neuroscience doctoral program at the University of Colorado Denver with a scholarship and Alexis was a Navy contractor with a 'secret' security clearance. Psychotropic drugs can lead to suicidal or homicidal behaviour, especially if they are discontinued abruptly after a certain period of use (Whitaker, 2002: 188–189; Breggin, 2013: 59). A terrorist group might kidnap regular people, administer them strong psychotropic drugs, subject them to propaganda to maximize their paranoia and then let them loose in a public place with plenty of weapons to cause chaos and mayhem.

It has been claimed that ISIS' super soldier drug captagon (an amphetamine-based drug also known as fenethylline) causes people at high doses to '"become violent and crazy, paranoid, unafraid of anything…They'll have a thirst for fighting and killing and will shoot at whatever they see. They lose any feeling or empathy for the people in front of them and can kill them without caring at all. They forget about their mother, father, and their families"' (S. Anderson, 2015). Not surprisingly, ISIS is easily amongst the most violent and savage terror groups in existence, conducting mass executions, including of women and children, by beheading, by crucifixion and by burning people alive.

9.2.4 Covert harassment

Tyrannical governments have always used harassment against dissidents and other enemies as a method of silencing dissent and of neutralising suspected enemy agents. In East Germany of the Cold War era the secret police, nicknamed 'Stasi', was taking psychological oppression to a new scientific level to 'exercise a "historically new form" of power' (Epstein, 2004: 324). Fearing negative Western press in a general climate of détente, the Stasi turned to covert harassment for keeping dissidents under control (Epstein, 2004: 324). In 1976, they developed a harassment protocol for what they called 'Zersetzung', which means literally disintegration. According to a Stasi dictionary,

> [*Zersetzung* is] an operational method of the Ministry for Security of State for the effective fight against subversive activities. With *Zersetzung*, across different operational political activities, one gains influence over hostile and negative persons, in particular over that which is hostile and negative in their dispositions and beliefs, in such a way that these would be shaken off and changed little by little, and, if applicable, the contradictions and differences between the hostile and negative forces would be provoked, exploited, and reinforced. The goal of *Zersetzung* is the fragmentation, paralysis, disorganization, and isolation of hostile and negative forces, in order to impede thereby, in a preventive manner, the hostile and negative doings, to limit them in large part, or to totally avert them, and if applicable to prepare grounds for a political and ideological reestablishment.
>
> (Suckut, 2001: 464)

In practice, this meant obvious surveillance such as gangstalking, various methods of discrediting people through damaging rumours, engineering professional failures to undermine self-confidence, damaging property of targeted individuals and even the poisoning of food, including with radioactive materials. There are claims that the Stasi also covertly used X-ray beam weapons on dissidents to induce cancer and to slowly kill them (Wensierski, 1999). What is most interesting about the Stasi harassment protocol is how much emphasis was put on secrecy: none of the harassment actions were meant to allow anybody to establish any proof of Stasi involvement. When the wall came down the Stasi tried to destroy all the evidence of its covert harassment program and was partially successful in that. So it is not known how extensive *Zersetzung* was and whether any psychotronic weapons were used against dissidents, although there are many allegations of this nature.

A high-profile case involving electromagnetic or psychotronic harassment is the Nepalese dissident Tek Nath's Rizal, who described his ordeal in his book *Killing Me Softly* (2009). Rizal alleged that he was psychologically tortured when imprisoned and that the torture continued after his release due to pressure from international human rights organizations. The Indian security analyst Indrajit Rai confirmed the existence of advanced torture techniques in the foreword to Rizal's book:

> Being a professor of War Studies, I found, during my military research, the mind-control technique applied to the war prisoners. It is an electromagnetic mind-control technique which can take full control of the person's body and mind permanently. It uses modulated microwave to produce audible voices in the person's head. It is in the form of subliminal hypnotic command and the victim can be hypnotically programmed for years without knowing...As a result, the mind works under hallucination that the victim sees different images in his mind which are implanted by the controller. It inflicts pain when he tries to divert the mind from the control. It causes breathing difficulties, terrible headaches, and high blood pressure, nose-bleeding and unbearable burning sensation while urinating. It makes one undergo deep hallucinations of dying, encountering ferocious tiger, eating flesh of one's own children and so on. Sometimes, he feels the food smelling noxious and it tastes like faeces which causes vomiting sensation and nausea.
>
> (Rizal, 2009)

In 2011, Rizal initiated a lawsuit in the US against the former king of Bhutan seeking $200 million in damages for torture (Himalayan Times, 2011).

Another credible story of psychotronic weapons being used for covert harassment comes from the ousted Honduran President Manual Zelaya, who filed a formal complaint to the Organization of American States in October 2009 about an attack he suffered while sheltering in the Brazilian embassy during the coup. When interviewed by Amy Goodman from *DemocracyNow!*, Zelaya explained: 'There are two kinds of unconventional weapons that have been used against us by the regime. There's a high-frequency pitch that has been used against protesters. And another weapon that has been used against us is an electronic device that issues microwaves, which is very harmful for your health. It causes headaches' (A. Goodman, 2009).

According to Zelaya, the Honduran military has such weapons in its arsenal and they have used them to force him out of the embassy. The problem of covert harassment is therefore not unique to the Eastern Bloc during the Cold War, but seems to occur in many countries. The Asian Human Rights Commission released a statement on neuroweapons in 2013: 'The threat is real...[there are] many indications that brain technology for neuroweapons is scientifically possible. Additionally, some say such technologies have been used systematically against select people in various jurisdictions' (AHRC, 2013).

9.2.5 Homemade 'ray guns'

DEW technology has been leaking to the public for quite some time and it is not beyond the capacity of an individual with modest electronics knowledge and a few hundred dollars to spare to build some primitive ray gun that can be used for the covert harassment of neighbours or for terrorist attacks. A guide published in 1996 explains how anybody can convert a regular 5000 W microwave oven into a Microwave Amplified Stimulated Emission of Radiation (MASER) weapon. The booklet claims that the '[m]icrowave's potential as a military weapon has been kept secret from the general public' and argues

> [w]hen maser energy hits a water molecule the molecule oscillates or vibrates extremely fast...Anything a maser beam hits that contains water will get hot very fast. The human body contains 67 percent water. A maser beam would atrophy and sear living heart muscle in seconds. Likewise brain tissue would be permanently impaired instantly. Any other organs would be similarly affected in a short period of time.
>
> (Gunn, 1996: 1–2)

After dealing with some theory about masers, the guide advises people how to take a microwave oven apart and modify it using readily available materials to turn it into a primitive ray gun. The final chapter describes some practical uses for the maser such as protecting a narrow entryway. It is suggested that a person could be incapacitated after 10 seconds exposure with a 2500 W magnetron at two feet away. It also suggests that the maser could disable vehicles if placed under a bridge as it would fry electronic components (Gunn, 1996: 23–24).

There are numerous companies in the US that offer custom-made electronics or schematics and parts for those, who can assemble RF/HPM weapons themselves. For example, Michael Maloof mentions in his book that he found on the Internet a company named *Information Unlimited* that offers an 'EMP/HERF/SHOCK Pulse Generator' that can disable vehicles driving at high speed. Also in the company's catalogue is an Electromagnetic EMP Blaster Gun, Gen. II that allegedly can shut down computers and blow up a gas station from 50 feet away (Maloof, 2013: 10–11). A company based in El Paso, Texas with the name *Lone Star Consulting* produces custom-made electronics and gives advice for countering mind control and electronic attack (www.lonestarconsultinginc.com).

Not surprisingly, there is an increasing number of court cases that deal with covert electronic attacks. A famous one is the case of James Walbert, who has sued his former business partner for attacking him with harmful electromagnetic radiation, which he described as feeling electroshocks and hearing noises in his head. Defence journalist David Hambling wrote in reference to the case: 'The UN is now taking the possibility of electromagnetic terrorism against people seriously. And for the first time this year's European Symposium on Non-lethal Weapons included a session on the social implications of non-lethal weapons, with specific reference to "privacy-invasive remote interrogation and behavioral influence applications"' (Hambling, 2009).

9.3 Towards a 'psychocivilized society'

Yale scientist Jose Delgado believed that brain manipulation holds the key to a future 'psychocivilized society', where the 'future psychocivilized human being [is] a less cruel, happier, and better man' (Delgado, 1969: 232). Delgado reduced the mind to brain processes that are mostly determined by sensory and other external stimuli: 'Thoughts and beliefs are necessarily dependent on neuro-physiological activity of the brain' (Delgado, 1969: 29). He claims that personality does not originate from some intangible soul, but is the product of genetics and environmental stimuli. The mind is thus a machine that can be completely understood and externally steered in desired directions, making freedom of will nothing more than an illusion: 'We cannot be free from parents, teachers, and society because they are extracerebral sources of our minds' (Delgado, 1969: 243). He therefore suggests: 'To discuss whether human behavior can or should be controlled is naïve and misleading. We should discuss what kinds of controls are ethical' (Delgado, 1969: 249).

Most shockingly, Delgado wanted to exercise psychophysical control of everybody's mind using brain chips to suppress behaviours that violate social norms. In a Congressional testimony from 1974 Delgado said:

> The individual may think that the most important reality is his own existence, but this is only his personal point of view. This lacks historical perspective. Man does not have the right to develop his own mind...We must electronically control the brain. Someday armies and generals will be controlled by electric stimulation of the brain.
>
> (Delgado, 1974)

Sadly, many technologically advanced societies are more and more moving in the direction spelled out so clearly by Delgado in the 1970s.

9.3.1 Huxley's 'Brave New World' dystopia

The writer Aldous Huxley developed in 1932 in his novel *Brave New World* a fascinating vision of a future world in which individuals are extremely controlled from birth to death without ever realising how enslaved they actually are. Instead of nations there is a world government that has taken complete control of reproduction and every human activity. A breeding program creates five major classes according to differences in intelligence. The individuals then get conditioned by the government to fulfil their role in society and to never think for themselves. All members are constantly drugged by an invented mystery drug Huxley called 'soma'. The drug has the ability to make people happy, induce hallucinations and suppress any strong emotions. People are encouraged to be promiscuous and engage in orgies, generally living a hedonistic life in the service of society until they reach the age of 60 when they automatically die. Huxley has to introduce an outsider, the savage, from the 'uncivilized' territories to show how inhumane, decadent and oppressive the brave new world actually is. The savage eventually leaves civilisation behind with disgust. Later Huxley reflected how realistic his vision was and how it could happen in *The Brave New World Revisited*. Writing in 1958, he suggested:

> The prophecies made in 1931 are coming true much sooner than I thought they would. The blessed interval between too little order and the nightmare of too much has not begun and shows no sign of beginning. In the West, it is true, individual men and women still enjoy a large measure of freedom. But even in those countries that have a tradition of democratic government, this freedom and even the desire for this freedom seem to be on the wane. In the rest of the world freedom for individuals has already gone, or is manifestly about to go. The nightmare of total organization, which I had situated in the seventh century After Ford, has emerged from the safe, remote future and is now awaiting us, just around the next corner.
>
> (Huxley, 2010: 238)

Huxley sees overpopulation as a key driving factor towards a world 'scientific dictatorship' as all resources and all aspects of society have to be controlled to avert disaster. He claims '[o]verpopulation leads to economic insecurity and social unrest. Unrest and insecurity lead to more control by central governments and an increase of their power' (Huxley, 2010: 245). Social engineers will mould society and move it into a desired direction of greater control. They rely on distracting the masses through media that create alternate fictional realities of constant entertainment so that they 'will find it hard to resist the encroachments of those who would manipulate and control' them (Huxley, 2010: 268). He discusses the use of propaganda and mind manipulation and suggests that the key to success will be to individually condition people. 'On the road to the Brave New World our rulers will have to rely on the transitional and provisional techniques of brainwashing' (Huxley, 2010: 295).

Successive chapters deal with chemical persuasion, subconscious persuasion (subliminals) and hypnosis. In a famous 1962 interview at UC Berkeley, he suggested there would be 'a method of control by which a people can be made to enjoy a state of affairs by which any decent standard they ought not to enjoy. This, the enjoyment of servitude, well this process is, as I say, has gone on for over the years, and I have become more and more interested in what is happening' (Huxley, 1962).

Looking at Western civilisation of the early twenty-first century, many of Huxley's predictions have come to pass. The masses are addicted to entertainment in the form of music, TV, Internet and movies, while there is an unprecedented assault on civil liberties. Americans consume digital media at an average of 15.5 hours per day and spend an average of 34 hours per week in front of the TV, which puts people in an addictive, trance-like and suggestible state of mind. Cognitive scientists have shown:

> When you watch TV, brain activity switches from the left to the right hemisphere. In fact, experiments conducted by researcher Herbert Krugman showed that while viewers are watching television, the right hemisphere is twice as active as the left, a neurological anomaly. The crossover from left to right releases a surge of the body's natural opiates: endorphins, which include beta-endorphins and enkephalins. Endorphins are *structurally identical* to opium and its derivatives (morphine, codeine, heroin, etc.).
>
> (Moore, 2001)

Mind manipulation techniques such as subliminals are now widely used in commercial advertising and have been discovered in political campaign ads. Western media are inundated with dark, satanic, violent and sexual subliminal stimuli as they seem to register better in subconscious minds. It is already a concern what such kind of permanent bombardment with negative subliminal messages could do to the mental health of a society in the long-term (Sharma 2015).

Particularly problematic for a democracy is the use of subliminals in political campaigns. For example, the George W. Bush campaign ran in 2000 an ad that had the word 'rats' shown for a fraction of a second over a picture of contender Al Gore (Borger, 2000). In Mexico City billboards with political candidates now contain hidden cameras to determine (through analysis of facial expression) the emotions of passer-bys, which makes it possible to tweak the message to get more positive responses (Randall, 2015). In short, elections are in danger of becoming little more than contests in sophisticated mass mind manipulation and hypnotic persuasion, as the real political issues fade into the background.

9.3.2 'Technocracy rising'

Aldous Huxley got many of his ideas for *Brave New World* from the technocracy movement of the 1920s that aimed to reorganize society in line with methods of scientific management, as first formulated by Frederick Taylor (Wood, 2015: XIII). To technocrats, genuine democracy is nothing more than a nuisance since changes in public opinion jeopardize their grand plan of a fully controlled society that operates most efficiently with little or no waste and no disturbance. If public opinion would matter and election outcomes could deliver real political change, then the technocratic project would be in danger. In other words, democracy must be minimized in order to realize the dream of turning society into a well-oiled cybernetic machine run by bureaucratic managers together with their scientific advisors.

As Patrick Wood can demonstrate in his book, there are extremely influential political groups, such as the Trilateral Commission and the Council on Foreign Relations, to move society along, often relying on hidden political manipulations and propaganda. He quotes Trilateralist and former National Security Advisor Zbigniew Brzezinski who wrote in his 1970 book *Between Two Ages*:

> [The technetronic era] involves the gradual appearance of a more controlled and directed society. Such a society would be dominated by an elite whose claim to power would rest on allegedly superior scientific know-how. Unhindered by the restraints of traditional liberal values, this elite would not hesitate to achieve its political ends by using the latest modern techniques for influencing public behavior and keeping society under close surveillance and control.
>
> (quoted from Wood, 2015: 41)

Since technocracy is not particularly appealing to the masses, especially since it would involve a radical transformation of the economy from capitalism to a planned economy totally based on energy (i.e. carbon permits) rather than currency. Such a move would inevitably lower living standards for the majority of people in Western societies. Highly restrictive environmental regulations that put severe limits on individual energy consumption and that restrict individual freedoms in many ways are not going to be favoured by the majority of the electorate. Therefore, Patrick Wood argues, the technocrats must offer the masses the carrot of transhumanism, which is the promise that everybody will have the chance to upgrade themselves to live better, happier, healthier and longer lives. Transhumanists promote the ideas of genetic modification, implantation and brain intervention to supposedly improve individual lives and society at large. The danger lies in the idea that people come to accept implants, prostheses and biological/ genetic modification as normal, which then makes it possible to exercise social control on some grand scale, if no legal and other safeguards are established ahead of time.

There are little doubts that Western governments are interested in the development of social engineering technologies for making sure that populations go along with government policies. On 15 September 2015 President Obama signed an Executive Order titled 'Using Behavioral Insights to Better Serve the American People', to 'improve how information is presented to consumers, borrowers, program beneficiaries, and other individuals' and to 'identify programs that offer choices and carefully consider how the presentation and structure of those choices... can most effectively promote public welfare' (US White House, 2015). In other words, the government wants to present information and structure individual choices in such a way as to influence individuals and 'nudge' them into compliance. While the executive order sounds benign enough, it is important to keep in mind that the methods that could be developed in this effort are likely to undermine individual choice and democracy in the long run. The similarities of Cass Sunstein's 'nudge' theory and the Soviet method of 'reflexive control' are very apparent and should provide some warning with respect to the potential for abuse.

9.3.3 Forced medication and medical intervention

What is most disturbing about the transhumanist movement is the apparent willingness of many of its outspoken proponents to force people to accept 'enhancements' and other modifications in the name of the greater good of society. Kurzweil essentially says that people will have little choice other than to become cyborgs: '[once] [t]here is ubiquitous use of neural-implant technology that provides enormous augmentation of human perceptual and cognitive abilities. Humans who do not utilize such implants are unable to meaningfully participate with those who do' (Kurzweil, 1998: X).

However, once an individual can be forced directly or through indirect pressure to take medications (e.g. genetics altering vaccines) or to accept BCIs, individual personality and freedom of will would be at stake. There are societal groups that are particularly vulnerable to forced medical intervention and that may lead the way towards a coming psychocivilized society, namely soldiers, criminals, certain professions (e.g. in the security and health sectors) and of course, those diagnosed with illness. Soldiers can be ordered to accept implanted microchips once they are no longer experimental. Indeed, DARPA has announced a project for the development of implantable nanosensors ('*In Vivo* Nanoplatforms' or IVNs) for monitoring health, including stress levels, detecting infectious agents and repairing soldiers on the battlefield (DARPA 2012). Presumably, the goal is to modify the bodies of all soldiers in this way.

It will not be long before criminals are implanted with chips that monitor and control their behaviour. Already in the 1960s, the psychologist Ralph Schwitzgiebel proposed 'an electronic device capable of tracking the wearer's location, transmitting information about his activities, communicating with him, and perhaps modifying his behaviour' (quoted from Fox, 1987: 131). At that time the technology was simply not available for pursuing the proposal, but today it would be fairly straightforward to implement the chipping of criminals, including microchips that monitor mental states and that release medication if, for example, the person gets agitated. This was actually proposed quite recently by a political candidate in the US:

> Brain implants able to manage out-of-control tempers and violent actions of prisons were suggested to minimize crime rates in the United States and as alternative for death penalty, according to Zoltan Istvan. The futurist and presidential candidate for the Transhumanist Party, Istvan

suggested that the technology could be a near-term alternative for criminals on death row and might be considered sufficient punishment.

(Malicdem, 2015)

The underlying assumption is that violence is not an expression of free will, but rather a disease that can be treated by technological means, so that violent offenders can be rendered harmless. NLW expert John Alexander discussed in a 2007 interview with Sharon Weinberger the prospect of 'neutering' potential terrorists. He said: 'Maybe I can fix you, or electronically neuter you, so it's safe to release you into society, so you won't come back and kill me'. According to Alexander, '[i]t's only a matter of time before technology allows that scenario to come true…We're now getting to where we can do that…Where does that fall in the ethics spectrum? That's a really tough question' (Weinberger, 2007a).

In a country that already has over a million people, including some toddlers, on terrorist watchlists, one can easily see how a government could abuse such a technology to deal with individuals that have 'extremist' views. The Soviet Union infamously put thousands of dissidents into mental institutions where they were subjected to forced medication with Haloperidol (or Haldol) and other dangerous drugs to cure their supposed illness (Gallo, 1982: 10). Obviously anybody who rejected the blessings of communism had to be mentally ill. In order to prevent such a scenario from happening in the free world, governments should not be allowed to forcibly medicate or microchip people in order to control their behaviour, except in very narrowly defined circumstances. Brain chips should be limited to the treatment of physiological illnesses and should not be used for bettering society, as the chances for abuse are far too great.

9.3.4 Towards a transhumanist hive mind

Neuroscientist Marios Kyriazis argues that '[t]he line between human brain function and digital information technologies is progressively becoming indistinct and less well-defined', which would enable 'the merging of the physical human brain abilities with virtual domains and automated web services'. The result would be a 'global brain' as 'a self-organizing system which encompasses all those humans who are connected with communication technologies, as well as the emergent properties of these connections' (Kyriazis, 2015: 2). In other words, the development of BCIs or brain chips might result in the creation of a global human/ virtual hive mind. Through BCIs humans could be constantly connected to a global computer network and thereby to each other, constantly communicating and thereby moving the collective brain or hive mind forward, transcending limitations of the individual human brain (Kyriazis, 2015: 1). According to an article on the futurist journal *IO9*, several scientists such as Kevin Warwick, Ramez Naam and Anders Sandberg believe that the technologies for creating a telepathic noosphere already exist or might exist soon (Dvorsky, 2013).

Of course there could be great benefits for humanity to be able to share thoughts, emotions, perceptions, dreams and memories in the way of wireless connectedness, but the risks are so severe that things like 'mind uploading' and the telepathic noosphere quickly start to look like an extremely bad idea. Humanity could face high-tech enslavement by an Artificial Super Intelligence. Humans would be turned into the Borg out of *Star Trek*. Unlike in earlier periods of history, where revolutions could overturn oppressive political orders and advance human freedom and development, there would be no resistance possible once everybody is connected to the hive mind, which could take control of every individual. Humanity as we know it would cease to exist.

9.4 Neurosecurity

Undoubtedly, brain intervention and a vastly improved understanding of the human psyche and human behaviour can result in serious dangers for individuals and society at large. Luckily, there are steps that can be taken to safeguard individual freedom and to protect nations against covert interference with minds of their populations. Jonathan Moreno has coined the term 'neurosecurity', which he defines in the following way: 'I use the term *neurosecurity* to refer both to the ways that science and technology targeted at the brain and nervous system should be managed for the public good, and the means that democratic states must develop to protect themselves from their adversaries' (Moreno, 2012: 185).

Neurosecurity would thus comprise two main components: a regulatory framework based on neuroethics and secondly, neurodefence or protective measures that need to be implemented for defending society. Moreno draws strong parallels between neurosecurity and biosecurity and suggests that many lessons learned from managing the threat of biological weapons would also apply to the newer field of neurosecurity. The argument makes a lot of sense, especially since the most powerful future neuroweapons, as indicated in the Russian literature, are likely to be of a biological nature (T. Thomas, 2014: 125). This section shall give some brief overview of what can and should be done in addressing some of the dangers discussed above.

9.4.1 Neurosecurity as neuroethics

The power to manipulate minds and to mentally coerce people seems to offer pretty obvious advantages in warfare and for the control of societies. However, before one can devise regulations based on neuroethical considerations one should determine what the actual threat of neuroweapons and neurowarfare is likely to be. Should neuroweapons be considered a new type of WMD and should there be heavy restrictions on their use? Moreno compared the threat posed by neuroweapons to nuclear weapons and argued that

The utility of neuroscience-based weapons, however, is mainly tactical, in that they might provide short-term and relatively targeted advantages such as disrupting an enemy patrol or disabling a terror cell. In that sense neuroweapons are much more manageable than nuclear weapons. Not weapons of mass destruction, they are better considered weapons of selective deception and manipulation.

(Moreno, 2012: 193)

In other words, Moreno thinks that neuroweapons are not a serious threat and their use may be preferable in ethical terms compared to other more lethal or crueller methods of war and interrogation. He argues that a 'no first use' principle should not apply to neuroweapons. This perspective overlooks the possibility that neuroweapons can be used strategically for manipulating the minds of enemy leaders, societal groups and indeed entire populations during peacetime. Secondly, Moreno is far too quick in dismissing the potential for the abuse of neuroweapons, which undeniably exists at least as a theoretical possibility.

Neuroweapons do raise some interesting ethical questions. For example, should it be permissible to use neuroweapons on non-combatants to pacify a city, e.g. by adding some calmatives to the drinking water? Should it be allowed to bombard enemies with subliminal messages or with messages through V2K that encourage them to commit suicide or to attack their comrades? What about degrading terrorist performance by biochemically neutering them or by changing their genetics? When can a government or corporation intrude the minds of individuals (including foreign political and business leaders) to obtain information on them for whatever reason? What about the ethics of deleting or implanting memories, or of changing key aspects of personality in order to make an individual a more effective fighter or to deceive or manipulate enemies? DARPA is already working on implants for suppressing traumatic memories of veterans – how long will it be before active service personnel is given the opportunity to erase memories at will? These questions will have to be addressed fairly soon.

One could argue that many of these things are already illegal under international law, so there would be no need for further regulation. The problem is that since virtually all conflicts in today's world are intrastate, often specific restrictions by international law do not apply. A government may claim that the use of calmatives on its population or segments thereof is 'law enforcement' or is 'therapeutic' and is thus outside of the scope of the CWC. Governments may change their interpretation of international law in the light of new nonlethal technology. For example, the FAS has pointed out that the US government recently changed its stance with respect to bioweapons. While previously supporting the view that all BW are prohibited, it now suggests that the BWC 'does not apply to non-lethal biological weapons' (FAS, 2013).

There are also no adequate protections with respect to the misuse of DEWs against combatants, as well as noncombatants. This is particularly disconcerting since terrorists have been redefined as 'unlawful' or 'unprivileged' combatants, while the terrorism label has been applied to 'extremist' individuals, who have not yet committed a terrorist act, but might do so at some unspecified time in the future.

Also more difficult to deal with than Moreno anticipates will be the issue of remote neural monitoring. It may be already too late in terms of technology to assume that the privacy of the mind could be adequately protected by the 5th Amendment (Moreno, 2012: 198). Mind reading in its most simple form of checking brainwave responses does not require putting anybody into a million dollar fMRI machine – it can in principle be done remotely and surreptitiously, as can many other biometric measurements. 'Lie detectors' that analyze stress in the voice and content of speech for deception or analyze facial expressions will become widely available. Mind reading can become as simple to use as installing an 'app' on a smart phone and, according Professor Nita Farahany from Duke University, there is not even a law against it (Dean, 2016). An article in the *London International Law Journal* argues that brainwave monitoring should be considered as a form of Signals Intelligence and that 'viewing brainwaves as an anatomical body part leaves us with no way of protecting the data and intellectual property as of today' (Rangarajan, 2014).

Furthermore, it is important to understand that just because certain methods of warfare are outlawed it does not mean that governments would not violate international norms, especially if they think that the violation can remain undetected. Covert action is, for example, a generally acknowledged tool of statecraft although it is clearly illegal under international law to interfere in the internal affairs of another nation. As outlined in the previous two chapters, neuroweapons are particularly useful in covert action and covert warfare since their effects are often not visible or easily detectable, which allows for some degree of deniability. It is the great secrecy surrounding neuroweapons and certain aspects of brain research that makes it difficult to even debate neuroethics from a position that realistically assesses existing or near-future capabilities. This has to change: governments have to be more transparent about neuroweapons research.

9.4.2 Neurodefence

It has been argued throughout the book that neuroweapons will have a strategic impact on future conflict. Nations that can exploit neuroscience for military purposes may gain a decisive advantage over other nations. Neurowarfare attacks might not be noticed until it is too late if no system for early warning and detection is established. Neurodefence should have the following components: (1) ending neuroweapons secrecy, (2) gathering intelligence on military neuroscience research of other countries, (3) taking surveillance measures for detecting neurowarfare attacks, (4) developing methods of correct attribution, (5) hardening for the protection of specific individuals and groups from attacks on their minds and finally (6) developing a program for civil defence that consists of education of the public about the threat and measures that everybody can take to prevent or minimize the danger of mind manipulation.

- *End neuroweapons secrecy*: Unlike in the case of other defence technologies, secrecy in methods of brain and mind manipulation cannot be tolerated at all since the stakes are too high. As pointed out by Lopatin and Tsyngankov, secrecy just makes neuroweapons (psychtronic weapons) arms races and the occurrence of neurowarfare (psi-warfare) more likely (Lopatin and Tsyngankov, 1999: 97). To begin with, secret weapons have no deterrence value. Even if they were announced by a government they cannot deter any aggressor unless they have been demonstrated first in a compelling way. The US could not coerce Japan with nuclear weapons until it actually dropped them that demonstrated without a shadow of a doubt that the weapons exist and can cause unacceptable damage. Short of causing the destruction of a foreign society through neurowarfare, there is only the possibility to make all principles and methods of neurowarfare public for everybody to see and evaluate in order to turn neuroweapons into an effective deterrent. Any neuroweapons/ mind control research should be therefore declassified and ideally all neuroweapons usage should be severely restricted under international law and domestic laws. Neurowarfare should only be allowed to exist for the purpose of deterrence similar to nuclear weapons. This, of course, would not rule out the use of neuroweapons against combatants on a tactical level in the context of an armed conflict, as suggested by Moreno.
- *Intelligence*: The 2008 NRC study recommended that the IC should carefully monitor technological development in the area of neuroscience. Acknowledging the inherent difficulties in providing technology warnings, it recommended:

Rapid advances in cognitive neuroscience, as in science and technology in general, represent a major challenge to the IC. The IC does not have the internal capacity to warn against scientific developments that could lead to major – even catastrophic – intelligence failures in the years ahead. An effective warning model must depend on continuous input from strong internal science and technology programs, strong interactive networks with outside scientific experts, and government decision makers who engage in the process and take it seriously as a driver of resources. All that remains a work in progress for the IC.

(NRC, 2008: 12)

Unlike other WMD arms control and counterproliferation monitoring there is the complication with neuroweapons research programs that they do not require any large-scale infrastructure and test sites like in the case of nuclear weapons development and that there are no relevant arms control inspection regimes that could discover military neuroscience research activities. Relevant military neuroscience research may be hidden within civilian, in particular medical research programs, as was the case with the Soviet Union's large-scale bioweapons program that was hidden within the civilian agency *Biopreparat* (Moreno, 2012: 189; Mangold and Goldberg 1999, 92–100). Furthermore, it can be assumed that neuroweapons will not be part of a country's official arsenal and would be employed covertly. The best chances for discovering such highly secret programs would be by monitoring and recruiting foreign neuroscientists, as well as the careful analysis of foreign published scientific literature.

- *Surveillance*: Neurodefence just like biodefence and cybersecurity must have a system of monitoring in place that can reliably detect hostile interference. This may necessitate the creation of a new organisation for neurodefence, specializing in defensive IW and neurowarfare. Since neuroweapons can take many forms, from biochemicals and microorganisms to electromagnetic radiation and computer code, there cannot be just one system for detection, but there has to be a combination of systems and methods that are employed. An important starting point would be to develop a concept of 'ecology of cognition' or 'neuroecology' (Lopatin and Tsygankov, 1999: 101). Giordano suggested defining it as 'the nature of neural functions in organisms embedded in, and responsive to, various conditions, effects, and cohorts within environment(s) occupied. These neural substrates of human ecology...function to influence a person's awareness, responses, decisions, and actions toward others; environmental conditions; and situations that may be regarded as positive, neutral, or negative' (Canna, 2013: 5). Since the brain connects in so many ways with the environment, most importantly through perception and the biological system, the surveillance of the neuro-ecology has to be comprehensive. For example, any imported foods, drugs and supplements need to be tested for hidden components, allowing for the possibility of binary and 'slow kill' weapons. Biodefence health monitoring should include the threat of nonlethal bioweapons and should not be limited to the classical biowarfare agents (anthrax, smallpox and botulinum). A difficulty is that symptoms may not be apparent for a long time, allowing nonlethal biowarfare agents to spread across society with fewer chances for early detection. Covert attacks on consciousness and a nation's value system that rely on gradualism are particularly difficult to detect and resist since changes occur over a longer period of time and might be falsely attributed to 'natural' causes or 'unintended' developments, such as 'progressive social morality'. This is not to suggest that any societal change has to be considered as negative, only that society has to be vigilant with respect to the possibility of harmful external influencing and social engineering. A particular worry relates to so-called 'ionospheric heaters' like HAARP and over-the-horizon radar systems, as they can be potentially used for remote brain entrainment (Becker and Seldon, 1985: 324). This is not only an idiosyncratic Russian fear, but a concern that has been voiced by other governments and Western scientists as well. For example, the European Parliament had an expert hearing on the potential environmental impacts of HAARP in 1999 and it concluded that they are 'a matter of global concern' and could have 'incalculable impact on human life' (Theorin, 1999). Any large-scale manipulation of the atmosphere using electromagnetic radiation by any state should be scientifically investigated for the possibility of brain interference. Automated software should be developed that can scan digital media for hidden subliminal messages, inaudible sounds and nefarious brain entrainment in real-time in order to protect the consumers of media and computer users from covert influence.

- *Develop methods for attribution:* Since neurowarfare attacks can be conducted covertly and remotely, attribution will be (very much like in cyber warfare) a major problem. Even if the attack is itself detected, it may not be trivial to figure out who is the sponsor of the attack. For example, a nefarious actor might create a 'mind virus' that is spread over the Internet and hacks the brains of people through neural devices and BCIs. It would not be straightforward to identify the origin of the attack because of the use of proxies and cut-outs, such as political front organizations, criminal syndicates, terrorist groups, etc. that can carry out attacks. The identity of the sponsor can be obscured even if a connection between certain groups and states is proven, as uncertainty could remain about the extent of control a state has over their proxies (UK MoD, 2010: 85). Obviously, without the credible capability to correctly attribute any attacks deterrence is bound to fail (Cohen, 2011: 8). This makes the establishment of an effective system for the surveillance of the neuroecology, including the development of new sensors, doubly important.
- *Hardening*: Robert McCreight suggested with respect to the emergence of neuroweaponry: '[t]he global battlespace will be dramatically altered and in light of this, it will be necessary to design and implement systems to protect humans from neural interference' (McCreight, 2014: 124). Methods and technologies should be developed for shielding personnel from remote brain monitoring and manipulation. The biggest potential vulnerability relates to future neural interfaces. They need to be designed from the very beginning with the security of the user in mind. According to an article by computer scientists Tamara Denning, Yoki Matsuoka and Tadayoshi Kohno:

Computer security and privacy is a field within computer science dedicated to the design and engineering of technologies so that they behave as intended, even in the presence of malicious third parties who seek to compromise the operations of the device. These malicious parties are often called hackers, attackers, or adversaries. Three of the standard goals in computer security are confidentiality, integrity, and availability: an attacker should private information (confidentiality); an attacker should not be able to change device settings or initiate unauthorized operations (integrity); and an attacker should not be able to disable a device altogether and render it ineffective (availability). We define neurosecurity as the protection of the confidentiality, integrity, and availability of neural devices from malicious parties with the goal of preserving the safety of a person's neural mechanisms, neural computation, and free will.

(Denning et al., 2009: 1–2)

Other hardening may take the form of shielding individuals from potentially harmful electromagnetic radiation and infrasound. For example, David Hambling suggested building 'a layer of conducting mesh into body armor' (Hambling, 2006). Soldiers could be also hardened psychologically against the stress of combat and potentially methods of covert influence – as the CIA desired to do as part of its MK ULTRA research. To some extent this is already the case, as some soldiers have to undergo so-called Survival, Evasion, Resistance, and Escape (SERE) training during which they learn how to resist harsh interrogation and brainwashing. The mental health of soldiers and other government personnel should be carefully monitored as part of early detection of and hardening against mind control.

- *Civil defence*: As neuroweapons can potentially target entire populations, it makes sense to think about developing a civil defence program that can strengthen the resilience of a population to any such covert attack. During the early Cold War US citizens were encouraged to build their own nuclear shelters and stock up on non-perishable food so that they had a better chance of surviving a nuclear attack. While it was unrealistic to assume that ordinary citizens could and would prepare for the end of civilization as we know it, the idea that average citizens could meaningfully contribute to the defence of a nation has again become relevant because of the threat of cyberwar. Peter Singer and Allan Friedman write in reference to cybersecurity '[t]he individual citizen must also play their part…When it comes to cybersecurity, most people are not being targeted by APTs [Advanced Persistent Threats], Stuxnet, or other high-end threats. We are, however, part of an ecosystem where we have responsibilities both to ourselves and to the broader community' (Singer and Friedman, 2014: 241). They essentially suggested that overall cybersecurity could be greatly improved if everybody would take sufficient steps for securing their own computers, i.e. use licensed software that is regularly updated, have a firewall installed, use and regularly change passwords and regularly run antivirus scans. Similarly, in neurowarfare much could be achieved if everybody would take measures to protect the health and integrity of the own brain and mind. This requires educating the public about methods of influencing and of encouraging everybody to monitor and support their own mental health (e.g. using novel neural devices and nootropics) and to resist irrational urges. It has to be acknowledged that there is such a thing as 'harmful information', which includes according to Lopatin and Tsyngankov the use of subliminals, hate speech and pornography, all of which can impact massively on the brain and a person's ability to make good judgments (Lopatin and Tsygankov, 1999: 104). However outdated it may sound, core liberal democratic values and overall morality must be strengthened in order to make the population more resilient to hostile ideologies and propaganda. A moral and free society is much harder to subvert and to corrupt. Of course, this would not help much if an individual was specifically targeted with a neuroweapon, but such measures can make it harder for a hostile power to manipulate a collective and to steer it into a direction that goes against the public interest.

9.5 The need for an ethics of mind manipulation

Neurowarfare could arrive much sooner than many analysts think, as the basic technologies necessary for neuro-enhancement and neuroweapons are largely available or may be ready within a decade or two, including brain chips and genetic weapons. Neurowarfare does raise a few novel ethical issues and might require rethinking some older problems in the ethics of warfare such as the targeting of civilians. This final section will outline some of the legal and ethical issues that could arise and will try to propose some solutions. It will be argued that a 'human right for mental self-determination' should be codified and that offensive neurowarfare (the deliberate destruction of another society by instigating internal chaos) should be clearly prohibited by international law.

9.5.1 Issues related to human decision-making

Neurowarfare can potentially weaken the capacity for human decision-making in two respects: 1) neurally-triggered weapons merely rely on the monitoring of subconscious brain processes and thereby bypass conscious human decision-making; and 2) the possibility for accurately predicting human behaviour will push militaries and other organisations engaged in a strategic competition (e.g. major corporations) to delegate more and more decision-making to AI, which may be less predictable and overall superior to human decision-making in terms of speed, the analysis of vast amounts of complex information and, most importantly, in achieving desired outcomes.

As pointed out by Stephen White, neurally triggered weapons create a plethora of legal dilemmas, especially with respect to assigning responsibility for war crimes (White, 2008). A neurally triggered weapon that accidentally opens fire on civilians raises the question of who would be responsible for such an atrocity. Although a human operator is involved in the decision-making process, no conscious human decision would be required. In effect, the human operator is downgraded from decision-maker to a mere functional component in a man-machine system, where the real control resides with the machine rather than the human component (Noll, 2014).

The human brain of the operator could theoretically be entirely simulated within a computer at some future point, in which case it would be possible to argue that the simulation of the human brain can be equated with actual human decision-making. The danger is that human judgment and human ethics may not be sufficiently replicated by machine intelligence, resulting in military conduct that is less restrained by human ethics and emotions such as empathy and may be less humane (Barrat, 2013: 18).

In the long run human brain simulations could give rise to self-aware strong AI and fully autonomous weapons systems. After all, the human brain can generate consciousness from which it would seem to follow that an accurate simulation of the brain would likely be the most promising approach for technologically replicating human consciousness in a machine. Of course, it remains speculative whether this can be done at all, but most AI researchers believe that it is going to happen sometime in the twenty-first century – a pretty short time period in historical terms (Barrat, 2013: 25). Numerous reputable scientists and engineers, including Bill Joy, Nick Bostrom, Stephen Hawking, Elon Musk and Steven Wozniak, have warned of a scenario of AI taking over and potentially destroying humanity in the process. It will thus be critical not to transfer important decision-making to machines and to keep humans in the loop, without turning humans into biological robots, controlled by brain chips, or into mere biological components in man-machine systems.

9.5.2 Issues related to human enhancement

Although human enhancement sounds desirable and overall positive, as it promises to improve humans and the human experience, it also raises some very daunting ethical questions. The first question regarding enhancement anybody should ask is: enhancement for what exactly? An assembly line worker might be 'enhanced' by artificially reducing natural curiosity and intelligence to make the work appear less boring, so that the worker is less distracted and achieves a greater output. Similarly, soldiers could be 'enhanced' by making them fearless and completely obedient to their superiors. In both cases natural human personality and capability would be altered to make them perform a particular job better. In this sense enhancement does not necessarily mean improvement of human capability across the board – it might not even mean any improvement that would be genuinely in the interest of the individual concerned. In other words, enhancement may downgrade humans from autonomous subjects into machines that can be modified and moulded for specific purposes. Admittedly, it could be beneficial for society as a whole to have individuals that are designed for fulfilling specific societal roles such as being ideal factory workers or soldiers, but it also clearly impacts massively on human freedom and the potential for human development.

The next big question is whether governments should try to steer human evolution by modifying the human genome and otherwise changing the physiology of their citizens. This is no longer a hypothetical question as the state of California has already enacted a law for forced vaccinations of school children with no more personal and religious exemptions (Willon and Mason, 2015). Vaccines are not simple medical treatments for a disease – they are arguably enhancements as they permanently alter the human immune system (Lin, 2012). This alteration is not necessarily beneficial for a vaccinated individual since vaccines have acknowledged health risks because of adjuvents and other additives and because of unpredictable immune system responses. Furthermore, vaccines may contain recombinant viruses that alter human DNA and thereby change a human being at a genetic level ('gene therapy' or 'gene doping': see chapter 3).

Even if individuals are comfortable with getting enhanced and having their genome and personality altered or with being perpetually connected to a noosphere or hive mind, it may have tremendous effects on the species as a whole. Enhancement could radically transform the human experience and there is no way of knowing whether the outcome is going to be a heaven or hell scenario (Garreau, 2005: 85–186). Governments should therefore tightly regulate certain enhancement technologies, such as brain chips and genetic modifications that could have unforeseeable negative consequences affecting the entire human species down the road.

9.5.3 Are neuroweapons legal?

The Hague Convention of 1907 famously declared: 'the right of Parties to the conflict to choose methods or means of warfare is not unlimited' (IV, Art. 22). Certain weapons and methods of warfare have been clearly outlawed, including chemical weapons, biological weapons, expanding bullets, blinding lasers and anti-personnel landmines because they are considered indiscriminate or cause 'superfluous injury and unnecessary suffering' (Green, 2000: 134–137). The question is: are certain neuroweapons and methods of neurowarfare inherently inhumane and should be clearly prohibited by international law?

Indeed, there have been already several attempts to outlaw psychotronic weapons, but none of them has succeeded. In 1999 a commission of the European Parliament called for a

> global ban on all research and development, whether military or civilian, which seeks to apply knowledge of the chemical, electrical, sound vibration or other functioning of the human brain to the development of weapons which might enable any form of manipulation of human beings, including a ban on any actual or possible deployment of such systems.
>
> (Theorin, 1999)

Similarly, Congressman Dennis Kucinich introduced the Space Preservation Act in 2001, which proposed the prohibition of space-based weapons and

> any other unacknowledged or as yet undeveloped means inflicting death or injury on, or damaging or destroying, a person (or biological life, bodily health, mental health, or physical and economic well-being of a person)…through the use of land-based, sea-based, or space-based systems using radiation, electromagnetic, psychotronic, sonic, laser, or other energies directed at individual persons or targeted populations for the purpose of information war, mood management, or mind control of such persons or populations.
>
> (H.R. 2977)

The bill received an '[u]nfavorable [e]xecutive [c]omment' via the House Armed Services Committee and the quoted passage was removed from the final bill. Presumably, DoD was/ is working on weapons described in Kucinich's bill and wanted to avoid a general prohibition. Otherwise, the unfavourable DoD comment makes no sense.

Although neuroweapons that only temporarily incapacitate or disperse combatants such as calmatives and pain rays could be used in full accordance with the laws and customs of war, there are also potential aspects of hypothetical neuroweapons that could make their use or certain types of uses appear decidedly inhumane. For example, if one could cause such a severe depression in enemy combatants that they commit suicide just to end their torment, then it seems like a method of psychological attack that amounts to torture and that represents excessive cruelty no different from physical suffering. It would certainly violate the prohibition of inflicting superfluous injury and unnecessary suffering (Rosenberg, 1994: 45).

Furthermore, it is easily imaginable that individual enemy combatants can be manipulated into attacking their comrades, e.g. by hacking their neural interfaces (Arizmendi and Diggins, 2012). Such manipulation would most certainly violate the customs of war. For example international law expert Leslie Green states: 'It is not illegal to use propaganda to incite the enemy civilians to rebel or enemy troops to mutiny or desert, but it is forbidden to compel nationals of the adverse party to take part in warlike operations against their own state' (Green, 2000: 148). Mind manipulation that mentally coerces an enemy combatant to open fire on his comrades would be thus already prohibited. More difficult is the matter of causing enemy civilians to riot.

9.5.4 *Targeting civilians with NLW*

A very big issue in relation to neurowarfare is going to be the question in what way and to what extent can noncombatants be broadly targeted. According to Green, '[o]ne of the oldest rules of the law of war provides for the protection of the civilian non-combatant population and forbids making civilians the direct object of attack' (Green, 2000: 229). The Geneva Conventions require belligerents to distinguish between combatants and noncombatants. Only combatants can be legally targeted in an armed conflict, while noncombatants are afforded immunity based on their status. Furthermore, '[t]he principle of discrimination relates not just to killing non-combatants, but to intentionally targeting non-combatants and/or engaging in indiscriminate attacks' and 'non-combatants have a right to self-defense', which would be negated by the use of NLW and which would amount to 'harm', since 'they can no longer exercise their autonomy and choice' (Kaurin, 2010: 105).

The ethicist Pauline Kaurin has also warned of the possibility that 'NLW being used in conjunction with conventional weaponry as a "force multiplier"' and that this 'could lead to the inflicting of unnecessary suffering, even if these weapons do not kill' (Kaurin, 2010: 106). Although there is the possibility that NLW can produce more humane outcomes, it is not necessarily so, if they are weapons used for softening up a target, so that other weapons are made more effective. Therefore, she suggests four principles that should be followed when using NLW in the context of armed conflict. NLW should be used:

(1) to provide the military with more flexible response times and options, allowing them more time and space to carefully make the strategic and ethical judgments necessary in war and to respond with appropriate and proportional force; (2) to reduce unnecessary suffering on the part of non-combatants; (3) to facilitate the eventual restoration of peace; and (4) to minimize combatant casualties.

(Kaurin, 2010: 102)

What follows is that noncombatants should only be targeted with NLW if their use is proportionate to a threat and if the effects are truly benign and temporary. NLW should not become compliance tools that enable authorities to force their will on populations, as some kind of 'soft repression'. In fact, any use of NLW against noncombatants should be subject to similar restrictions as the use of lethal force within the limitations suggested by Kaurin.

9.5.5 Time to rethink the ethics of propaganda?

Propaganda is a modern phenomenon that is intrinsically linked to the existence of mass media (Ellul, 1965: 89). It is present in all societies, including democratic ones, where propaganda has to be more subtle and sophisticated since there is media competition, unlike in dictatorships, where all media are controlled by the government. Most generally, governments can use several channels for disseminating propaganda: government PR and official statements, deliberate leaks to the private media, 'strategic communication' with foreign audiences and disinformation planted on the Internet or in foreign media by proxies. White propaganda that is correctly attributed to a government is most common and least controversial, while unattributed or falsely attributed grey and black propaganda, sometimes containing disinformation for purposes of deception and confusion, is the domain of intelligence services and military PSYOPS.

Any attempt by a supposedly democratic government to exercise politically motivated censorship of information, or the deliberate dissemination of disinformation (false information coming from trusted sources), has to be considered an attempt of undermining democracy, since it makes it impossible for the electorate to understand the extent to which the government's actions serves the interest of the people. Furthermore, even the use of propaganda that includes disinformation directed against foreign audiences is problematic for democratic states, since it can be expected that there will be 'blowback' in the manner that the disinformation gets picked up and is disseminated by domestic media. It is just not possible to merely lie to foreigners and tell the truth to the own people: one has to deceive foreign *and* domestic audiences for any deception to work.

Currently, democratic states like the US do not have any clear rules for how propaganda is to be used for advancing foreign policy and domestic policy objectives. The Smith-Mundt Act, which restricted the dissemination of State Department media designed for foreign audiences to domestic audiences, has been repealed. Besides the Pentagon, which spends billions a year on 'strategic communication', was never covered by this restriction (Hudson, 2013). The biggest concern is that governments could use propaganda to spread subversive messages designed to incite hate or to cause civil war, which is not even illegal under international law. Only recently have there been international efforts to criminalize at least hate speech and incitement to genocide in relation to the Rwandan genocide of 1994 (Timmermann, 2005: 257). Back then 800,000 Rwandans perished in a killing frenzy that lasted a hundred days, in which ordinary Rwandans incited by years of 'vicious hate propaganda' through the radio began to attack Tutsis and moderate Hutu with machetes (Timmermann, 2005: 258).

There have been also efforts to regulate government propaganda directed at foreign audiences. For example, the Russian government has proposed a 'Convention on International Information Security' in 2011, which not only covers the issue of cyber security, but also what the Russians call 'information-psychological' warfare. The proposed convention defines IW as:

> conflict between two or more States in information space with the goal of inflicting damage to information systems, processes, and resources, as well as to critically important structures and other structures; undermining political, economic, and social systems; carrying out mass psychological campaigns against the population of a State in order to destabilize society and the government; as well as forcing a State to make decisions in the interests of their opponents. Information weapon – information technology, means, and methods intended for use in information warfare.
>
> (quoted from T. Thomas, 2014: 112)

The proposed convention suggests to 'take action aimed at limiting the proliferation of "information weapons" and the technology for their creation' (Russian Federation, 2011). Information weapons include in Russian terminology also the mass media and 'psychotronic weapons' (Lopating and Tsygankov, 1999: 93). Admittedly, the language of the proposed convention is far too vague to be acceptable for most states. However, it seems like a step in the right direction, as it would obligate governments to use 'information weapons' more responsibly.

It seems obvious that, unless internationally regulated, there is a huge danger that IW waged by intelligence services against foreign societies will produce blowback that eventually jeopardises the future of democracy everywhere, as democratic states will feel compelled to conduct arbitrary censorship and curtail free speech to counter the new threat. The only solution is to have some standards regarding the dissemination of information to which governments and corporations can be held. Wilfully lying and exploiting psychological weaknesses in mass audiences in order to destabilize countries or bankrupt companies should not be permissible, especially if the propaganda is enhanced by mind manipulation techniques. An international body could be established within the UN or another relevant international organisation or body to monitor and investigate cases of particularly malicious information that is deliberately disseminated to cause harm.

9.5.6 The need for a 'human right for mental self-determination'

In today's world, every individual's faculty of critical thinking is under constant attack. Advertisers, politicians and PSYOPS planners are continuously manipulating people into changing their perceptions of reality and making choices that ultimately do not benefit them. The situation is bound to get worse, as knowledge of the human brain and human behaviour advances and with it the capability of influencing or controlling it. Currently, there are not many legal protections for people against mind reading, nefarious influencing, or even mind control. Western legal systems are based on the concept of the unassailability of the freedom of will. Courts rarely accept the argument that behaviours of an individual did not originate in themselves, but were caused by some form of external manipulation of the individual's mind.

As incredulous as it may sound, brainwashing or psychologically manipulating people into making choices opposed to their own genuine self-interest and well-being is not actually illegal. As a matter of fact, American courts have often rejected claims of brainwashing or 'thought reform' as a legal defence. This is complicated by the tendency that victims of brainwashing often themselves claim that they have never been brainwashed and that everything they have done was the result of their own free will.

A very famous case that illustrates this problem very well is the Patty Hearst kidnapping by members of the Symbionese Liberation Army (SLA) in February 1974 (K. Taylor, 2004: 10–12). After two months of systematic abuse and rape, she had been brainwashed by the SLA leader Donald DeFreeze to join the group and to rob a bank with them. For 19 months, she was on the run with the terrorists until she was finally apprehended in September 1975. Her lawyer based her defence on the circumstance of brainwashing, although Hearst claimed otherwise during the trial. MK ULTRA veteran, psychiatrist Jolyon West, testified on Hearst's behalf that she was indeed brainwashed by DeFreeze since she suffered from a 'non-specific form of dissociative identity disorder' (C. Ross, 2006: 199). She was eventually convicted to 35 years of prison for bank robbery, but later pardoned by President Carter. In 2002, almost 30 years later, she said in an interview with Larry King: 'Most of the time I was with them, my mind was going through doing exactly what I was supposed to do… I had no freewill' (Hearst, 2002).

It is time to take the concept of brainwashing and other mind manipulation seriously in light of neuroscience research that has proven that minds can be changed and indeed controlled by methods such as psychiatric drugs and invasive and non-invasive brain stimulation. Effective legal protections need to be created that put clear limitations on the permissible methods for influencing others. The legal scholars Jan Christoph Bubliz and Reinhard Merkel have proposed making severe interventions into other's minds a criminal offense and recognising a 'human right to mental self-determination' (Bublitz and Merkel, 2012). They argue that mental injuries are different and independent from bodily injuries – one does not necessitate the other.

For example, a poker player may covertly spray oxytocin to manipulate the other players into trusting him. Since oxytocin is a harmless and naturally-occurring biochemical, there would be no grounds for claiming bodily injury. Similarly, marketers might surreptitiously monitor the emotional responses of a potential customer to various products to present them with subliminal stimuli that most likely will cause them to buy the product the customer's brain responded to most positively. In both cases subtle mental manipulation reduced the capacity of people to make the best decisions in their own interest. But at what point should mental interference be considered a crime? Should governments have the right to subject citizens to sophisticated brainwashing or 're-education', in which their thought patterns are radically changed to remove tendencies towards violence or to simply make them more compliant citizens? One has to only look at the success of authoritarian regimes like Russia, where President Putin enjoys an almost 90 per cent approval rating thanks to his ability to control all major media in his country, to understand that techniques of mass mind manipulation have become already very sophisticated. Ultimately, mind manipulation enabled by advances in neuroscience is a human rights issue, as they 'will open the way for abuses such as invasion of personal liberty, control of behaviour and brainwashing' (Butler, 1998: 316).

According to Bublitz and Merkel, '[i]t appears evident that states must be barred from invading the inner sphere of persons, from accessing their thoughts, modulating their emotions or manipulating their personal preferences. At the very least, such measures are in grave need of justification' (Bublitz and Merkel, 2012: 61). Bublitz and Merkel suggest distinguishing between indirect and direct interventions in the brain of which only direct interventions should be criminalized. Indirect intervention through conscious perception and other external stimuli does not undermine the faculty of critical thinking and decision-making in contrast to direct intervention through subconscious cues and direct alteration of the brain, e.g. through mind-altering biochemicals. Severe mind manipulation also violates human dignity as it undermines the moral autonomy of a person. Furthermore, freedom of thought and freedom of conscience are the cornerstones of democracy – if they are not protected and governments and other third parties are allowed to manipulate minds as they please, the result can only be either a totalitarian dictatorship or a lawless chaos, none of which is desirable.

9.6 Conclusion

Military neuroscience applications go beyond the realm of military operations and traditional warfare. Future conflicts may be invisible and will take place at the level of society, bypassing not only the military, but even the state. These technologies may be exploited by criminals, terrorists and other nefarious groups. It is fair to say that neuroscience and neurotechnologies do create many new dangers to individual freedom and to society at large, if governments and the public remain complacent about the very real threats of remote neural monitoring and mind control. It is also important to see military neuroscience in the much larger context of 'macro ideologies', such as socialism, technocracy and transhumanism.

While technocracy promises efficiency in government through social engineering, transhumanism promises individual perfection and immortality. In the end, both ideologies are two sides of the same coin: they ultimately aim for the establishment of a global system for the cybernetic integration and control of all human beings into a societal mega-machine. This would effectively end all human conflict, but also humanity as we know it. It would lead to the creation of a 'scientific dictatorship' or a 'psychocivilized' society as Jose Delgado called it: a society where every individual can be made to conform to societal norms with no possibility for deviation or resistance. Social progress and two thousand years of growth in individual rights and freedom would end.

It is therefore imperative that a public and political debate over neurosecurity and neuroethics commences, so that measures can be taken to protect society and individual freedoms. Most importantly, a right to 'mental self-determination' should be established and safeguarded by democratic governments in order to prevent nefarious actors from taking advantage over others using sophisticated influence technologies. International treaties should be negotiated that restrict IW, or what has been described in here as 'neurowarfare'. Only by strengthening the freedom of the mind and by empowering individuals, rather than subjecting them to overreaching government control, can society become resilient against attacks on the individual and collective national consciousness and prosper.

Bibliography

Academy of Medical Sciences (2011) 'Animals Containing Human Material', Working Group Report (July), [Online], Available: www.acmedsci.ac.uk/viewFile/ publicationDownloads/Animalsc.pdf [Accessed 4 December 2015].

ACHRE (1995) 'Final Report', Washington, DC: Advisory Committee on Human Radiation Experimentation, [Online], Available: biotech.law.lsu.edu/research/reports/ ACHRE/index.htm [Accessed 28 August 2015].

Adams, T.K. (2001) 'Future Warfare and the Decline of Human Decision-Making', *Parameters* (Winter), pp. 57–71.

Adey, R. (1975) 'The Influences of Impressed Electrical Fields at EEG Frequencies on Brain and Behavior, in Behavior and Brain Electrical Activity', in: Burch, N. and Altshuler, H.I. (eds.), *Behavior and Brain Electrical Activity*, New York: Plenum Press.

Aftergood, S. (1994) 'The Soft-Kill Fallacy', *The Bulletin of American Scientists* (September/ October), pp. 40–45.

Ahmed, N. (2013) 'Pentagon Bracing for Public Dissent Over Climate and Energy Shocks', *The Guardian*, 14 June, [Online], Available: www.theguardian.com/ environment/earth-insight/2013/jun/14/climate-change-energy-shocks-nsa-prism [Accessed 23 January 2016].

——— (2014) 'Pentagon Preparing for Mass Civil Breakdown', *The Guardian*, 12 June, [Online], Available: www.theguardian.com/environment/earth-insight/2014/ jun/12/pentagon-mass-civil-breakdown [Accessed 22 January 2016].

Ainscough, M.J. (2002) 'Next Generation Bioweapons: The Technology of Genetic Engineering Applied to Biowarfare and Bioterrorism', Air War College, The Counterproliferation Papers, Future Warfare Series, no. 14, [Online], Available: fas. org/irp/threat/cbw/nextgen.pdf [Accessed 7 December 2015].

Albarelli, H.P. (2009) *A Terrible Mistake: The Murder of Frank Olson and the CIA's Cold War Experiments*, Walterville, OR: TrineDay Publishing.

Alder, K. (2007) *The Lie Detectors: The History of an American Obsession*, New York: The Free Press.

Alexander, J. (1980) 'The New Mental Battlefield – Beam Me Up, Spock', *Military Review* (December), pp. 47–54.

——— (1999) *Future War: Non-Lethal Weapons in Modern Warfare*, New York: Thomas Dunne Books.

Alibek, K. (1999) 'Biological Weapons in the Former Soviet Union: An Interview with Kenneth Alibek', *The Nonproliferation Review* (Spring-Summer), pp. 1–10.

Alibek, K. and Handelman, S. (1999) *Biohazard: The Chilling True Story of the Largest Covert Biological Weapons Program in the World – Told from Inside by the Man Who Ran It*, New York: Dell Publishing.

Altmann, J. (1999) 'Acoustic Weapons – A Prospective Assessment: Source, Propagation, and Effects of Strong Sounds', *Cornell University Occasional Paper* 22.

––––––– (2001) 'Non-lethal' Weapons Technologies – the Case for Independent Scientific Analysis', *Medicine, Conflict and Survival*, vol. 17, no. 3, pp. 234–47.

Ames, D.L. and Fiske, S.T. (2010) 'Cultural Neuroscience', *Asian Journal of Social Psychology*, vol. 13, pp. 72–82.

Amos, J. (2003) 'Organ Music "Instills Religious Feelings"', *BBC News*, 8 September, [Online], Available: news.bbc.co.uk/2/hi/science/nature/3087674.stm [Accessed 25 June 2015].

Amnesty International (2013) 'China's "Reeducation Through Labor" Camps: Replacing One System of Repression with Another?', *Press Release*, 17 December, [Online], Available: www.amnesty.org/en/latest/news/2013/12/china-s-re-education-through-labour-camps-replacing-one-system-repression-another/ [Accessed 20 November 2015].

Anderson, J. (1972) ' "Brainwash" Attempt by Russians?', *The Washington Post*, 10 May, p. B15.

––––––– (1981) 'Yes, Psychic Warfare Is Part of the Game', *The Washington Post*, 5 February, p. 11.

––––––– (1985) 'Radio Waves Studied for Arms Potential', *Washington Post*, 31 July, p. C9.

Anderson, S. (2015) 'These Are the People Making Captagon, the Drug ISIS Fighters Take to Be Invincible', *New York Magazine*, 9 December, [Online], Available: nymag.com/daily/intelligencer/2015/12/men-making-captagon-the-drug-fueling-isis.html# [Accessed 23 January 2016].

Anibogu, M.C. (1998) 'The Future of Electromagnetic Field Litigation', *Pace Environmental Law Review*, vol. 15, no. 2, pp. 2–75.

Annas, G., Andrews, L. and Isai, R. (2002) 'Protecting the Endangered Human: Toward an International Treaty Prohibiting Cloning and Inheritable Alterations', *American Journal of Law and Medicine*, vol. 28, no. 2–3, pp. 151–178.

Antal, J. (2011) 'Phasers on Stun: A Status Report on Directed Energy Weapons Programmes', *Military Technology* (July), pp. 66–73.

Anthony, S. (2012a) 'DARPA Combines Human Brains and 120-Megapixel Cameras to Create the Ultimate Military Threat Detection System', *ExtremeTech*, 19 September, [Online], Available: www.extremetech.com/extreme/136446-darpa-combines-human-brains-and-120-megapixel-cameras-for-the-ultimate-military-threat-detection-system [Accessed 5 May 2015].

––––––– (2012b) 'Canadian Camouflage Company Claims to Have Created Perfect Invisibility Cloak, US Military Soon to Be Invisible', *ExtremeTech*, 14 December, [Online], Available: www.extremetech.com/extreme/143353-canadian-camouflage-company-claims-to-have-created-perfect-invisibility-cloak-us-military-soon-to-be-invisible [Accessed 15 July 2015].

Aquino, M. (1980) 'From PSYOPS to MindWar: The Psychology of Victory', San Francisco: 7th Psychological Operations Group, [Online], Available: archive.org/details/pdfy-Mv-q4qGq8_TBPcwL [Accessed 28 August 2015].

Araújo, H. R. C., Carvalho, D. O., Ioshino, R. S., Costa-da-Silva, A. L. and Capurro, M. L. (2015) 'Aedes aegypti Control Strategies in Brazil: Incorporation of New Technologies to Overcome the Persistence of Dengue Epidemics', *Insects*, vol. 6, no. 2, pp. 576–594.

Asian Human Rights Commission (2013) 'Neuroweapons, Neuroscience, and "Brain Circuit Manipulation": Inside Story of Obama's Mind Control Project', *GlobalResearch.Ca*, 8 October, [Online], Available: www.globalresearch.ca/neuro weapons-neuroscience-and-brain-circuit-manipulation-inside-story-of-obamas-mind-control-project/5353451 [Accessed 8 July 2015].

Associated Press (2015) 'US Missiles to Counter Russia in Europe? Options Weighed After Alleged Nuclear Treaty Breach', *FoxNews.com*, 4 June, [Online], Available: www.foxnews.com/us/2015/06/04/us-missiles-to-counter-russia-in-europe-options-weighed-after-alleged-nuclear/ [Accessed 15 July 2015].

Averre, D. and Davies, L. (2015) 'Russia, Humanitarian Intervention and the Responsibility to Protect: the Case of Syria', *International Affairs*, vol. 91, no. 4, pp. 813–834.

Axe, D. (2010) 'Military One Step Closer to Battlefield Holograms', *Wired.com*, 6 December, [Online], Available: www.wired.com/2010/12/military-one-step-closer-to-battlefield-holograms/ [Accessed 24 July 2015].

———— (2012) 'This Scientist Wants Tomorrow's Troops to Be Mutant-Powered', *Wired.com*, 26 December, [Online], Available: www.wired.com/2012/12/andrew-herr/ [Accessed 4 June 2015].

Baard, M. (2007) 'Sentient World: War Games at the Grandest Scale', *The Register*, 23 June, [Online], Available: www.theregister.co.uk/2007/06/23/sentient_worlds/ [Accessed 31 July 2015].

Babacek, M. (2013) 'Psychotronic and Electromagnetic Weapons: Remote Control of the Human Nervous System', *GlobalResearch.ca*, 31 January, [Online], Available: www.globalresearch.ca/psychotronic-and-electromagnetic-weapons-remote-control-of-the-human-nervous-system/5319111 [Accessed 13 January 2016].

Bahrami, B., Lavie, N. and Rees, G. (2007) 'Attentional Load Modulates Responses of Human Primary Visual Cortex to Invisible Stimuli', *Current Biology*, vol. 17, no. 6, pp. 509–513.

Baldwin, M., Bach, S.A. and Lewis, S.A. (1960) 'Effects of Radio-Frequency Energy on Primate Cerebral Activity', *Neurology*, vol. 10, pp. 178–187.

Balko, R. (2013) *Rise of the Warrior Cop: The Militarization of America's Police Forces*, New York: Public Affairs.

Bamford, J. (2012) 'The NSA Is Building the Country's Biggest Spy Center (Watch What You Say)', *Wired.com*, 15 March, [Online], Available: www.wired.com/2012/03/ff_nsadatacenter/ [Accessed 30 June 2015].

Barranski, J.V. and Pigeau, R.A. (1997) 'Self-monitoring Cognitive Performance During Sleep Deprivation: Effects of Modafinil, d-Amphetamine and Placebo', *Journal of Sleep Research*, no. 6, pp. 84–91.

Barrat, J. (2013) *Our Final Invention: Artificial Intelligence and the End of the Human Era*, New York: St. Martin's Press.

Basick, R.L. (2015) 'Social and Neurological Construction of Martyrdom Project Selected to Receive $3.4 Million from Department of Defense', *University of Chicago website*, [Online], Available: socialsciences.uchicago.edu/story/uchicago-study-explores-how-isis-lights-brains-recruits [Accessed 26 November 2015].

Bast, F. (2015) 'Extraordinary Tales: Parasites Hijacking the Minds of Hosts', *Resonance*, vol. 20, no. 10, pp. 893–902.

BBC (2005) 'US Military Pondered Love Not War', *BBC News*, 15 January, [Online], Available: news.bbc.co.uk/2/hi/4174519.stm [Accessed 20 June 2015].

——— (2014) 'Russian TV Uses 25th Frame Effect in Reports on Ukraine – Security Service', *BBC Monitoring Service*, 20 May.

——— (2015) 'Russia Reveals Giant Nuclear Torpedo in State TV "Leak"', *BBC News*, 12 November, [Online], Available: www.bbc.com/news/world-europe-34797252 [Accessed 20 December 2015].

Beck, R.C. (1986) 'Mood Modification with ELF Magnetic Fields: A Preliminary Exploration', *ARCHAEUS*, vol. 4, pp. 47–53.

Becker, R. O. and Selden, G. (1985) *The Body Electric: Electromagnetism and the Foundation of Life*, New York: William Morrow.

Becker, R. O. (1990) *Cross Currents: The Perils of Electropollution and the Promise of Electromedicine*, New York: Jeremy Tarcher/ Penguin.

Beckett, B. (1983) *Weapons of Tomorrow*, New York: Plenum Press.

Beckhusen, R. (2012) 'Fight or Flight: DARPA Explores the Neuroscience of Threat Response', *Wired.com*, 24 August, [Online], Available: www.wired.com/2012/08/movement-control/ [Accessed 10 June 2015].

Begich, N. (2006) *Controlling the Human Mind: The Technologies of Political Control or Tools for Peak Performance*, Anchorage, AK: Earthpulse Press.

Belfiore, M. (2009) *The Department of Mad Scientists: How DARPA Is Remaking Our World, from the Internet to Artificial Limbs*, New York: Harper.

Bell, L. (2013) 'Why the Heck Is DHS Buying More Than a Billion Bullets Plus Thousands of Guns and Mine-Resistant Armored Vehicles?', *Forbes*, 10 March, [Online], Available: www.forbes.com/sites/larrybell/2013/03/10/why-the-heck-is-dhs-buying-more-than-a-billion-bullets-plus-thousands-of-guns-and-mine-resistant-armored-vehicles/ [Accessed 13 June 2015].

Bellamy, C. (1990) *The Evolution of Modern Land Warfare*, London: Routledge.

Belous, V. (2009) 'Weapons of the 21st Century', *International Affairs*, vol. 55, no. 2, pp. 64–82.

Ben Ouargham-Gormley, S. (2013) 'Dissuading Biological Weapons Proliferation', *Contemporary Security Policy*, vol. 34, no. 3, pp. 473–500.

Benson, K. and Weber, J. (2012) 'Full Spectrum Operations in the Homeland: A "Vision" of the Future', *Small Wars Journal*, 25 July, [Online], Available: smallwarsjournal.com/jrnl/art/full-spectrum-operations-in-the-homeland-a-%E2%80%9Cvision%E2%80%9D-of-the-future?page=1 [Accessed 12 July 2015].

Bergstein, B. (2014) 'Hacking the Soul: Q&A Joseph LeDoux', *MIT Technology Review*, 17 June, [Online], Available: www.technologyreview.com/qa/528156/the-promise-and-perils-of-manipulating-memory/ [Accessed 5 June 2015].

Berk, M., Dodd, S. and Henry, M. (2006) 'Do Ambient Electromagnetic Fields Affect Behavior? A Demonstration of the Relationship Between Geomagnetic Storm Activity and Suicide', *Bioelectromagnetics*, vol. 27, no. 2, pp. 151–155.

Berns, G., Bell, E., Capra, C.M., Prietula, M.J., Moore, S., Anderson, B., Ginges, J. and Atran, S. (2012) 'The Price of Your Soul: Neural Evidence for the Non-Utilitarian Representation of Sacred Values', *Philosophical Transactions of the Royal Society*, vol. 367, pp. 754–762.

Berry, S. (2001) 'Decoding Minds, Foiling Adversaries', *SIGNAL Magazine* (October), p. 55.

Bezmenov, Y. (1984) 'Soviet Subversion of the Free World Press: A Conversation with Yuri Bezmenov', Westlake Village, CA: American Media, [Online], Available: www.youtube.com/watch?v=y3qkf3bajd4 [Accessed 28 August 2015].

Bhonsle, R. (2013) 'Future Warfare: Men and Machines at War', *Global Policy*, vol. 4, no. 2, pp. 213–215.

Bimmerle, G. (1993) '"Truth Drugs" in Interrogation', *Studies in Intelligence*, vol. 5, no. 2, [Online], Available: www.cia.gov/library/center-for-the-study-of-intelligence/kent-csi/vol5no2/html/v05i2a09p_0001.htm [Accessed 29 August 2015].

Binhi, V. N. (2008) 'Electromagnetic Aspects of Mind Control: A Scientific View', [Online], Available: waves.lima-city.de/dok/Electromagnetic%20aspect%20of%20mind%20control%20-%20V.N.%20Binhi.pdf [Accessed 7 July 2014].

———— (2010) *Electromagnetic Mind Control: Fact or Fiction? A Scientific View*, New York: Nova Science Publishers.

Binney, W. (2012) 'Whistleblower: The NSA Is Lying – The U.S. Government Has Copies of Most of Your Emails', *Democracy Now!* [Online], Available: www.democracynow.org/2012/4/20/whistleblower_the_nsa_is_lying_us!, [Accessed 20 April 2015].

Blackwell, J. (2007) 'The Cognitive Domain of War: The Role of Science in Approaching the Enemy?', *International Economy* (Summer), pp. 33–35.

Blanchfield, A., Hardy, J. and Marcora, S. (2014) 'Non-conscious Visual Cues Related to Affect and Action Alter Perception of Effort and Endurance Performance', *Frontiers in Human Neuroscience*, vol. 8, no. 967, pp. 1–16.

Bland, E. (2008) 'Army Developing "Synthetic Telepathy"', *NBC News*, 13 October, [Online], Available: www.nbcnews.com/id/27162401/ns/technology_and_science-science/t/army-developing-synthetic-telepathy/#.VYG_uflVhBc [Accessed 1 June 2015].

———— (2010) 'Zapping the Brain Improves Maths Skills', *Discovery News*, 4 November, [Online], Available: news.discovery.com/human/psychology/brain-electricity-math.htm [Accessed 3 July 2014].

Blank, R. H. (2013) *Intervention in the Brain: Politics, Policy, and Ethics*, Cambridge, MA: MIT Press.

Blank, S. (2013) 'Russian Information Warfare as Domestic Counterinsurgency'. *American Foreign Policy Interests*, vol. 35, no. 1, pp. 31–44.

Block, S.M. (2001) 'The Growing Threat of Biological Weapons', *American Scientist*, vol. 89, pp. 28–37.

Bogue, R. (2010) 'Recent Developments in Miniature Flying Robots', *Industrial Robot*, vol. 37, no. 1, pp. 17–22.

Bok, S. (1989) *Secrets: On the Ethics of Concealment and Revelation*, New York: Vintage Books.

Bonné, J. (2003) '"Go Pills": A War on Drugs?', *NBC News*, 9 January, [Online], Available: www.nbcnews.com/id/3071789/ns/us_news-only/t/go-pills-war-drugs/#.VZl3PflWLCs [Accessed 5 May 2015].

Boot, M., Kirkpatrick, J.J. and Doran, M. (2013) 'Political Warfare', Policy Innovation Memorandum No. 33, *Council on Foreign Relations*, [Online], Available: www.cfr.org/wars-and-warfare/political-warfare/p30894 [Accessed 1 January 2016].

Borger, J. (2000) 'Dirty Rats Leave Gore a Subliminal Message', *The Guardian*, 12 September, [Online], Available: www.theguardian.com/world/2000/sep/13/uselections2000.usa. [Accessed 4 August 2015].

Bostrom, N. (2005) 'Transhumanist Values', *Review of Contemporary Philosophy*, vol. 4 (May), [Online], Available: www.nickbostrom.com/ethics/values.html [Accessed 29 August 2015].

Boult, A. (2015) 'Silicon Valley Professionals Are Taking LSD at Work to Increase Productivity', *The Telegraph*, 26 November, [Online], Available: www.telegraph.co.uk/news/newstopics/howaboutthat/12019140/Silicon-Valley-professionals-are-taking-LSD-at-work-to-increase-productivity.html [Accessed 3 December 2015].

Bowart, W. (1978) *Operation Mind Control*, Glasgow, UK: Fontana.

——— (1990) 'The Invisible Third World War', *Unfinished Manuscript*, [Online], Available: www.whale.to/b/bowart4.html [Accessed 29 August 2015].

Bowen, J. (2009) 'Light and Sound Experiments Show Non-Lethal Applications', *Joint Non-Lethal Weapons Program*, [Online], Available: jnlwp.defense.gov/PressRoom/InTheNews/tabid/4777/Article/577760/light-and-sound-experiments-show-non-lethal-applications.aspx [Accessed 15 December 2015].

Brady, J. (2015) 'Revolutionizing Soldier Communication', *Army Technology*, vol. 3, no. 3, pp. 6–8.

Brait, E. (2015) 'Poll Finds Almost a Third of Americans Support a Military Coup', *The Guardian*, 12 September, [Online], Available: www.theguardian.com/us-news/2015/sep/11/military-coup-some-americans-would-vote-yes [Accessed 28 December 2015].

Breggin, Peter (1975) 'Psychosurgery for Political Purposes', *Duquesne Law Review*, vol. 13, pp. 841–862.

——— (2013) *Psychiatric Drug Withdrawal: A Guide for Prescribers, Therapists, Patients, and Their Families*, New York: Springer.

Brewer, J. (2014) 'Mindfulness in the Military', *American Journal of Psychiatry*, vol. 171, no. 8, pp. 803–806.

Brodeur, P. (1977) *The Zapping of America: Microwaves, Their Deadly Risk, and the Cover-Up*, New York: W.W. Norton & Co.

Brown, T., Johnson, R. and Milavetz, G (2013) 'Identifying Periods of Drowsy Driving Using EEG', *Annals of Advances in Automotive Medicine*, vol. 57, pp. 99–108.

Browne, L. (2012) 'When Microorganisms Control the Mind', *ABC News*, 12 December, [Online], Available: abcnews.go.com/blogs/health/2012/12/06/when-microorganisms-control-the-mind/ [Accessed 10 June 2015].

Brunderman, J.A. (1999) *High Power Radio Frequency Weapons: A Potential Counter to U.S. Stealth and Cruise Missile Technology*, Montgomery, AL: Maxwell AFB, Air War College.

Brzezianska, E., Domanska, D. and Jegier, A. (2014) 'Gene Doping in Sport: Perspectives and Risks', *Biology of Sport*, vol. 31, no. 4, pp. 251–259.

Bublitz, J.C. and Merkel, R. (2014), 'Crimes Against Minds: On Mental Manipulations, Harms and a Human Right to Mental Self-Determination', *Criminal Law and Philosophy*, vol. 8, pp. 51–77.

Bulletin of Atomic Scientists (1994) 'The "Soft Kill" Solution', *The Bulletin of Atomic Scientists*, vol. 50, no. 2, pp. 4–5.

Bunker, R. (1997) *Nonlethal Weapons: Terms and References*, U.S. Air Force Academy, CO: Institute for National Security Studies.

——— (2002) 'Radio Frequency Weapons: Issues and Potentials', *Journal of California Law*, vol. 36, no. 1, pp. 6–17.

Burnam-Fink, M. (2011) 'The Rise and Decline of Military Human Enhancement', *Science Progress*, 7 January, [Online], Available: scienceprogress.org/2011/01/the-rise-and-decline-of-military-human-enhancement/ [Accessed 25 November 2015].

Burrell, I. (2013) 'Inside Google HQ: What Does the Future Hold for the Company Whose Visionary Plans Include Implanting a Chip in Our Brains?', *The Independent*, 20 July, p. 18.

Bushnell, D.M. (2001) 'Future Strategic Issues/ Future Warfare [circa 2025]', *NASA Langley Research Center* (July), [Online], Available: www.fas.org/man/eprint/FutureWarfare.ppt. [Accessed 10 July 2015].

Butler, D. (1998) 'Advances in Neuroscience "May Threaten Human Rights"', *Nature*, vol. 391, no. 6665, p. 316.

Byman, D. (2005) *Deadly Connections: States that Sponsor Terrorism*, Cambridge: Cambridge University Press.

Canepari, Z., Cooper, D. and Cott, E. (2015) 'Prosthetic Limbs, Controlled by Thought', *The New York Times*, 20 May, [Online], Available: www.nytimes.com/2015/05/21/technology/a-bionic-approach-to-prosthetics-controlled-by-thought.html?_r=0 [Accessed 25 May 2015].

Canli, T., Brandon, S., Casebeer, W., Crowley, P.J., DuRouseau, D., Greely, H.T. and Pascual-Leone, A. (2007) 'Neuroethics and National Security', *The American Journal of Bioethics*, vol. 7, no. 5, pp. 3–13.

Canna, S. (2013) 'Leveraging Neuroscientific and Neurotechnological Developments with a Focus on Influence and Deterrence in a Networked World', *NSI Strategic Multi-Layer Assessment Workshop Summary*, 18 October.

Cannon, M. (1992) 'Mind Control and the American Government', *Lobster*, vol. 23, pp. 3–24.

Carlton, J. (2001) 'Early Warning: The U.S. Military and Secret Tests on the American Public – Years Ago, It Sprayed Microbes Thought to Be Safe Over Major Cities', *Asian Wall Street Journal*, 23 October, p. 1.

Carr, N. (2011) *The Shallows: What the Internet Is Doing to Our Brains*, New York: W.W. Norton & Co.

Carroll, M.C. (2005) *Lab 257: The Disturbing Story of the Government's Secret Germ Laboratory*, New York: Harper Collins.

CDC (2011) 'Fentanyl: Incapacitating Agent', *Centers for Disease Control*, [Online], Available: www.cdc.gov/niosh/ershdb/EmergencyResponseCard_29750022.html [Accessed 10 July 2015].

Cerny, P. G. (1998) 'Neomedievalism, Civil War and the New Security Dilemma: Globalisation as Durable Disorder', *Civil Wars*, vol. 1., no. 1, pp. 36–64.

Cerry, T. and Chaturvedi, A. (2006) 'Sentient World Simulation: A Continuously Running Model of the Real World', W. Lafayette, IN: Purdue University, 22 August, [Online], Available: www.krannert.purdue.edu/academics/mis/workshop/ac2_100606.pdf [Accessed 20 June 2015].

Cha, A.U. (2015) 'Why DARPA Is Paying People to Watch Alfred Hitchcock Cliffhangers', *The Washington Post*, 28 July.

Chambers, C. (2013) 'Neuro-Enhancement in the Military: Far-Fetched or an Inevitable Future?', *The Guardian*, 7 October, [Online], Available: www.theguardian.com/science/head-quarters/2013/oct/07/neuroscience-psychology [Accessed 27 July 2015].

Chandler, D.L. (2015) 'Magnetic Brain Stimulation', *MIT Technology Review*, 12 March, [Online], Available: newsoffice.mit.edu/2015/magnetic-brain-stimulation-0312 [Accessed 11 May 2015].

Chandler, R. (2008) *Shadow World: Resurgent Russia, the Global New Left, and Radical Islam*, Washington, DC: Regnery Publishing.

Chekinov, S.G. and Bogdanov, S.A. (2013) 'The Nature and Content of a New-Generation War', *Military Thought*, vol. 4, pp. 12–23.

Chen, A. (1990) 'Expert Discusses the Effects of Subliminal Advertising', *The Tech*, vol. 110, no. 7, pp. 2, 23.

Cherry, N. (2002) 'Actual or Potential Effects of ELF and RF/MW Radiation on Enhancing Violence and Homicide, and Accelerating Aging of Human, Animal or Plant Cells', Lincoln University, Canterbury, New Zealand, 30 August, [Online], Available: researcharchive.lincoln.ac.nz/handle/10182/4006 [Accessed 23 May 2015].

Chicurel, M. (2002) 'Magnetic Mind Games', *Nature*, vol. 417, no. 6885, pp. 114–116.

Chotikul, D. (1986) *The Soviet Theory of Reflexive Control in Historical and Psychocultural Perspective: A Preliminary Study*, Monterrey, CA: Naval Postgraduate School (July).

CIA (1951) 'Special Research, Bluebird'.

CIA (1952a) 'Memorandum', (January).

CIA (1954) 'Hypnotic Experimentation and Research', 10 February.

CIA (1961a) 'Memorandum: For the Record: MK ULTRA Subproject 136', 23 August.

CIA (1961b) '"Guided Animal" Studies', Memorandum for the Deputy Director of Plans, 21 April.

CIA (1962) 'MK ULTRA Subproject 142 Invoice', 25 May.

CIA (1963) 'Memorandum for Director of Central Intelligence; Subject: Report of Inspection of MK ULTRA', 26 July.

CIA (1975) 'Memorandum: For the Record: Project Artichoke', 31 January.

Cioffi-Revilla, C. (2000) 'War and Warfare: Scales of Conflict in Long-Range Analysis', in: Denemark, R.A., Friedman, J., Gills, B.K., Modelski, G. (eds.) *World System History: The Social Science of Long-Term Change*, London: Routledge, pp. 253–272.

Clarke, R.A. and Knake, R. (2010) *Cyber War: The Next Threat to National Security and What to Do About It*, New York: Harper Collins.

Coghlan, A. (2014) 'The Smart Mouse with the Half-Human Brain', *New Scientist*, 1 December, [Online], Available: www.newscientist.com/article/dn26639-the-smart-mouse-with-the-half-human-brain/ [Accessed 4 December 2015].

Cohen, F. (2011) 'Influence Operations', *Fred Cohen & Associates*, [Online], Available: all.net/journal/deception/CyberWar-InfluenceOperations.pdf [Accessed 13 January 2016].

Cole, L.A. (1988) *Clouds of Secrecy: The Army's Germ Warfare Tests over Populated Areas*, Totowa, NJ: Rowman & Littlefield.

Collins, N. (2012) 'It's Nature, Not Nurture: Personality Lies in Genes, Twins Study Shows', *The Telegraph*, 16 May, [Online], Available: www.telegraph.co.uk/news/science/science-news/9267147/Its-nature-not-nurture-personality-lies-in-genes-twins-study-shows.html [Accessed 12 June 2015].

Condon, J.B. (2014) 'Illegal Secrets', *Washington University Law Review*, vol. 91, no. 5, pp. 1099–1168.

Condon, R. (1959) *The Manchurian Candidate*, New York: New American Library.

Constantine, A. (1995) *Psychic Dictatorship in the USA*, Portland, OR: Feral House.

Cordesman, A. (2014) 'Russia and the "Color Revolution": A Russian Military View on a World Destabilized by the US and the West', *Center for Strategic and International Studies*, 28 May.

Corman, S. (2012) 'Arizona State University Narrative Networks (N2) – Pre-Award Progress, Status and Management Report Quarterly Progress Report', *Arizona State University*, 15 July, [Online], Available: documents.theblackvault.com/documents/mindcontrol/13-F-1207Doc3.pdf [10 July 2015].

Costanzo, M.A. and Gerrity, E. (2009) 'The Effects and Effectiveness of Using Torture as an Interrogation Device: Using Research to Inform the Policy Debate', *Social Issues and Policy Review*, vol. 3, no. 1, pp. 179–210.

Coupland, R.M. (1996) 'The Effect of Weapons: Defining Superfluous Injury and Unnecessary Suffering', *Medicine & Global Survival*, vol. 3.

Crick, F. and Koch, C. (1998) 'Consciousness and Neuroscience', *Cerebral Cortex*, vol. 8, no. 2, pp. 97–107.

Croody, E., Perez-Armendariz, C. and Hart, J. (2002) *Chemical and Biological Warfare: A Comprehensive Survey for the Concerned Citizen*, New York: Copernicus Books.

Crowley, M. (2009a) 'Dangerous Ambiguities: Regulation of Riot Control Agents and Incapacitants under the Chemical Weapons Convention', *University of Bradford*, Bradford Non-Lethal Weapons Research Program, [Online], Available: www.omegaresearchfoundation.org/assets/downloads/publications/BNLWRP Dangerous1.pdf [Accessed October 2015].

———— (2009b) 'Dangers Of Incapacitating Chemical Weapons And Widespread Misuse Of Riot Control Agents', *Science Daily*, 9 November, [Online], Available: www.sciencedaily.com/releases/2009/10/091029161809.htm [Accessed 21 June 2015].

Cruceanu, V.D. and Rotarescu, V.S. (2013) 'Alpha Brainwave Entrainment as a Cognitive Performance Activator', *Cognition, Brain, Behavior*, vol. 27, no. 3, pp. 249–261.

Cutler, W.B. (2000) 'Human Sex Attractant Pheromones: Discovery, Research, and Development', *Psychoneuroendocrinology*, vol. 25, no. 1, p. 65.

Dando, M. (2005) 'The Malign Misuse of Neuroscience', *Disarmament Forum*, no. 1, pp. 17–24.

Darczewska, J. (2015) 'The Devil Is in the Details: Information Warfare in the Light of Russia's Military Doctrine', *Center for Eastern Studies Point of View*, no. 50.

DARPA (2011) 'Broad Agency Announcement Narrative Networks', *Defense Sciences Office*, 7 October, [Online], Available: www.slideshare.net/UnitB166ER/darpa-baa1203 [Accessed 1 May 2015].

———— (2012) 'In Vivo Nanoplatforms for Diagnostics: Broad Agency Announcement', *Microsystems Technology Office* [Accessed 15 March 2015].

———— (undated), 'Narrative Networks: Dr. Justin Sanchez', [Online], Available: www.darpa.mil/program/narrative-networks [Accessed 21 July 2015].

———— (2013) 'Advanced Tools for Mammalian Genome Engineering', *Project Solicitation* ST13B-001, [Online], Available: sbirsource.com/grantiq#/topics/88854 [Accessed 31 July 2015].

———— (2015) 'Targeted Electrical Stimulation of the Brain Shows Promise as a Memory Aid', *Press Release*, 11 September, [Online], Available: www.darpa.mil/news-events/2015-09-11a [Accessed 25 January 2016].

———— (2016) 'Bridging the Bio-Electronic Divide', *Press Release*, 19 January, [Online], Available: www.darpa.mil/news-events/2015-01-19 [Accessed 25 January 2016].

Davison, N. (2006) 'The Early History of "Non-Lethal" Weapons', Bradford Nonlethal Weapons Project Occasional Paper 1 (December).

———— (2009) *Non-Lethal Weapons*, New York: Palgrave Macmillan.

Dean, J. (2016) 'Robots Could Soon Read Your Minds and There Are No Laws Against It', *Mirror*, 23 January, [Online], Available: www.mirror.co.uk/news/weird-news/robots-could-soon-read-your-7230366 [Accessible 23 January 2016].

DeCamp, J. (1996) *The Franklin Cover-Up: Child Abuse, Satanism, and Murder in Nebraska*, Lincoln, NE: AWT Inc.

De Caro, C. (1985) 'Special Assignment Weapons of War: Is There an RF Gap?', *CNN News Broadcast*, November, [Online], Available: www.youtube.com/watch?v=EK7K1Li71YY [Accessed 30 August 2015].

—— (1987) 'The Modern-Day Death Ray', *The Washington Post*, 21 June, p. B03.

De Cordoba, J. (1995) 'Drugged in Colombia: Street Thugs Dope Unwitting Victims', *Wall Street Journal*, 3 July, p. A1.

DeGrazia, D. (2007) 'Human-Animal Chimeras: Human Dignity, Moral Status, and Species Prejudice', *Metaphilosophy*, vol. 38, no. 2–3, pp. 309–329.

Delgado, J. (1969) *Physical Control of the Mind: Toward a Psychocivilized Society*, New York: Harper & Row.

—— (1974) 'Jose Delgado Testimony', *Congressional Record*, vol. 26, no. 118.

Deshere, E.F. (1960) 'Hypnosis in Interrogation', *Studies in Intelligence*, vol. 4, no. 1, pp. 51–64.

Dhami, M. (2011) 'Behavioral Support for JTRIG's Effects and Online HUMINT Operations', *GCHQ*, 10 March, [Online], Available: firstlook.org/theintercept/document/2015/06/22/behavioural-science-support-jtrig/. [Accessed 4 November 2015].

Diggins, C. and Arizmendi, C. (2012) 'Hacking the Human Brain: The Next Domain of Warfare', *Wired.com*, 11 December, [Online], Available: www.wired.com/2012/12/the-next-warfare-domain-is-your-brain [Accessed 16 December 2015].

Dillow, C. (2010) 'DARPA Wants to Install Transcranial Ultrasound Mind Control Devices in Soldiers' Helmets', *Popular Science*, 9 September, [Online], Available: www.popsci.com/technology/article/2010-09/darpa-wants-mind-control-keep-soldiers-sharp-smart-and-safe [Accessed 3 May 2015].

Dixon, R. (2002) 'Abusing the Power of Suggestion in Russian Ads', *LA Times*, 25 August, p. A5.

Dobbins, J. (2012) 'War with China', *Survival*, vol. 54, no. 4, pp. 7–24.

Doherty, T.J. and Clayton, S. (2011) 'The Psychological Impacts of Climate Change', *American Psychologist*, vol. 66, no. 4, pp. 265–276.

Doswald-Beck, L. and Cauderey, G.C. (1990) 'The Development of New Anti-Personnel Weapons', *International Review of the Red Cross*, vol. 279, pp. 565–577.

Douglass, J.D. (1999) *Red Cocaine: The Drugging of America and the West*, London: Edward Harle Ltd.

—— (2001) 'Influencing Behavior and Mental Processes in Covert Operations', *Medical Sentinel*, vol. 6, no. 4, pp. 130–133.

Drummond, K. (2012) 'Military's "Luke Skywalker" Binoculars Use Brain Waves to Detect Threats', *Forbes.com*, 18 September, [Online], Available: www.forbes.com/sites/katiedrummond/2012/09/18/darpa-threat-recognition/ [Accessed 4 December 2015].

Dunnigan, J.F. (1996) *Digital Soldiers: The Evolution of High-Tech Weaponry and Tomorrow's Brave New Battlefield*, New York: St. Martin's Press.

Dupree, C. (2004) 'Cushioning Hard Memories', *Harvard Magazine* (July/ August), [Online], Available: harvardmagazine.com/2004/07/cushioning-hard-memories.html [Accessed 10 May 2015].

Durnell, L. (2014) 'Train Your Brain Using the Navy SEAL Mental Toughness Program', *The Huffington Post*, 25 June, [Online], Available: www.huffingtonpost.com/linda-durnell/train-your-brain-using-th_b_5527152.html [Accessed 15 August 2015].

Dvorsky, G. (2014) 'How the Military Could Turn Your Mind Into the Next Battlefield', *IO9*, 19 December, [Online], Available: io9.com/how-the-military-could-turn-your-mind-into-the-next-bat-1673214050 [Accessed 12 February 2015].

Economist (2008) 'How to Disappear', *The Economist*, [Online], Available: www.economist.com/node/11999355 [Accessed 6 September 2015].

Edmonds, D. and Eidinow, J. (2004) *Bobby Fischer Goes to War: How the Soviets Lost the Most Extraordinary Chess Match of all Time*, New York: Harper Collins.

Ehrenberg, R. (2015) 'Brain Scans Pinpoint Individuals from a Crowd', *Nature.com*, 12 October, [Online], Available: www.nature.com/news/brain-scans-pinpoint-individuals-from-a-crowd-1.18541 [Accessed 3 December 2015].

Eidelson, E. (2013) 'Neuroscience, Special Forces and Yale', *CounterPunch*, 6 March, [Online], Available: www.counterpunch.org/2013/03/06/neuroscience-special-forces-and-yale/ [Accessed 10 January 2014].

Ellis, J.D. (2015) 'Directed Energy Weapons: Promise and Prospects', *Center for a New American Security* (April) [Online], Available: www.cnas.org/sites/default/files/publications-pdf/CNAS_Directed_Energy_Weapons_April-2015.pdf [Accessed 4 November 2015].

Ellul, Jacques (1965) *Propaganda: The Formation of Men's Attitudes*, New York: Vintage Books.

Engelhardt, T. (2014) *Shadow Government: Surveillance, Secret Wars, and a Global Security State in a Single-Superpower World*, Chicago, IL: Haymarket Books.

Entous, A. (2015) 'In Depth: Covert CIA Mission to Arm Syrian Rebels Goes Awry', *The Wall Street Journal*, 28 January, p. 10.

Epstein, C. (2004) 'The Stasi: 'New Research on the East German Ministry of State Security', *Kritika: Explorations in Russian and Eurasian History*, vol. 5, no. 2, pp. 321–348.

European Parliament (2000) 'Crowd Control Technologies (An Appraisal of Technologies for Crowd Control Technology Options for the European Union)', Final Report to STOA from the Omega Foundation (EP/1V/B/STOA/99/14/01).

Evans, G. (2013) 'Brain-Computer Interfacing: A Big Step Towards Military Mind Control', *Army-Technology.com*, 17 July, [Online], Available: www.army-technology.com/features/featurebrain-computer-interfacing-military-mind-control/ [Accessed 3 May 2015].

Farren, M. (2010) *Speed-Speed-Speedfreak: A Fast History of Amphetamine*, Port Townsend, WA: Feral House.

Farwell, L.A., Richardson, D.C. and Richardson, G.M. (2013) 'Brain Fingerprinting Field Studies Comparing P300-MERMER and P300 Brainwave Responses in the Detection of Concealed Information', *Cognitive Neurodynamics*, vol. 7, pp. 263–299.

Federation of American Scientists (2002) 'Russian Parliament Concerned About US Plans to Develop New Weapon', [Online], Available: fas.org/irp/program/collect/haarp-duma.htm [Accessed 13 January 2016].

——— (2013) 'Introduction to Biological Weapons', [Online], Available: fas.org/programs/bio/bwintro.html [Accessed 10 August 2015].

Ferguson, C.D. and Potter, W.C. (2005) *Four Faces of Nuclear Terrorism*, New York: Routledge.

Fernandez, C. (2015) 'Could "Supercharged Genes" Be Used by Terrorists? Technique to Genetically Modify Insects Could Spread Lethal Diseases', *Daily Mail*, 4 August, p. 21.

Festinger, L. (1957) *Cognitive Dissonance Theory*, Stanford, CA: Stanford University Press.

Fidler, D.P. (2005) 'The Meaning of Moscow: "Non-Lethal Weapons" and International Law in the 21st Century', *International Review of the Red Cross*, vol. 87, no. 859, pp. 525–552.

Fielding, N. and Cobain, I. (2011) 'Revealed: US Spy Operation That Manipulates Social Media', *The Guardian*, 17 March [Online], Available: www.theguardian.com/technology/2011/mar/17/us-spy-operation-social-networks. [Accessed 15 June 2015].

Fitz, N. (2015) 'Economic Inequality: It's Far Worse Than You Think', *Scientific American*, 31 March, [Online], Available: www.scientificamerican.com/article/economic-inequality-it-s-far-worse-than-you-think/ [Accessed 27 May 2015].

Fitzgerald, M. (1996) 'Russian Views on Electronic and Information Warfare: Volume I', *Hudson Institute* (December).

Ford, K. and Glymour, C. (2014) 'The Enhanced Warfighter', *The Bulletin of Atomic Scientists*, vol. 70, no. 1, pp. 43–53.

Foreign Policy (2008) 'This Is Your Brain on War', *Foreign Policy*, no. 169, p. 29.

Foucault, M. (1983) 'The Subject and Power', in: H. Dreyfus and P. Rabinow (eds.), *Michel Foucault: Beyond Structuralism and Hermeneutics*, Chicago, IL: University of Chicago Press, pp. 208–226.

——— (1988) 'Technologies of the Self', in: Martin, L.H., Gutman, H. and Hutton, P.H. (eds.) *Technologies of the Self: A Seminar with Michel Foucault*, Boston, MA: University of Massachusetts Press, pp. 16–49.

——— (1997) *Society Must Be Defended: Lectures at the Collège de France, 1975–1976*, New York: Picador.

Fox, R.G. (1987) 'Dr Schwitzgiebel's Machine Revisited: Electronic Monitoring of Offenders', *Australia & New Zealand Journal of Criminology*, vol. 20, pp. 131–147.

Frederickson, S. (2011) 'Brain Fingerprint or Lie Detector: Does Canada's Polygraph Jurisprudence Apply to Emerging Forensic Neuroscience Technologies?', *Information & Communications Technology Law*, vol. 20, no. 2, pp. 115–132.

Freedman, L. (1996) 'Vietnam and the Disillusioned Strategist', *International Affairs*, vol. 72, no. 1, pp. 133–151.

Frégnac, Y. and Laurent, G. (2014) 'Neuroscience: Where Is the Brain In the Human Brain Project?', *Nature*, vol. 513, no. 7516, pp. 27–29.

Freier, N. (2008) '*Known Unknowns: Unconventional "Strategic Shocks" in Defense Strategy Development*', U.S. Army War College, Strategic Studies Institute (November).

——— (2009) 'Hybrid Vs. Compound War', *Armed Forces Journal*, 1 October, [Online], Available: www.armedforcesjournal.com/hybrid-vs-compound-war/ [Accessed 5 May 2014].

Frey, A.H. (1962) 'Human Auditory System Response to Modulated Electromagnetic Energy', *Journal of Applied Physiology*, vol. 17, no. 4, pp. 689–692.

Frey, C.B. and Osborne, M.A. (2013) 'The Future of Employment: How Susceptible Are Jobs to Computerisation?', Oxford Martin School Programme on the Future of Technology, 17 September.

Friedman, R.A. (2013) 'A Dry Pipeline for Psychiatric Drugs', *The New York Times*, 20 August, p. D3.

——— (2016) 'A Drug to Cure Fear', *The New York Times*, 22 January, [Online], Available: www.nytimes.com/2016/01/24/opinion/sunday/a-drug-to-cure-fear.html?_r=0 [Accessed 25 January 2016].

Friman, H. (1999) 'Perception Warfare: Perspectives for the Future', Discussion Paper, Stockholm: *Swedish National Defense College*.

Frost, D. (1978) 'An Interview with Richard Helms', *Studies in Intelligence*, vol. 44, no. 4, pp.107–136.

Gafford, R. (1995) 'The Operational Potential of Subliminal Perception', *Studies in Intelligence*, vol. 2, no. 2, [Online], Available: www.cia.gov/library/center-for-the-study-of-intelligence/kent-csi/vol2no2/html/v02i2a07p_0001.htm [Accessed 30 August 2015].

Gallagher, Robert (1987) 'How Russia's Radiofrequency Weapons Can Kill', *EIR Science & Technology*, vol. 14, no. 28, pp. 18–23.

Gallagher, Ryan (2014) 'The Surveillance Engine: How the NSA Built Its Own Secret Google', *The Intercept*, 25 August, [Online], Available: firstlook.org/theintercept/2014/08/25/icreach-nsa-cia-secret-google-crisscross-proton/ [Accessed 7 June 2015].

Gallo, C. (1982) 'Why Cover Up Torture of Soviet Dissidents', *Human Events*, vol. 42, no. 11, p. 10.

Garamone, J. (2000) 'Soldiers, Set Your Weapons on Stun', *DoD News: American Forces Press Service*, 3 January, [Online], Available: www.defense.gov/news/newsarticle.aspx?id=44311 [Accessed 20 May 2015].

Gareev, M.A. (1998) *If War Comes Tomorrow? The Contours of Future Armed Conflict*, London: Frank Cass.

Garreau, J. (2005) *Radical Evolution: The Promise and Perils of Enhancing Our Minds, Our Bodies – and What It Means to Be Human*, New York: Broadway Books.

Gayle, D. (2012) 'Army of the Future: Soldiers Will Be Able to Run at Olympic Speed and Won't Need Food or Sleep with Gene Technology', *Mail Online*, 12 August, [Online], Available: www.dailymail.co.uk/sciencetech/article-2187276/U-S-Army-Soldiers-able-run-Olympic-speed-wont-need-food-sleep-gene-technology.html [Accessed 10 June 2015].

Gimbel, J. (1986) 'U.S. Policy and German Scientists: The Early Cold War', *Political Science Quarterly*, vol. 101, no. 3, pp. 433–451.

Giordano, J. (ed.) (2012) *Neurotechnology: Premises, Promises, and Problems*, Boca Raton, FL: CRC Press.

——— (ed.) (2014) *Neurotechnology in National Security and Defense: Practical Considerations, Neuroethical Concerns*, Boca Raton, FL: CRC Press.

Giordano, J. and Wurzman, R. (2011) 'Neurotechnologies as Weapons in National Intelligence and Defense – An Overview', *Synesis: A Journal of Science, Technology, Ethics, and Policy*, pp. 55–71.

Giri, D.V. (2004) *High-Power Electromagnetic Radiators: Nonlethal Weapons and Other Applications*, Cambridge, MA: Harvard University Press.

Goldsmith, J.R. (1996) 'Balancing the Interests of Patients, Science and Employees: Case Study of RF (Microwave) Exposures of US Embassy Staff in Eastern European Posts', *The Science of the Total Environment*, vol. 184, pp. 83–89.

Goldstein, F.L. and Findley, B.F. (1996) *Psychological Operations: Principles and Case Studies*, Montgomery, AL: Air University Press.

Golitsyn, A. (1984) *New Lies for Old*, New York: Dodd & Mead.

Goodman, A. (2009) 'Ousted Honduran President Manuel Zelaya Speaks from the Brazilian Embassy in Tegucigalpa', *DemocracyNow!*, [Online], Available: www.democracynow.org/2009/10/5/ousted_honduran_president_manuel_zelaya_speaks [Accessed 20 May 2015].

Goodman, M. (2015) *Future Crimes: Everything Is Connected, Everything Is Vulnerable and What We Can Do About It*, New York: DoubleDay.

Gorman, J. (2013) 'Agency Initiative Will Focus on Deep Brain Stimulation', *The New York Times*, 24 October, p. A16.

Graimann, B., Allison, B. and Pfurtscher, G. (2010) 'Brain-Computer Interfaces a Gentle Introduction', in: Graimann e.a. (eds.), *Brain-Computer Interfaces*, Heidelberg: Springer Verlag.

Granqvist, P., Fredriksen, M., Unge, P., Hagenfeldt, A., Valind, S., Larhammar, D. and Larsson, M. (2004) 'Sensed Presence and Mystical Experiences Are Predicted by Suggestibility, Not by the Application of Transcranial Weak Complex Magnetic Fields', *Neuroscience Letters*, vol. 379, no. 1, pp. 1–6.

Grau, C., Ginhoux, R., Riera, A., Nguyen, T.L., Chauvat, H., Berg, M., Amengual, J.L., Pascual-Leone, A. and Ruffini, G. (2014) 'Conscious Brain-to-Brain Communication in Humans Using Non-Invasive Technologies', *PLOS One* vol. 10 pp.1371.

Green, L.C. (2000) *The Contemporary Law of Armed Conflict*, Manchester: Manchester University Press.

Greenwald, G. (2013) 'Washington Gets Explicit: Its "War on Terror" Is Permanent', *The Guardian*, 17 May [Online], Available: www.theguardian.com/commentisfree/2013/may/17/endless-war-on-terror-obama [Accessed 10 December 2015].

Gregory, S., ffytche, D., Simmons, A., Kumari, V., Howard, M., Hodgins, S. and Blackwood, N. (2012) 'The Antisocial Brain: Psychopathy Matters', *Archives of General Psychiatry*, vol. 49, no. 9, pp. 962–972.

Griffin, A. (2015) 'Patents for Technology to Read People's Minds Hugely Increasing', *The Independent*, 8 May, [Online], Available: www.independent.co.uk/life-style/gadgets-and-tech/news/patents-for-technology-to-read-peoples-minds-hugely-increasing-10236211.html [Accessed 13 September 2015].

Gunn, D. (1996) *The Poor Man's Ray Gun*, El Dorado, AZ: Desert Publications.

Guo, J.W. (2005) 'Ultramicro, Nonlethal, and Reversible: Looking Ahead to Military Biotechnology', *Military Review*, vol. 85, no. 4, pp. 75–78.

Gusterson, H. (2008) 'Project Minerva Revisited', *The Bulletin of Atomic Scientists*, 5 August, [Online], Available: thebulletin.org/project-minerva-revisited [Accessed 15 January 2016].

———— (2009) 'Project Minerva and the Militarization of Anthropology', *Radical Teacher*, vol. 86, pp. 4–16.

Gutmann, A. (2011a) 'Transcript: Meeting 4, Session 10', Presidential Commission for the Study of Bioethical Issues,1 March, [Online], Available: bioethics.gov/node/203 [Accessed 5 May 2015].

———— (2011b) 'Moral Science: Protecting Participants in Human Subjects Research', Presidential Commission for the Study of Bioethical Issues (December), [Online], Available: bioethics.gov/sites/default/files/Moral%20Science%20June%202012.pdf [Accessed 5 August 2016].

Guyatt, D. (1996) 'Some Aspects of Antipersonnel Electromagnetic Weapons', Synopsis of a Lecture Given at a Symposium of the International Committee of the Red Cross: *The Medical Profession and the Effects of Weapons* in Montreaux, Schweiz, [Online], Available: www.deepblacklies.co.uk/some_aspects_of_a-pem_weapons.htm [Accessed 5 August 2016].

Hall, J. (2009) *A New Breed: Satellite Terrorism in America*, New York: Strategic Book Publishing.

Hamblin, J.D. (2013) *Arming Mother Nature: The Birth of Catastrophic Environmentalism*, Oxford: Oxford University Press.

Hamblin, J. (2013) 'The Neuroscience of War', *The Atlantic*, 31 August, [Online], Available: www.theatlantic.com/health/archive/2013/08/the-neuroscience-of-war/279242/ [Accessed 20 July 2015].

Hambling, D. (2006) 'Moscow's Remote Controlled Heart Attacks', *DefenseTech.com*, 14 February, [Online], Available: defensetech.org/2006/02/14/moscows-remote-controlled-heart-attacks/ [Accessed 11 February 2014].

——— (2009) 'Court to Defendant: Stop Blasting That Man's Mind!', *Wired.com*, 1 July, [Online], Available: www.wired.com/2009/07/court-to-defendant-stop-blasting-that-mans-mind/ [Accessed 13 August 2015].

Harkinson, J. (2015) 'Scores of Scientists Raise Alarm About Long-Term Dangers of Cell Phones', *Mother Jones*, 11 May, [Online], Available: www.motherjones.com/environment/2015/05/cellphone-emf-wifi-health-risks-scientists-letter [Accessed 12 August 2015].

Harland, C.J., Clark, T.D. and Prance, R.J. (2002) 'Remote Detection of Human Electroencephalograms Using Ultrahigh Input Impedance Electric Potential Sensors', *Applied Physics Letters*, vol. 81, no. 17, pp. 3284–3286.

Harris, M. (2010) 'MRI Lie Detectors', *IEEE Spectrum*, 30 July, [Online], Available: spectrum.ieee.org/biomedical/imaging/mri-lie-detectors [Accessed 16 August 2015].

Haseltine, E. (2015) 'Terrorism in Paris: New Neuroscience Tells Us How to Respond', *Psychology Today*, 14 November, [Online], Available: www.psychologytoday.com/blog/long-fuse-big-bang/201511/terrorism-in-paris-new-neuroscience-tells-us-how-respond [Accessed 15 November 2015].

Hassan, R. (2009) 'What Motivates the Suicide Bombers: Study of a Comprehensive Database Gives a Surprising Answer', *Yale Global*, 3 September, [Online], Available: yaleglobal.yale.edu/content/what-motivates-suicide-bombers-0 [Accessed 8 May 2015].

Hauck, D.J. (2005) 'Pandora's Box Opened Wide: UAVs Carrying Genetic Bioweapons', *Air War College*, Center for Strategy and Technology, [Online], Available: www.au.af.mil/au/awc/awcgate/cst/bugs_ch08.pdf [Accessed 9 December 2015].

He, B., Gao, S., Yuan, H. and Wolpaw, J.R. 'Brain-Computer Interfaces', in: He, B. (ed.) (2013) *Neural Engineering*, New York: Springer, pp. 87–151.

Hearn, K. (2008) 'Neuroscience Wake-Up Call: U.S. Lags in Ability to Monitor Iran and China', *The Washington Times*, 2 October, p. B1.

Hearst, P. (2002) 'Interview with Patty Hearst', *CNN: Larry King Live*, [Online], Available: www.cnn.com/TRANSCRIPTS/0201/22/lkl.00.html [Accessed 22 January 2016].

Heger, M. (2008) 'Why Microwave Auditory Effect Crowd-Control Gun Won't Work', *IEEE Spectrum*, 1 July, [Online], Available: spectrum.ieee.org/biomedical/devices/why-microwave-auditory-effect-crowdcontrol-gun-wont-work [Accessed 18 December 2015].

Heidenrich, J.G. (2007) 'The State of Strategic Intelligence', *Studies in Intelligence*, vol. 51, no. 2.

Heikerö, R. (2010) 'Emerging Cyber Threats and Russian Views on Information Warfare and Information Operations', Stockholm: *Swedish Defence Research Agency* (March).

Heuer, R.J. (1999) *Psychology of Intelligence Analysis*, Langley, VA: Center for the Study of Intelligence.

Himalayan Times (2011) 'Diaspora Mulls Torture Suits against Bhutan's Ex-King', *Himalayan Times*, 26 April.

Hochberg, L.R., Serruya, M.D., Friehs, G.M., Mukand, J.A., Saleh, M., Caplan, A.H., Branner, A., Chen, D., Penn, R.D. and Donoghue, J.P. (2006) 'Neuronal Ensemble Control of Prosthetic Devices by a Human with Tetraplegia', *Nature*, vol.: 442, pp. 164–171.

Hoffman, B. (1998) *Inside Terrorism*, New York: Columbia University Press.

Hoffman, F.G. (2015) 'The Contemporary Spectrum of Conflict: Protracted, Gray Zone, Ambiguous, and Hybrid Modes of War', *The Heritage Foundation*, 2016 Index of U.S. Military Strength, pp. 25–36.

Hoffman, M. (2016) 'Telepathy: "Mind Reading" Computer Deciphers Words from Brainwaves', *Science World Report*, 6 January, [Online], Available: www.scienceworldreport.com/articles/35517/20160106/telepathy-mind-reading-computer-deciphers-words-brainwaves.htm [Accessed 7 January 2016].

Hofmann, A. (1996) 'LSD: Completely Personal', *Newsletter of the Multidisciplinary Association for Psychedelic Studies*, vol. 6, no. 3.

Holley, P. (2015) 'The Tiny Pill Fueling Syria's War and Turning Fighters into Superhuman Soldiers', *The Washington Post*, 19 November, [Online], Available: www.washingtonpost.com/news/worldviews/wp/2015/11/19/the-tiny-pill-fueling-syrias-war-and-turning-fighters-into-super-human-soldiers/ [Accessed 18 December 2015].

Hollister, S. (2013) 'Could the NSA Use Microsoft's Xbox One to Spy on You?', *The Verge*, 16 July, [Online], Available: www.theverge.com/2013/7/16/4526770/will-the-nsa-use-the-xbox-one-to-spy-on-your-family [Accessed 20 August 2015].

Holsboer, F. (2008) 'How Can We Realize the Promise of Personalized Antidepressant Medicines?', *Nature Reviews Neuroscience*, no. 9, pp. 638–646.

Horgan, J. (2004) 'The Myth of Mind Control', *Discover*, vol. 25, no. 10, pp. 40–47.

——— (2005) 'The Forgotten Era of Brain Chips', *The Scientific American* (October), pp. 66–73.

House, T.J., Near, J.B, Shields, W.B., Celentano, R.J., Husband, D.M., Mercer, A.E. and Pugh, J.E. (1996) 'Weather as a Force Multiplier: Owning the Weather by 2025', *U.S. Air Force* (August), [Online], Available: csat.au.af.mil/2025/volume3/vol3ch15.pdf [Accessed 28 December 2015].

Hudson, J. (2013) 'U.S. Repeals Propaganda Ban, Spreads Government-Made News to Americans', *Foreign Policy*, 14 July, [Online], Available: foreignpolicy.com/2013/07/14/u-s-repeals-propaganda-ban-spreads-government-made-news-to-americans/ [Accessed 15 January 2016].

Hughes, D. (2007) 'Honeywell Uses Brainwaves to Spot Targets for DARPA', *Aerospace Daily & Defense Report*, vol. 224, no. 24, p. 5.

Hulnick, A. (2004) 'Espionage: Does It Have a Future in the 21st Century', *Brown Journal of World Affairs*, vol. 11, no. 1, pp. 165–173.

Hunter, E. (1956) *Brainwashing: The Story of the Men That Defied It*, New York: Farrar, Giroux & Cudahy.

Hurd, M.D., Martorell, P., Delavande, A., Mullen, K.J. and Langa, K.M. (2013) *Monetary Costs of Dementia in the United States*, Santa Monica, CA: RAND.

Hurley, D. (2014) 'US Military Leads Quest for Futuristic Ways to Boost IQ', *Newsweek*, 5 March, [Online], Available: www.newsweek.com/2014/03/14/us-military-leads-quest-futuristic-ways-boost-iq-247945.html [Accessed 18 July 2015].

Huxley, A. (1962) 'Interview with Aldous Huxley', Berkeley Language Center, 20 March, [Online], Available: www.informationclearinghouse.info/article31319.htm [Accessed 31 August 2015].

——— (2010) *Brave New World with the Essay Brave New World Revisited*, New York: Harper.

Insel, T.R., Landis, S.C. and Collins, F.S. (2013) 'The BRAIN Initiative', *Science*, vol. 340, no. 6133, pp. 687–688.

Isaacson, B. (2015) 'Silicon Valley Is Trying to Make Humans Immortal – and Finding Some Success', *Newsweek*, 5 March, [Online], Available: www.newsweek.com/2015/03/13/silicon-valley-trying-make-humans-immortal-and-finding-some-success-311402.html [Accessed 16 January 2016].

Isachenkov, V. (2015) 'Russia Steps Up Military Modernization. But at What Cost?', *Associated Press*, 12 June, [Online], Available: www.military.com/daily-news/2015/06/12/russia-steps-up-military-modernization-but-at-what-cost.html [Accessed 26 August 2015].

Jacobson, A. (2014) *Operation Paperclip: The Secret Intelligence Program that Brought Nazi Scientists to America*, New York: Little, Brown & Co.

――― (2015) *DARPA: The Pentagon's Brain*, New York: Little Brown & Company.

JASON (2008) *Human Performance*, McLean, VA: The MITRE Group (March).

Jefferson, C. (2012) 'Neuroscience, Conflict and Security: A Dual-Use Dilemma', *The Royal Society*, 8 February, [Online], Available: blogs.royalsociety.org/in-verba/2012/02/08/neuroscience-conflict-and-security-a-dual-use-dilemma/ [Accessed 7 December 2015].

Jervis, R. (2010) *Why Intelligence Fails: Lessons from the Iranian Revolution and the Iraq War*, Ithaca, NY: Cornell University.

Jha, A. (2007) 'Brain Absorbs Subliminal Messages – If Not Too Busy', *The Guardian*, 9 March, p. 14.

Jogalekar, A. (2014) 'Why Drugs Are Expensive: It's the Science, Stupid', *The Scientific American*, 6 January, [Online], Available: blogs.scientificamerican.com/the-curious-wavefunction/why-drugs-are-expensive-ite28099s-the-science-stupid/ [Accessed 3 December 2015].

Johnson, L.K. (2010) 'Evaluating "HUMINT": The Role of Foreign Agents in U.S. Security', *Comparative Strategy*, vol. 29, no. 4, pp. 308–332.

Jones, D. (1984) 'Opening Pandora's Box', *BBC Special Channel 4*, [Online], Available: mindjustice.org/BBCposta2012.htm [Accessed 23 July 2015].

Jouvenal, J. (2016) 'The New Way the Police Are Surveilling You: Calculating Your "Threat Score"', *Washington Post*, 10 January, [Online], Available: www.washingtonpost.com/local/public-safety/the-new-way-police-are-surveilling-you-calculating-your-threat-score/2016/01/10/e42bccac-8e15-11e5-baf4-bdf37355da0c_story.html [Accessed 26 January 2016].

Justesen, D. (1975) 'Microwaves and Behavior', *American Psychologist*, vol. 30, no. 3, pp. 391–401.

Kaldor, M. (1999) *Old and New Wars: Organized Violence in a Global Era*, Stanford, CA: Stanford University Press.

――― (2013) 'In Defence of New Wars', *Stability*, vol. 2, no. 1, pp. 1–16.

Kamath, M. (2015) '"Bulk Data Failure": NSA Has Been Collecting so Much Data that Now It Can't Sift Through It', *TechWorm.com*, 8 May, [Online], Available: www.techworm.net/2015/05/bulk-data-failure-nsa-has-been-collecting-so-much-data-that-now-it-cant-sift-through-it.html [Accessed 16 January 2016].

Kaplan, R. (1999) *The Coming Anarchy: Shattering the Dreams of the Post Cold War*, New York: Vintage Books.

Kattenburg, D. (1987) 'Zapping the Movement', *Peace Magazine* (June-July), p. 9.

Kaurin, P. (2010) 'With Fear and Trembling: An Ethical Framework for Non-Lethal Weapons', *Journal of Military Ethics*, vol. 9, no. 1, pp. 100–114.

Kavanagh, E. (2007) 'A Proposal for a Decade of the Mind Initiative', *Science*, vol. 317, p. 1321.

Kay, D. (2003) 'Genetically Engineered Bioweapons', [Online], Available: www.aaas. org/sites/default/files/migrate/uploads/ch17.pdf [Accessed 4 December 2015].

Keeler, A. (2008) 'Remote Mind Control Technology', in: Keith, J. (ed.) *Secret and Suppressed: Banned Ideas and Hidden History*, Port Townsend, WA: Feral House.

Keiper, A. (2006) 'The Age of Neuroelectronics', *The New Atlantis* (Winter), pp. 4–41.

Kellman, B. (2008) 'Bioviolence: A Growing Threat', *The Futurist* (May-June), pp. 25–30.

Kernbach, S. (2013) 'Unconventional Research in USSR and Russia: Short Overview', *Cybertronica Research*, Research Center of Advanced Robotics and Environmental Science, Stuttgart, [Online], Available: arxiv.org/pdf/1312.1148v2.pdf. [Accessed 14 July 2015].

Khatchadourian, R. (2012) 'Operation Delirium', *The New Yorker*, 17 December, [Online], Available: www.newyorker.com/magazine/2012/12/17/operation-delirium [Accessed 12 January 2016].

Kiernan, V. (1994) 'Set Lasers on "Stun"', *The Vancouver Sun*, 19 February.

Kime, C. and Lyons, D. (2007) 'CT2WS: A Military User's Perspective', *DARPA*, 13 March, [Online], Available: www.wired.com/images_blogs/dangerroom/files/ltcol_carl_kime_ct2ws_31307.pdf [Accessed 15 July 2015].

Kintner, W.R. and Kornfeder, J.Z. (1962) *The New Frontier of War: Political Warfare, Present and Future*, Chicago, IL: Henry Regnery Company.

Kintsch, E. (2010) *Hack the Planet: Science's Best Hope – Or Worst Nightmare – For Averting Climate Catastrophe*, Hoboken, NJ: John Wiley & Sons.

Koivuniemi, A. and Ott, K. (2014) 'When "Altering Brain Functions" Becomes "Mind Control"', *Frontiers in Systems Neuroscience*, vol. 8, no. 202, pp. 1–6.

Koplow, D. A. (2006) *Non-Lethal Weapons: The Law and Policy of Revolutionary Technologies for the Military and Law Enforcement*, Cambridge: Cambridge University Press.

Korybko, A. (2015) *Hybrid Wars: The Indirect Adaptive Approach to Regime Change*, Moscow: People's Friendship University of Russia, Project of the Institute for Strategic Studies and Predictions.

Kouzminov, A. (2005) *Biological Espionage: Special Operations of the Soviet and Russian Foreign Intelligence Services in the West*, London: Greenhill Books.

Krepinevich, A. (2009) *Seven Deadly Scenarios: A Military Futurist Explores War in the 21st Century*, New York: Bantam Books.

Kress, K. A. (1977) 'Parapsychology in Intelligence: A Personal Review and Conclusions', *Studies in Intelligence*, vol. 21, pp. 7–17.

Kristensen, H. M. and Norris, R.S. (2014) 'Slowing Nuclear Weapon Reductions and Endless Nuclear Weapon Modernizations: A Challenge to the NPT', *Bulletin of Atomic Scientists*, vol. 70, no. 4, pp. 94–107.

Kumar, T., Jha, R.K. and Ray, S.M. (2012) 'Mind Control Using Psychotronics', *International Journal of Engineering and Computer Science*, vol. 1, no. 1, pp. 42–47.

Kurzweil, R. (1998) *In the Age of Spiritual Machines*, New York: Viking.

——— (2005) *The Singularity Is Near!*, New York: Viking.

——— (2012) *How to Create a Mind: The Secret of Human Thought Revealed*, New York: Viking.

Kuzmin, A. (2015) 'Russian Media Take Climate Cue from Skeptical Putin', *Reuters*, 29 October, [Online], Available: www.reuters.com/article/us-climatechange-summit-russia-media-idUSKCN0SN1GI20151029 [Accessed 13 January 2016].

Kyriazis, M. (2015) 'Systems Neuroscience in Focus: From a Human Brain to a Global Brain?', *Frontiers in Neuroscience*, vol. 9, pp. 1–4.

Lakoski, J.M., Bosseau Murray, W. and Kenny, J.M. (2000) 'The Advantages and Limitations of Calmatives for Use as a Non-Lethal Technique', Pennsylvania State University, College of Medicine, Applied Physics Laboratory, 3 October.

Landon-Murray (2013) 'Thinking in 140 Characters: The Internet, Neuroplasticity, and Intelligence Analysis', *Journal of Strategic Security*, vol. 6, no. 3, pp. 73–82.

Lango, J. (2010) 'Nonlethal Weapons, Noncombatant Immunity, and Combatant Nonimmunity: A Study of Just War Theory', *Philosophia*, vol. 38, pp. 475–497.

Laytner, R. (1998) 'Weapons that Won't Kill: A New Arsenal of Weapons Can Capture Gunmen in a Net, Halt Attackers with a Sticky Goo and Disable Aircraft with a Shock Wave. Welcome to a New World', *The Gazette*, 4 January.

Leake, C. and Stuart, W. (2012) 'Putin Targets Dissidents with "Zombie" Which Attacks Victims' Nervous System', *Daily Mail*, 31 March, [Online], Available: www.dailymail.co.uk/news/article-2123415/Putin-targets-foes-zombie-gun-attack-victims-central-nervous-system.html [Accessed 15 January 2016].

Lee, D. (2014) ' "Remote Control" Contraceptive Chip Available "By 2018" ', *BBC.com*, 7 July, [Online], Available: www.bbc.com/news/technology-28193720 [Accessed 3 December 2015].

Lee, J. (2001) 'An Audio Spotlight Creates a Personal Wall of Sound', *New York Times*, 15 May, p. F4.

Lee, M.A. and Davidson, B. (1982) 'Mad, Mad, Mad War: BZ: The U.S. Has Enough of this Chemical Weapon to Turn Everyone on Earth into a Stark, Raving Lunatic', *Mother Jones* (May), pp. 14–22.

Leiby, R. (2015) ' "Brain Zapping": Veterans Say Experimental PTSD Treatment Has Changed Their Lives', *The Washington Post*, 12 January, [Online], Available: www.washingtonpost.com/lifestyle/style/brain-zapping-veterans-say-experimental-ptsd-treatment-has-changed-their-lives/2015/01/12/2fc8b3ca-58aa-11e4-8264-deed989ae9a2_story.html [Accessed 21 May 2015].

Leitenberg, M. (2003) 'Distinguishing Offensive from Defensive Biological Weapons Research', *Critical Reviews in Microbiology*, vol. 29, no. 3, pp. 223–257.

Levitt, B.B. and Lai, H. (2010) 'Biological Effects from Exposure of Electromagnetic Radiation Emitted by Cell Tower Base Stations and Other Antenna Arrays', *Environmental Reviews*, vol. 18, pp. 369–395.

Lewer, N. and Schofield, S. (1997), *Non-Lethal Weapons: A Fatal Attraction?*, London: Zed Books.

Libet, B. (2002) 'The Timing of Mental Events: Libet's Experimental Findings and Their Implications', *Consciousness and Cognition*, Vol. 11, No. 2, pp. 291–299.

Libicki, M. (2012), 'The Specter of Non-Obvious Warfare', *Strategic Studies Quarterly* (Fall), pp. 88–101.

Lilly, J. C. (2004) *Programming the Human Biocomputer*, Oakland, CA: Ronin Publishing.

Lim, D. (2011) 'DARPA Wants to Master the Science of Propaganda', *Wired.com*, 18 October, [Online], Available: www.wired.com/2011/10/darpa-science-propaganda/ [Accessed 15 January 2016].

Lin, P. (2010), 'Ethical Blowback from Emerging Technologies', *Journal of Military Ethics*, vol. 9, no. 4, pp. 313–331.

——— (2012) 'More Than Human? The Ethics of Biologically Enhancing Soldiers', *The Atlantic*, 16 February, [Online], Available: www.theatlantic.com/technology/archive/2012/02/more-than-human-the-ethics-of-biologically-enhancing-soldiers/253217/ [Accessed 15 January 2016].

———— (2013) 'Could Human Enhancement Turn Soldiers Into Weapons that Violate International Law? Yes', *The Atlantic*, 4 January, [Online], Available: www.theatlantic. com/technology/archive/2013/01/could-human-enhancement-turn-soldiers-into-weapons-that-violate-international-law-yes/266732/ [Accessed 14 February 2014].

Lin, P., Mehlman, M.J. and Abney, K. (2013) 'Enhanced Warfighters: Risks, Ethics, and Politics', Research Report, *The Greenwall Foundation*, 1 January.

Lind, W.S., Nightengale, K., Schmitt, J.F., Sutton, J.W. and Wilson, G.I. (1989) 'The Changing Face of War: Into the Fourth Generation', *Marine Corps Gazette*, vol. 73, no. 10, pp. 22–26.

Linebarger, Paul M.A. (1972) *Psychological Warfare*, New York: Arno Press.

Lockwood, N. (2011) 'How the Soviet Union Transformed Terrorism', *The Atlantic*, 23 December, [Online], Available: www.theatlantic.com/international/archive/2011/12/how-the-soviet-union-transformed-terrorism/250433/ [Accessed 26 December 2015].

Lopatin, V.N. and Tsygankov, V.D. (1999) *Psychotronic War and the Security of Russia* [Psichotronnoje oružie i bezopasnost Rossii], Moscow: SINTEG.

Los Angeles Times (1994) 'Autopsy of Researcher Given LSD by CIA Proves Inconclusive', *Los Angeles Times*, 29 November, p. 17.

Love, D. (2014) 'Future Mind-Altering Drugs Could Make Prisoners Think They're in Jail for 1,000 Years', *Business Insider*, 19 August, [Online], Available: www. businessinsider.com/drugs-alter-prisoners-perception-of-time-2014-8 [Accessed 29 December 2015].

Lyell, Lord (1997) 'Non-Lethal Weapons: General Rapporteur (United Kingdom)', *Committee of the North Atlantic Assembly*, 1 September.

Lynch, Z. (2009) *The Neuro Revolution: How Brain Science Is Changing Our World*, New York: St. Martin's Press.

MacGregor (1970) *A Brief Survey of the Literature Relating to the Influence of Low Energy Microwaves on Nervous Function*, Santa Monica, CA: RAND.

Malech, R. (1976) 'Apparatus and Method for Remotely Monitoring and Altering Brainwaves', *U.S. Patent: 3951134 A*, filed on 5 August.

Malicdem, D. (2015) 'Futurist Suggests Replacing Death Penalty with Brain Implants That Control Prisoners' Minds, Behavior', *International Business Times*, 27 July, [Online], Available: www.ibtimes.com.au/futurist-suggests-replacing-death-penalty-brain-implants-control-prisoners-mind-behaviour-1457786 [Accessed 15 September 2015].

Maloney, J.M., Uhland, B.F.P., Sheppard, N.F., Pelta, C.M. and Santini, J.T. (2005) 'Electrothermally Activated Microchips for Implantable Drug Delivery and Biosensing', *Journal of Controlled Release*, vol. 109, No. 1–3, pp. 244–255.

Maloof, M. (2013) *A Nation Forsaken: EMP: The Escalating Threat of an American Catastrophe*, Washington, DC: WND Books.

Malykh, A.G. and Reza, S.M. (2010) 'Piracetam and Piracetam-Like Drugs: From Basic Science to Novel Clinical Applications to CNS Disorders', *Drugs*, vol. 70.3, pp. 287–312.

Mangold, T. and Goldberg, J. (1999) *Plague Wars: The Terrifying Reality of Biological Warfare*, New York: St. Martin's Press.

Manwaring, M.G. (ed.) (1993) *Gray Area Phenomena: Confronting the New World Disorder*, Boulder, CO: Westview Press.

Marchant, G.E. and Gaudet, L.M. (2014) 'Neuroscience, National Security, and the Reverse Dual-Use Dilemma', in: Giordano, J. (ed.) *Neurotechnology in National Security and Defense: Practical Considerations, Neuroethical Concerns*, Boca Raton, FL: CRC Press.

Marcus, G. and Koch, C. (2014) 'The Future of Brain Implants', *Wall Street Journal*, 14 March, [Online], Available: www.wsj.com/articles/SB10001424052702304914904 579435592981780528 [Accessed 4 July 2015].

Margolis, J. (2013) *The Secret Life of Uri Geller*, London: Watkins Publishing.

Markou, A., Chiamulera, C., Geyer, M.A., Tricklebank, M. and Steckler, T. (2009) 'Removing Obstacles in Neuroscience Drug Discovery: The Future Path for Animal Models', *Neuropsychopharmacology*, vol. 34, no. 1, pp. 74–89.

Markram, H. (2012) 'The Human Brain Project', *A Report to the European Commission*, [Online], Available: www.humanbrainproject.eu/documents/10180/1298661/The+ HBP+Report/ddae4bfc-31eb-448e-8974-f5882f926451 [Accessed 5 August 2016].

Marks, J. (1979) *The Search for the Manchurian Candidate: The CIA and Mind Control: The Secret History of Behavioral Sciences*, New York: W.W. Norton & Co.

Marks, J. H. (2010) 'A Neuroskeptic's Guide to Neuroethics and National Security', *AJOB Neuroscience*, vol. 1, no. 2, pp. 4–12.

Marrin, S. (2012) 'Is Intelligence Analysis an Art or a Science?', *International Journal of Intelligence and Counterintelligence*, vol. 25, no. 3, pp. 529–545.

Matheson, R. (2015) 'Deal Reached to Commercialize Microchip Drug-Delivery Implant', *Tech Swarm*, 29 June, [Online], Available: www.techswarm.com/2015/06/ deal-reached-to-commercialize-microchip.html [Accessed 10 August 2015].

Matyszczyk, R. (2015) 'Google Exec.: Humans Will Be Hybrids by 2030', *CNET. com*, 4 June, [Online], Available: www.cnet.com/news/google-exec-humans-will-be-hybrids-by-2030/ [Accessed 4 December 2015].

May, E.C. (1996) 'The American Institutes for Research Review of the Department of Defense's STAR GATE Program: A Commentary', *The Journal of Parapsychology*, vol. 60, pp. 3–23.

May, L. (2007) *War Crimes and Just War*, Cambridge: Cambridge University Press.

McCoy, A. (2006) *A Question of Torture: CIA Interrogation from the Cold War to the War on Terror*, New York: Metropolitan Books.

McCreight, R. (2014) 'Brain Brinkmanship: Devising Neuroweapons Looking at Battlespace, Doctrine, and Strategy', in: James Giordano (ed.) *Neurotechnology in National Security and Defense: Practical Considerations, Neuroethical Concerns*, Boca Raton, FL: CRC Press, pp. 115–132.

McFate, M. (2005) 'The Military Utility of Understanding Adversary Culture', *Joint Forces Quarterly*, vol. 38, no. 3, pp. 42–48.

McFate, M. and Jackson, A. (2005) 'An Organizational Solution to DoD's Cultural Knowledge Needs', *Military Review* (July-August), pp. 18–21.

McGreal, C. (2015) 'Vladimir Putin's "Misinformation" Offensive Prompts US to Deploy Its Cold War Propaganda Tools', *The Guardian*, 25 April, [Online], Available: www.theguardian.com/world/2015/apr/25/us-set-to-revive-propaganda-war-as-putin-pr-machine-undermines-baltic-states [Accessed 13 January 2016].

McKale, D. (1998) *War by Revolution: Germany and Great Britain in the Middle East in the Era of World War I*, Kent, OH: Kent State University Press.

McKendrick, R., Parasuraman, R. and Ayaz, H. (2015) 'Wearable Functional Near-Infrared Spectroscopy (fNIRS) and transcranial Direct Current Stimulation (tDCS): Expanding Vistas for Neurocognitive Augmentation', *Frontiers in Neuroscience*, vol. 9, pp. 1–14.

Mehlman, M, Lin, P. and Abney, K. (2013) 'Enhanced Warfighters: A Policy Framework', *Military Medical Ethics for the 21st Century*, pp. 113–126.

Melley, T. (2011) 'Brain Warfare: The Covert Sphere, Terrorism, and the Legacy of the Cold War', *Grey Room*, vol. 45, pp. 19–40.

Merloo, J. A.M. (2009) *The Rape of the Mind: The Psychology of Thought Control, Menticide, and Brainwashing*, Joshua Tree, CA: Progressive Press.

Messing, P. (2010) 'Did the CIA Test LSD in the New York Subway System?', *New York Post*, 14 March, p. 28.

Metz, S. and Kievit, J. (1994) *The Revolution in Military Affairs and Conflict Short of War*, U.S. Army War College, Strategic Studies Institute.

Miller, K.J., Schalk, G., Hermes, D., Ojemann, J.G. and Rao, R. (2016) 'Spontaneous Decoding of the Timing and the Content of Human Object Perception from Cortical Surface Recordings Reveals Complimentary Information in Event-Related Potential and Broadbent Spectral Change', *PLoS Computational Biology*, vol. 12, no. 1, pp. 1–20.

Minghui (2014) 'Summary Report: The Role of China's Brainwashing Centers in the Torture and Deaths of Falun Gong Practitioners', *Minghui.org*, 19 March, [Online], Available: www.clearwisdom.net/html/articles/2014/3/19/145983.html [Accessed 15 January 2016].

Miranda, R.A., Casebeer, W.D., Hein, A.M., Judy, J.W., Krotkov, E.P., Laabs, T.L., Manzo, J.E., Pankratz, K.J., Pratt, G.A., Sanchez, J.C., Weber, D.J., Wheeler, T.L. and Ling, G.S.F. (2015) 'DARPA-funded Efforts in the Development of Novel Brain–Computer Interface Technologies', *Journal of Neuroscience Methods*, vol. 244, pp. 52–69.

MIT (2013) 'How Smart Dust Could Spy on Your Brain', *MIT Technology Review*, 16 July, [Online], Available: www.technologyreview.com/view/517091/how-smart-dust-could-spy-on-your-brain/ [Accessed 4 May 2015].

Moeser, W. (1962) 'Whiz Kid, Hands Down', *Life*, 14 September, pp. 69–72.

Mohr, C.L. and Gordon, J.E. (2001) *Tulane: The Emergence of a Modern University, 1945–1980*, Baton Rouge, LA: Louisiana State University Press.

Moore, B.E. (2013) *The Brain Computer Interface Future: Time for a Strategy*, Montgomery, AL: Air War College.

Moore, W. (2001) 'Television: Opiate of the Masses', *The Journal of Cognitive Liberties*, vol. 2, no. 2, pp. 59–66.

Moravec, H. (1999) *Robot: Mere Machine to Transcendent Mind*, Oxford: Oxford University Press.

Moreno, J. (2000) *Undue Risk: Secret State Experiments on Humans*, New York: W.H. Freeman & Co.

——— (2004) 'DARPA on Your Mind', *University of Pennsylvania Scholarly Commons*, 25 November, [Online], Available: repository.upenn.edu/cgi/viewcontent.cgi?article=1029&context=neuroethics_pubs [Accessed 16 January 2016].

——— (2006a) *Mind Wars: Brain Research and National Defense*, New York: Dana Press.

——— (2006b) 'Military Mind Wars', *The Scientist* vol. 26, no. 11, p. 25.

——— (2012) *Mind Wars: Brain Science and the Military in the 21st Century*, New York: Bellevue Literary Press.

MTHR (2012) 'Mobile Telecommunications and Health Research Programme Report', *MTHR Programme Management Committee*, [Online], Available: www.mthr.org.uk/documents/MTHRreport2012.pdf [Accessed 10 July 2014].

Mumford, M.D., Rose, A.M. and Goslin, D.A. (1995) 'An Evaluation of Remote Viewing: Research and Applications', *The American Institutes for Research*, 29 September.

Münkler, H. (2004) *The New Wars*, Cambridge: Polity.

Murdock, C., Crotty, R. and Weaver, A. (2014) *Building the 2021 Affordable Military*, *Center for Strategic and International Studies* (June), Lanham, MD: Rowman & Littlefield.

Murphy, K. (2013) 'Jump Starter Kits for the Brain', *The New York Times*, 29 October, p. D3.

Naish, J. (2012) 'Genetically Modified Athletes: Forget Drugs: There Are Even Suggestions Some Chinese Athletes' Genes Are Altered to Make Them Stronger', *The Daily Mail*, 31 July, [Online], Available: www.dailymail.co.uk/news/article-2181873/Genetically-modified-athletes-Forget-drugs-There-suggestions-Chinese-athletes-genes-altered-make-stronger.html [Accessed 8 July 2015].

Narula, S. (2004) 'Psychological Operations (PSYOPS): A Conceptual Overview', *Strategic Analysis*, vol. 28, no. 1, pp. 177–192.

National Institutes of Health (2015) 'Brain Research through Advancing Innovative Neurotechnologies', [Online], Available: braininitiative.nih.gov/ [Accessed 17 June 2015].

National Intelligence Council (2012) 'Global Trends 2030: Alternative Worlds', [Online], www.dni.gov/index.php/about/organization/global-trends-2030?highlight=WyJnbG9 iYWwiLCJ0cmVuZHMiLDIwMzAsImdsb2JhbCB0cmVuZHMiLCJnbG9iYWwgd HJlbmRzIDIwMzAiLCJ0cmVuZHMgMjAzMCJd [Accessed 5 August 2016].

National Research Council (2003) *An Assessment of Nonlethal Weapons: Science and Technology*, Washington, DC: National Academies Press.

———— (2008), *Emerging Cognitive Neuroscience and Related Technologies*, Washington, DC: National Academies Press.

———— (2009) *Opportunities in Neuroscience for Future Army Applications*, Washington, DC: National Academies Press.

NATO (2012) 'Deterrence and Defense Posture Review', *Press Release*, 20 May, [Online], Available: www.nato.int/cps/en/natolive/official_texts_87597.htm [Accessed 29 December 2015].

Nature (2012) 'Secret Weapons', *Nature*, vol. 489, no. 7415, pp. 177–178.

Newman, S.A. (2003) 'Averting the Clone Age: Prospects and Perils of Human Developmental Manipulation', *Journal of Contemporary Health and Law*, vol. 19, no. 1, pp. 431–463.

New York Times Staff (2015) 'Prosthetic Limbs, Controlled by Thought', *New York Times*, 21 May, p. B1.

Nisbet, R.A. (1966) 'Project Camelot: An Autopsy', *Public Interest*, vol. 5, no. 5, pp. 45–69.

Nitsche, M.A. and Paulus, W. (2011) 'Transcranial Direct Current Stimulation – an Update', *Restorative Neurology and Neuroscience*, vol. 29, pp. 463–492.

Nitze, P. H. (1994) 'A Conventional Approach', *Proceedings*, vol. 120, no. 1, pp. 46–51.

Noll, G. (2014) 'Weaponising Neurotechnology: International Humanitarian Law and the Loss of Language', *London Review of International Law*, vol. 2, no. 2, pp. 201–231.

Nutley, E.L. (2003) 'Non-Lethal Weapons: Setting Phasers on Stun? Potential Strategic Blessings and Curses of Non-Lethal Weapons on the Battlefield', *Air War College*, Occasional Paper No. 34 (August).

O'Brien, N. (2012) 'US Use of Truth Drug Revealed', *The Sydney Morning Herald*, [Online], Available: www.smh.com.au/national/us-use-of-truth-drug-revealed-20120929-26sja.html [Accessed 30 September 2015].

Odierno, R. (2012) 'The Army in a Time of Transition: Building a Flexible Force', *Foreign Affairs* (May-June).

Office of the Director of National Intelligence (2012) 'National Intelligence Program: FY 2013 Congressional Budget Justification', *Community Management Account Volume VII* (February).

O'Mara, S. (2009) 'Torturing the Brain: On the Folk Psychology and Neurobiology Motivating "Enhanced and Coercive Interrogation Techniques"', *Trends in Cognitive Sciences*, vol. 13, no. 12, pp. 497–500.

Osborne, W.B., Bethel, S.A., Chew, N.R., Nostrand, P.M. and Whitehead, Y.G. (1996) 'Information Operations: A New Warfighting Capability', A Research Paper to Be Presented at Air Force 2025 (August).

Ostrander, S. and Schroeder, L. (1997) *Psychic Discoveries behind the Iron Curtain*, New York: Marlowe & Co.

Oye, K.A., Esvelt, K., Appleton, E., Catteruccia, F., Chruch, G., Kuiken, T., Lightfood, S.B., McNamara, J., Smidler, A. and Collins, J.P. (2015) 'Regulating Gene Drives', *Science*, vol. 345, no. 6197, pp. 626–628.

Pacepa, I.M. (2006) 'Russian Footprints', *The National Review*, 24 August, [Online], Available: www.nationalreview.com/article/218533/russian-footprints-ion-mihai-pacepa [Accessed 26 December 2015].

Packard, V. (1957) *The Hidden Persuaders*, New York: D. McKay Co.

Paltrow, S.J. (2013) 'Special Report: The Pentagon's Doctored Ledgers Conceal Epic Waste', *Reuters*, 18 November [Online], Available: www.reuters.com/article/2013/11/18/us-usa-pentagon-waste-specialreport-idUSBRE9AH0LQ20131118 [Accessed 20 July 2015].

Parashar, S. (2011) 'Neuroscientific Technologies in Security and Defense Strategies', *Penn Bioethics Journal*, pp. 23–27.

Parnell, B. (2013) 'Forget Invisibility Kittens, Now Tanks Draped in Invisibility Cloak', *The Register*, 13 November, [Online], Available: www.theregister.co.uk/2013/11/13/antenna_invisibility_cloak/ [Accessed 10 August 2015].

Pasternak, D. (1997) 'Wonder Weapons: The Pentagon's Quest for Nonlethal Weapons is Amazing. But Is It Smart?', *U.S. News and World Report*, 7 July, [Online], Available: www.randomcollection.info/mcf/us-news07-07-97.htm [Accessed 5 May 2015].

——— (2000), 'John Norseen', *U.S. News and World Report*, 10 January, vol. 128, no. 1.

Patrick, S.M. (2010) 'Are "Ungoverned Spaces" a Threat?', *Council on Foreign Relations Expert Brief*, 11 January, [Online], Available: www.cfr.org/somalia/ungoverned-spaces-threat/p21165 [Accessed 28 December 2015].

Paul, C. (2011) *Strategic Communication: Origins, Concepts, and Current Debates*, Santa Barbara, CA: Praeger.

Pearson, A.M., Chevrier, M.I. and Wheelis, M. (2007) *Incapacitating Biochemical Weapons*, Plymouth, UK: Lexington Books.

Pelletier, D. (2013) 'Tomorrow's Wars: Bio-Weapons, Mind-Control; Is Nothing Sacred?', *Ethical Technology*, 13 November, [Online], Available: ieet.org/index.php/IEET/more/pelletier20131113 [Accessed 7 December 2015].

Perlmutter, D. and Loberg, K. (2013) *The Grain Brain: The Surprising Truth about Wheat, Carbs, and Sugar- Your Brains Silent Killers*, New York: Little, Brown and Co.

Perry, B. (2015) 'Non-Linear Warfare in the Ukraine: The Critical Role of Information Operations and Special Operations', *Small Wars Journal* (August), pp. 1–30.

Persinger, M.A. (1995) 'On the Possibility of Remotely Accessing Every Human Brain by Electromagnetic Induction of Fundamental Algorithms', *Perceptual and Motor Skills*, vol. 80, pp. 791–799.

——— (2003) 'The Sensed Presence Within Experimental Settings: Implications for the Male and Female Concept of Self', *The Journal of Psychology*, vol. 137, no. 1, pp. 5–16.

Peterson, A. (2013) 'Yes, Terrorists Could Have Hacked Dick Cheney's Heart: Security Vulnerabilities Are a Major Problem in Medical Software and Devices', *The Washington Post*, 21 October, [Online], Available: www.washingtonpost.com/news/the-switch/wp/2013/10/21/yes-terrorists-could-have-hacked-dick-cheneys-heart/ [Accessed 15 January 2016].

Petit, B. (2012) 'Social Media and Unconventional Warfare', *Special Warfare*, vol. 25, no. 2, pp. 21–28.

Petro, J.B., Plasse, T.R. and McNulty, J.A. (2003) 'Biotechnology: Impact on Biological Warfare and Biodefense', *Biosecurity and Bioterrorism: Biodefense Strategy, Practice, and Science*, vol. 1, no. 3, pp. 161–168.

Phan, D. (2015) 'The Brain Wave: Optogenetics May Be the Future of Mind Control', *Johns Hopkins University Newsletter*, 10 September, [Online], Available: nlonthedl.wordpress.com/2015/09/10/the-brain-wave-optogenetics-may-be-the-future-of-mind-control/ [Accessed 3 December 2015].

Pilkington, E. (2009) 'CIA Doctors Face Human Experimentation Claims', *The Guardian*, 2 September, [Online], Available: www.theguardian.com/world/2009/sep/02/cia-usa [Accessed 12 August 2015].

Pilkington, M. (2004) 'Life: Death Rays: Far Out', *The Guardian*, 12 August, p. 10.

Pines, M. (1973) *The Brain Changers: Scientists and the New Mind Control*, New York: Harcourt Brace Jovanovich Inc.

Pollack, H. (1979) 'Epidemiological Data on American Personnel in the Moscow Embassy', *Bulletin of the New York Academy of Medicine*, vol. 55, no. 11, pp. 1182–1186.

Pomerantsev, P. (2014) 'Russia and the Menace of Unreality', *The Atlantic*, 9 September, [Online], Available: www.theatlantic.com/international/archive/2014/09/russia-putin-revolutionizing-information-warfare/379880/ [Accessed 10 April 2015].

——— (2015) 'The Kremlin's Hall of Mirrors', *The Guardian*, 12 May, p. 25.

Presidential Commission for the Study of Bioethical Issues (2011) 'Moral Science: Protecting Participants in Human Subject Research', Washington, DC (December).

Price, R. (1995) 'A Genealogy of the Chemical Weapons Taboo', *International Organization*, vol. 49, no. 1, pp. 73–103.

Pyle, C.H. (2009) *Getting Away with Torture*, Washington, DC: Potomac Books.

Randall, K. (2015) 'Neuropolitics, Where Campaigns Try to Read Your Mind', *The New York Times*, 3 November, [Online], Available: www.nytimes.com/2015/11/04/world/americas/neuropolitics-where-campaigns-try-to-read-your-mind.html?_r=0 [Accessed 15 January 2016].

Ranelagh, J. (1986) *The Agency: The Rise and Decline of the CIA From Wild Bill Donovan to William Casey*, New York: Simon & Schuster.

Rangarajan, P. (2014) 'Protecting Brainwaves Considering Signals Intelligence Advances', *International Law Journal of London*, 25 November, [Online], Available: www.internationallawjournaloflondon.com/protecting-brainwaves-considering-signal-intelligence-technology-advances.html [Accessed 10 July 2015].

Rao, R.P. (2013) *Brain-Computer Interfacing: An Introduction*, Cambridge: Cambridge University Press.

Rao, R.P.N., Stocco, A., Bryan, M., Sarma, D., Youngquist, T.M., Wu, J. and Prat, C.S. (2014) 'A Direct Brain-to-Brain Interface in Humans', *PLOS One* vol. 10, pp.137.

Reisman, W.M. and Armstrong, A. (2006) 'The Past and the Future of the Claim of Preemptive Self-Defense', *The American Journal of International Law*, vol. 100, no. 3, pp. 525–550.

Reno, J. (2014) '"Medicating Our Troops Into Oblivion": Prescription Drugs Said to Be Endangering U.S. Soldiers', *International Business Times*, 19 April, [Online],

Available: www.ibtimes.com/medicating-our-troops-oblivion-prescription-drugs-said-be-endangering-us-soldiers-1572217 [Accessed 22 July 2015].

Repantis, D., Schlattmann, P., Laisney, O. and Heuser, I. (2010) 'Modafinil and Methylphenidate for Neuroenhancement in Healthy Individuals: A Systematic Review', *Pharmacological Research*, no. 26, pp. 187–206.

Requarth, T. (2015) 'This Is Your Brain. This Is Your Brain as a Weapon', *Foreign Policy*, 14 September, [Online], Available: foreignpolicy.com/2015/09/14/this-is-your-brain-this-is-your-brain-as-a-weapon-darpa-dual-use-neuroscience/ [Accessed 25 November 2015].

Resnik, D. B. (2007) 'Neuroethics, National Security and Secrecy', *The American Journal of Bioethics*, vol. 7, no. 5, pp. 14–15.

Reuters (2012a) 'Internet Search Yields Bogus Arms Parts from China', *Reuters*, 26 March, [Online], Available: www.reuters.com/article/usa-china-weapons-idUSL2E8EQBM020120326 [Accessed 20 December 2015].

――― (2012b) 'Obama Authorizes Secret Support to Syrian Rebels', *Reuters*, 1 August, [Online], Available: www.reuters.com/article/us-usa-syria-obama-order-idUSBRE8701OK20120802 [Accessed 24 December 2015].

Richards, B. (1977) 'CIA Project Eyed Lobotomy, Electric Shock Therapy', *The Washington Post*, 29 October, p. A2.

――― (1979) 'Book Disputes CIA Chief on Mind-Control Efforts', *The Washington Post*, 29 January, p. A2.

Richelson, J. (2001) *The Wizards of Langley: Inside the CIA's Directorate of Science and Technology*, Boulder, CO: Westview Press.

Rickards, J. (2014) *The Death of Money: The Coming Collapse of the International Monetary System*, New York: Penguin.

Rid, T. (2013) *Cyber War Will Not Take Place*, Oxford: Oxford University Press.

Rifat, T. (2002) *Remote Viewing: What It Is, Who Uses It and How to Do It*, London: Vision Books.

Rizal, T. N. (2009) *Torture Killing Me Softly: Bhutan Through the Eyes of Mind Control Victim*, Hattiban, Lalitpur: Jagadamba Press.

Rhine, J.B. (1966) *Extra-sensory Perception*, Boston: Bruce Humphries.

Roache, R. (2014) 'Enhanced Punishment: Can Technology Make Life Sentences Longer?', [Online], Available: blog.practicalethics.ox.ac.uk/2013/08/enhanced-punishment-can-technology-make-life-sentences-longer/ [Accessed 29 December 2015].

Roberts, D., Ackerman, S., Siddique, H. and Chrisafis, A. (2013) '"We Have Our Plans": Vladimir Putin Warns US Against Syria Military Action', *The Guardian*, 4 September, [Online], Available: www.theguardian.com/world/2013/sep/04/putin-warns-military-action-syria [Accessed 20 December 2015].

Robinson, R. (2009) 'Exploring the "Global Workspace" of Consciousness', *PLOS Biology*, vol. 7, no. 3, p. E1000066.

Robock, A. (2011) 'Nuclear Winter Is a Real and Present Danger: Models Show that even a "Small" Nuclear War Would Cause Catastrophic Climate Change', *Nature*, vol. 473, no. 7347, pp. 275–277.

――― (2015) 'The CIA Asked Me About Controlling the Climate – This is Why We Should Worry', *The Guardian*, 17 February, [Online], Available: www.theguardian.com/commentisfree/2015/feb/17/cia-controlling-climate-geoengineering-climate-change [Accessed 28 December 2015].

Robock, A. and Toon, O. B. (2006) 'Self-Assured Destruction: The Climate Impacts of Nuclear War', *Bulletin of Atomic Scientists*, vol. 68, no. 5, pp. 66–74.

Ronanki, R. and Steier, D. (2014) 'Cognitive Analytics: Tech Trends 2014', *Deloitte University Press*, 21 February, [Online], Available: dupress.com/articles/2014-tech-trends-cognitive-analytics/ [Accessed 25 August 2015].

Ronson, J. (2004) *The Men Who Stare at Goats*, New York: Simon & Schuster.

Rose, S. (2006) 'We Are Moving Ever Closer to the Era of Mind Control', *The Observer*, 5 February, p. 31.

Roseifelde, S. and Hlouskova, R. (2007) 'Why Russia Is Not a Democracy', *Comparative Strategy*, vol. 26, no. 3, pp. 215–229.

Rosenberg, B.H. (1994) 'Non-lethal Weapons May Violate Treaties', *Bulletin of Atomic Scientists*, September-October, pp. 44–45.

Ross, C. (2006) *The C.I.A. Doctors: Human Rights Violations by American Psychiatrists*, Richardson, TX: Manitou.

——— (2007) 'Ethics of CIA and Military Contracting by Psychiatrists and Psychologists', *Ethical and Human Psychology and Psychiatry*, vol. 9, no. 1, pp. 25–34.

Ross, V. (2011) ' "StunRay": A Light Weapon that Overstimulates the Brain', *Discover Magazine*, 5 April, [Online], Available: blogs.discovermagazine.com/sciencenotfiction/2011/04/05/stunray-a-light-weapon-that-overstimulates-the-brain/#.Vapd2flViko [Accessed 7 January 2016].

Rothman, P. (2015) 'Biology Is Technology – DARPA Is Back in the Game with a Big Vision and It Is H+' *H+ Magazine*, 15 February, [Online], Available: hplusmagazine.com/2015/02/15/biology-technology-darpa-back-game-big-vision-h/ [Accessed 16 January 2016].

Rothstein, L. (1998) 'All in the (Russian) Mind?', *Bulletin of the Atomic Scientists*, vol. 54, no. 4, p. 11.

Royal Society (2012) 'Brain Waves Module 3: Neuroscience, Conflict and Security', *The Royal Society RS Policy Document 06/11* (February), [Online], Available: https://royalsociety.org/~/media/Royal_Society_Content/policy/projects/brain-waves/2012-02-06-BW3.pdf. [Accessed 5 August 2016].

Rozsa, L. (2009) 'Drugs as Weapons: A Psychochemical Weapon Considered by the Warsaw Pact', *Substance Use & Misuse*, vol. 44, pp. 172–178.

Rusconi, E., Scott-Brown, K.C. and Szymkoviak (2014) 'Neuroscience Perspectives on Security', *Frontiers in Human Neuroscience*, vol. 8, no. 996, pp. 1–2.

Russian Federation (2010) 'The Military Doctrine of the Russian Federation Approved by the Russian Federation Presidential Edict on 5 February 2010', [Online], Available: www.sras.org/military_doctrine_russian_federation_2010 [Accessed 9 January 2016].

——— (2011) 'Convention on International Information Security', The Ministry of Foreign Affairs, [Online], Available: cryptome.org/2014/05/ru-international-infosec.htm [Accessed 15 January 2016].

Russon, M.A. (2014) 'Mind-Controlled UAV Drones Being Developed for the Military in Texas', *International Business Times*, 4 September, [Online], Available: www.ibtimes.co.uk/mind-controlled-uav-drones-being-developed-texas-us-military-1463980 [Accessed 16 January 2016].

Sanders, L. (2001) 'Brains May Be Wars' Battleground', *Science News*, vol. 180, no. 13, p. 14.

Schneier, B. (2015) 'What Is Next in Government Surveillance', *The Atlantic*, 2 March, [Online], Available: www.theatlantic.com/international/archive/2015/03/whats-next-in-government-surveillance/385667/ [Accessed 16 January 2016].

Schrag, P. (1978) *Mind Control*, New York: Pantheon Books.

Schuman, T. (1985) *World Thought Police*, Los Angeles, CA: Almanac.

Science Daily (2008) 'Logo Can Make You "Think Different"', *Science Daily*, 30 March, [Online], Available: www.sciencedaily.com/releases/2008/03/080328085918. htm [Accessed 20 November 2015].

Segel, L. (2002) 'Operation Midnight Climax: Among the Greatest Medical Conspiracies of all Time Would Have to Be Listed the CIA's Odd Experiments with LSD', *Medical Post*, vol. 38, no. 33, p. 27.

Sejna, J. (1982) *We Will Bury You: The Soviet Plan for the Subversion of the West by the Highest Ranking Communist Ever to Defect*, London: Sidgwick & Jackson.

Shachtman, N. (2007) 'Be More Than You Can Be', *Wired Magazine* 15.3 (March), [Online], Available: archive.wired.com/wired/archive/15.03/bemore.html [Accessed 10 January 2014].

———— (2010a), 'Pain Ray Recalled from Afghanistan', *Wired.com*, 20 July, [Online], Available: www.wired.com/2010/07/pain-ray-recalled-from-afghanistan/ [Accessed 16 January 2016].

———— (2010b) 'Air Force Wants Neuroweapons to Overwhelm Enemy Minds', *Wired. com*, 2 November, [Online], Available: www.wired.com/2010/11/air-force-looks-to-artificially-overwhelm-enemy-cognitive-capabilities/ [Accessed 16 January 2016].

———— (2012) 'DARPA's Magic Plan: "Battlefield Illusions" to Mess with Enemy Minds', *Wired.com*, 14 February, [Online], Available: www.wired.com/2012/02/darpa-magic/ [Accessed 16 January 2016].

Sharkey, N. (2012) 'The Evitability of Autonomous Robot Warfare', *International Review of the Red Cross*, vol. 94, no. 886, pp. 787–799.

Sharma, A. (2015) 'Subliminal Perception: Conceptional Analysis and Its Rampant Usage in Advertisements and Music Industry', *Indian Journal of Health and Wellbeing*, vol. 6, no. 6, pp. 640–643.

Shea, T.C. (2002) 'Post-Soviet Maskirovka, Cold War Nostalgia, and Peacetime Engagement', *Military Review*, vol. 82, no. 3, pp. 63–67.

Shen, H. (2013) 'Zapping the Brain Can Help Spot-Clean Nasty Memories', *Nature. com*, 22 December, [Online], Available: www.nature.com/news/zapping-the-brain-can-help-to-spot-clean-nasty-memories-1.14431 [Accessed 6 July 2015].

Sher, L. (2000) 'The Effects of Natural and Man-Made Electromagnetic Fields on Mood and Behavior: The Role of Sleep Disturbances', *Medical Hypotheses*, vol. 54, no. 4, pp. 630–633.

Shermer, M. (2016) 'Can Our Minds Live Forever?', *Scientific American*, 1 February, [Online], Available: www.scientificamerican.com/article/can-our-minds-live-forever/ [Accessed 5 February 2016].

Shukman, D. (1996) *Tomorrow's War: The Threat of High-Technology Weapons*, New York: Harcourt Brace & Co.

Silva, M. (2014) 'Intelligence is for the SHARP', *Bloomberg*, 27 January, [Online], Available: go.bloomberg.com/political-capital/2014-01-27/intelligence-is-for-the-sharp/ [Accessed 10 July 2015].

Singer, P.W. and Friedman, A. (2014) *Cybersecurity and Cyberwar: What Everyone Needs to Know*, Oxford: Oxford University Press.

Sirén, T. (2013) *Winning Wars Before They Emerge: From Kinetic Warfare to Strategic Communications as a Proactive Mind-Centric Paradigm of the Art of War*, Boca Raton, FL: Universal Publishers.

Slater, L. (2005) 'Who Holds the Clicker?', *Mother Jones* (November), [Online], Available: www.motherjones.com/politics/2005/11/who-holds-clicker [Accessed 20 August 2015].

Smith, C.L. (1988) 'Soviet Maskirovko', *Airpower Journal* (Spring).

Smith, K. (2013) 'Mind-Reading Technology Speeds Ahead', *Nature*, vol. 502 (24 October), pp. 428–430.

Snegovaya, M. (2015) '*Putin's Information Warfare in Ukraine: Soviet Origins of Russia's Hybrid Warfare*', Washington, DC: Institute for the Study of War (September).

Squires, S. (1988) 'The Military's Twilight Zone: Uncle Sam's Jedi Warriors', *The Washington Post*, 17 April, p. 47.

Stanton Candlin, A.H. (1974) *Psycho-Chemical Warfare: The Chinese Communist Drug Offensive Against the West*, New York: Arlington House.

Stewart, H. (2013) 'How Do Filmmakers Manipulate Our Emotions with Music?', *BBC Arts and Culture*, 13 September, [Online], Available: www.bbc.co.uk/arts/0/24083243 [Accessed 12 December 2015].

Stix, G. (2008) 'Can fMRI Really Tell When You Are Lying?', *Scientific American* (August), [Online], Available: www.scientificamerican.com/article/new-lie-detector/ [Accessed 10 July 2015].

Stock, G. (2002) *Redesigning Humans: Our Inevitable Genetic Future*, New York: Houghton Mifflin.

Streatfield, D. (2001) *Cocaine: An Unauthorized Biography*, New York: St. Martin's Press.

—— (2008) *Brainwash: The Secret History of Mind Control*, New York: Picador.

Suckut, S. (2001) *Das Wörterbuch der Staatssicherheit. Definitionen zur politisch-operativen Arbeit*, Berlin: Links Verlag.

Sullivan, J.F. (2007) *Gatekeeper: Memoirs of a CIA Polygraph Examiner*, Washington, DC: Potomac Books.

Sunstein, C. and Vermeule, A. (2009) 'Conspiracy Theories: Causes and Cures', *The Journal of Political Philosophy*, vol. 17, no. 2, pp. 202–227.

Suvorov, V. (1988) *Spetsnaz: The Inside Story of Soviet Special Forces*, New York: W.W. Norton & Co.

Szafranski, R. (1994) 'Neocortical Warfare? The Acme of Skill', *Military Review* (November), pp. 41–55.

Tancredi, L. (2005) *Hardwired Behavior: What Neuroscience Reveals About Morality*, Cambridge: Cambridge University Press.

Tanielian, T. and Jaycox, L.H. (2008) *Invisible Wounds of War: Psychological and Cognitive Injuries, Their Consequences, and Services to Assist Recovery*, Santa Monica, CA: RAND.Taylor, E. (1990) *Subliminal Communication: The Emperor's New Clothes*, Las Vegas, NV: Just Another Reality Publishing.

—— (2007) 'Subliminal Information Theory Revisited: Casting Light on a Controversy', *Annals of the American Psychotherapy Association* (fall), pp. 28–33.

—— (2009) *Mind Programming: From Persuasion and Brainwashing to Self-Help and Practical Metaphysics*, Carlsbad, CA: Hay House.

Taylor, K. (2004) *Brainwashing: The Science of Thought Control*, Oxford: Oxford University Press.

Taylor, S. (1992) 'A History of Secret CIA Mind Control Research', *Nexus* (April/May), [Online], Available: all.net/journal/deception/MKULTRA/www.profreedom.free4all.co.uk/skeletons_1.html [Accessed 10 July 2015].

Tennison, M.N. and Moreno, J. (2012) 'Neuroscience, Ethics, and National Security: The State of the Art', *PLOS Biology*, vol. 10, no. 3, pp. 1–4.

Terhune, D.B., Russo, S., Near, J., Stagg, C.J. and Kadosh, R.C. (2014) 'GABA Predicts Time Perception', *Journal of Neuroscience*, vol. 34, no. 12, pp. 4364–4370.

The Week (2012) 'How Future Criminals Could Hack Your Brain and Steal Your PIN', *The Week*, 28 August, [Online], Available: theweek.com/articles/472864/how-future-criminals-could-hack-brain-steal-pin [Accessed 5 May 2015].

Theorin, B. (1999) 'On the Environment, Security, and Foreign Policy', *European Parliament Committee on Foreign Affairs, Security and Defence Policy*, 14 January, [Online], Available: www.europarl.europa.eu/sides/getDoc.do?pubRef=-//EP//TEXT+REPORT+A4-1999-0005+0+DOC+XML+V0//EN [Accessed 3 May 2015].

Thomas, G. (1990) *Journey Into Madness: The True Story of Secret CIA Mind Control and Medical Abuse*, New York: Bantam.

——— (2008) *Secrets & Lies: A History of CIA Mind Control & Germ Warfare*, London: JR Books.

Thomas, M.R. (1998) *Non-Lethal Weaponry: A Framework for Future Integration*, Montgomery, AL: Air Command and Staff College.

Thomas, T.L. (1997) 'Russian Information-Psychological Actions: Implications for U.S. PSYOPS', *Special Warfare*, vol. 10, no. 1, pp. 12–19.

——— (1998) 'The Mind Has No Firewall', *Parameters* (Spring), pp. 84–92.

——— (1999) 'Human Network Attacks', *Military Review* (September/ October), pp. 23–33.

——— (2000) 'The Russian View of Information Warfare', Foreign Military Studies Office, Fort Leavenworth, KA, [Online], Available: fmso.leavenworth.army.mil/documents/Russianvuiw.htm [Accessed 13 January 2016].

——— (2014) 'Russia's Information Warfare Strategy: Can the Nation Cope in Future Conflicts?', *Journal of Slavic Military Studies*, no. 27, pp. 101–130.

——— (2015) 'Psycho Viruses and Reflexive Control: Russian Theories of Information-Psychological War', in: *Information War: From China's Three Warfares to NATO's Narratives*, Legatum Institute (September), pp. 16–21.

Thompson, R. F. (2012) *Das Gehirn: Von der Nervenzelle zur Verhaltenssteuerung*, Heidelberg: Spektrum: Akademischer Verlag.

Tilghman, A. and Pawlyk, O. (2015) 'U.S. vs. Russia: What a War Would Look Like Between the World's Most Fearsome Militaries', *Military Times*, 5 October, [Online], Available: www.militarytimes.com/story/military/2015/10/05/us-russia-vladimir-putin-syria-ukraine-american-military-plans/73147344/ [Accessed 12 January 2016].

Time (1975) 'Of Dart Guns and Poisons', Time Magazine, vol. 106, no. 13.

——— (1976) 'Foreign Relations: The Microwave Furor', Time Magazine, vol. 107, no. 12, p. 19.

——— (1979) 'Secret Voices: Messages that Manipulate', *Time Magazine* (10 September).

Timmermann, W.K. (2005) 'The Relationship Between Hate Propaganda and the Incitement of Genocide: A New Trend in International Law Towards Criminalization of Hate Propaganda?', *Leiden Journal of International Law*, vol. 18, no. 2, pp. 257–282.

Toffler, A. and Toffler, H. (1993) *War and Anti-War: Survival at the Dawn of the 21st Century*, Boston, MA: Little Brown & Co.

Tracey, I. and Flower, R. (2014) 'The Warrior in the Machine: Neuroscience Goes to War', *Nature*, vol. 15, pp. 825–834.

Trower, B. (2001) 'Confidential Report TETRA Strictly for the Police Federation of England and Wales', [Online], Available: www.tetrawatch.net/papers/trower_report.pdf [Accessed 26 August 2015].

Tucker, J.B. (2010) 'The Future of Chemical Weapons', *The New Atlantis* (Winter), pp. 3–29.

Tufail, Y., Matyushov, A., Baldwin, N., Tauchmann, M.L., Georges, J., Yoshihiro, A., Helms Tillery, S.I. and Tyler, W.J. (2010) 'Transcranial Pulsed Ultrasound Stimulates Intact Brain Circuits', *Neuro*, vol. 66, no. 5, pp. 681–694.

Tversky, A. and Kahneman, D. (1974) 'Judgment Under Uncertainty: Heuristics and Biases', *Science*, vol. 185, no. 4157, pp. 1124–1131.

Tyler, P. E. (1986) 'The Electromagnetic Spectrum in Low-Intensity Conflict', in: Dean, D.J. (ed.) *Low-Intensity Conflict and Modern Technology*, Montgomery, AL: Air University Press, pp. 249–260.

UK Ministry of Defence (2010) 'Global Strategic Trends – Out to 2040', London, UK Ministry of Defence, [Online], Available: www.gov.uk/government/uploads/system/uploads/attachment_data/file/33717/GST4_v9_Feb10.pdf [Accessed 30 August 2015].

——— (2014) 'Global Strategic Trends – Out to 2045', London, UK Ministry of Defence, [Online], Available: www.gov.uk/government/uploads/system/uploads/attachment_data/file/348164/20140821_DCDC_GST_5_Web_Secured.pdf [Accessed 30 August 2015].

Ullman, H. and Wade, J.P. (1996) *Shock and Awe: Achieving Rapid Dominance*, Washington, DC: National Defense University.

United Nations (1978) 'Convention on the Prohibition of Military or Any Other Hostile Use of Environmental Modification Techniques (ENMOD)', *United Nations Office for Disarmament Affairs*, [Online], Available: disarmament.un.org/treaties/t/enmod/text [Accessed 15 July 2015].

United Nations Institute for Disarmament Research (2014) 'The Weaponization of Increasingly Autonomous Technologies: Considering How Meaningful Human Control Might Move the Discussion Forward', *UNIDIR*, [Online], Available: www.unidir.org/files/publications/pdfs/considering-how-meaningful-human-control-might-move-the-discussion-forward-en-615.pdf [Accessed 25 November 2015].

U.S. Air Force (1994) 'New World Vistas for Air and Space Power in the 21st Century', Proceedings of the 50th Anniversary Symposium of the USAF Scientific Advisory Board (10 November).

——— (2003) '5.6. Airborne Holographic Projector', [Online], Available: www.stealthskater.com/Documents/Holography_1.pdf [Accessed 20 July 2015].

U.S. Army (2005) 'China: Medical Research on Bio-Effects of Electromagnetic Pulse and High-Power Microwave Radiation', National Ground Intelligence Center (17 August).

——— (2014) 'Counter-Unconventional Warfare', Army Special Operations Command, White Paper (26 September).

U.S. Congress (1963) 'Communist Goals', Extension of Remarks of Hon. A.S. Herlong, JR. of Florida in the House of Representatives (10 January).

——— (1975a) 'Report to the President by the Commission on CIA Activities within the United States', U.S. Senate (June).

——— (1975b) 'Alleged Assassination Plots Involving Foreign Leaders: An Interim Report of the Select Committee of the Study of Governmental Operations with Respect to Intelligence Activities', U.S. Senate, 20 November.

——— (1977) 'Joint Hearing Before the Select Committee on Intelligence and the Subcommittee on Health and Scientific Research', Washington, DC: U.S. Senate, 3 August.

———— (1997) 'Report of the Commission on Protecting and Reducing Government Secrecy', Washington, DC: U.S. Senate, 31 December.

———— (2002) 'Authorization for the Use of Military Force Against Iraq Resolution', H.J.Res 114, 107th Congress (2001–2002).

———— (2007) 'Final Report of the U.S. Senate Committee on the Use of Brain Fingerprinting Technology', U.S. Senate, Senate Research Office.

———— (2011) 'The Committee's Investigation in Counterfeit Electronics Parts in the Department of Defense Supply Chain', Senate Armed Services Committee Hearing, 8 November.

U.S. Department of Defense (1972) 'Controlled Offensive Behavior – USSR', Defense Intelligence Agency, Medical Intelligence Office (July).

———— (1976) 'Biological Effects of Electromagnetic Radiation (Radio Waves and Microwaves) – Eurasian Communist Countries', Defense Intelligence Agency: Army Medical Intelligence and Information Agency (March).

———— (1977) 'Experimentation Programs Conducted by the Department of Defense That Had CIA Sponsorship or Participation and That Involved the Administration to Human Subjects of Drugs Intended for Mind-control or Behavior-modification Purposes', Memorandum to the Secretary of Defense, 20 September.

———— (1986) 'Low Intensity Conflict: American Dilemma', 15 January, [Online], Available: usacac.army.mil/cac2/csi/docs/Gorman/06_Retired/01_Retired_1985_90/02_86_LIC_16Jan.pdf [Accessed 18 December 2015].

———— (1996) 'Policy for Non-Lethal Weapons', Department of Defense Directive No. 3000.3 (9 July).

———— (1998) 'Bioeffects of Selected Nonlethal Weapons', [Online], Available: www.metatexte.net/docs/index-6.html [Accessed 29 August 2015].

———— (2003) 'Psychological Operations, Tactics, Techniques, and Procedures', FM 3–05.301, 31 December.

———— (2007) 'Defense Science Board Task Force on Directed Energy Weapons', Washington, DC: Office of the Under-Secretary of Defense for Acquisition, Technology, and Logistics.

———— (2009) 'Strategic Communication Science and Technology Plan', Washington, DC (April).

———— (2014) 'Counter-Unconventional Warfare: White Paper', U.S. Special Operations Command, 26 September.

———— (2015) 'The National Military Strategy of the United States 2015', Washington, DC: Department of Defense (June).

U.S. Department of Homeland Security (2008) 'Privacy Impact Assessment for the Future Attribute Screening Technology (FAST) Project', Chief Privacy Officer, Department of Homeland Security, 15 December.

———— (2013) 'Transportation Security Administration's Screening of Passengers by Observation Techniques', Office of the Inspector General, 29 May.

U.S. Department of State (2011) 'Colombia 2011 Crime and Safety Report', *Bureau of Diplomatic Security*, [Online], Available: www.osac.gov/pages/contentreportdetails.aspx?cid=10559 [Accessed 16 January 2016].

U.S. General Accounting Office (1994) 'Human Experimentation: An Overview of Cold War Era Programs', Testimony before the Legislation and National Security Subcommittee, Committee on Government Operations, House of Representatives, 28 September.

U.S. National Security Council (2002) 'National Strategy to Combat Weapons of Mass Destruction', December, [Online], Available: www.state.gov/documents/organization/16092.pdf [Accessed 26 November 2015].

———— (2009) 'National Strategy for Countering Biological Threats', November [Online], Available: www.whitehouse.gov/sites/default/files/National_Strategy_for_Countering_BioThreats.pdf [Accessed 9 December 2015].

U.S. White House (2015) 'Executive Order: Using Behavioral Science to Better Serve the American People', [Online], Available: www.whitehouse.gov/the-press-office/2015/09/15/executive-order-using-behavioral-science-insights-better-serve-american [Accessed 15 January 2016].

Van Aken, J. and Hammond, E. (2003) 'Genetic Engineering and Biological Weapons', *EMBO Report*, vol. 4, no. 1, pp. S57-S60.

Van Creveld, M. (1991) *The Transformation of War*, New York: The Free Press.

Van Herpen, M. (2015) *Putin's Propaganda Machine: Soft Power and Russian Foreign Policy*, Lanham, MD: Rowman & Littlefield.

Victorian, A. (1996) 'United States, Canada, and Britain: Partners in Mind-Control Operations', *MindNet Journal* (July), [Online], Available: wikispooks.com/wiki/Document:United_States,_Canada,_Britain:_Partners_in_mind-control_operations [Accessed 10 May 2015].

———— (1999) *The Mind Controllers*, London: Vision.

Viegas, J. (2012) 'Brain in a Dish Flies Plane', *Discovery News*, 27 November, [Online], Available: news.discovery.com/tech/robotics/brain-dish-flies-plane-041022.htm [Accessed 16 January 2016].

Vielhaber, D. (2013) 'The Stasi-Meinhof Complex', *Studies in Conflict and Terrorism*, vol. 36, no. 7, pp. 533–546.

Volcler, J. (2013), *Extremely Loud: Sound as a Weapon*, New York: The New Press.

Wadhwa, T. (2012) 'Yes, You Can Hack a Pacemaker', *Forbes*, 6 December, [Online], Available: www.forbes.com/sites/singularity/2012/12/06/yes-you-can-hack-a-pacemaker-and-other-medical-devices-too/ [Accessed 12 August 2015].

Walker, M. (2009) 'Ship of Fools: Why Transhumanism Is the Best Bet to Prevent the Extinction of Civilization', *MetaNexus*, 5 February, [Online], Available: www.metanexus.net/essay/h-ship-fools-why-transhumanism-best-bet-prevent-extinction-civilization [Accessed 4 December 2015].

Wall, J. (1998) 'Military Use of Mind Control Weapons', *Nexus* (October/ November), pp. 13–18.

———— (1999) 'Aerial Mind Control: The Threat to Civil Liberties', *Nexus* (October/ November).

Wall Street Journal (2013) '90 Million Americans Not Working', *Wall Street Journal*, 24 October, p. 12.

Wallace, R., Melton, K. and Schlesinger, R. (2009) *Spycraft: Inside the CIA's Top Secret Spy Lab*, London: Bantam Press.

Warmflash, D. (2015) 'Brain fingerprints: Will semantic memory identification replace fingerprints and passwords?', *Genetic Literacy Project*, 3 June, [Online], Available: www.geneticliteracyproject.org/2015/06/03/brain-fingerprints-will-semantic-memory-identification-replace-fingerprints-and-passwords/ [Accessed 20 November 2015].

Weberman, A.J. (1980) 'Mind Control: The Story of Mankind Research Unlimited, Inc.', *Covert Action Quarterly*, no. 9 (June), pp. 15–21.

Weinberger, S. (2007a) 'Mind Games', *The Washington Post*, 14 January, [Online], Available: www.washingtonpost.com/wp-dyn/content/article/2007/01/10/AR2007011001399.html [Accessed 10 May 2015].

——— (2007b) 'The Other Medusa: A Microwave Sound Weapon', *Wired.com*, 16 August, [Online], Available: www.wired.com/2007/08/the-other-medus/ [Accessed 16 January 2016].

——— (2007c) 'The Weird Russian Mind Control Behind a DHS Contract', *Wired. com*, 20 September, [Online], Available: archive.wired.com/politics/security/news/ 2007/09/mind_reading?currentPage=all [Accessed 16 January 2016].

——— (2007d) 'Light + Sound = New Weapon', *Wired.com*, 12 December, [Online], Available: www.wired.com/2007/12/light-sound-new/ [Accessed 16 January 2016].

——— (2007e) *Imaginary Weapons: A Journey Through the Pentagon's Scientific Underworld*, New York: Nation Books.

——— (2008a) *Scary Things That Don't Exist: Separating Myth From Reality*, Muscatine, IA: The Stanley Foundation (June).

——— (2008b) 'Army Yanks "Voice-to-Skull-Devices" Website', *Wired.com*, 9 May, [Online], Available: www.wired.com/2008/05/army-removes-pa/ [Accessed 16 January 2016].

——— (2010) "Intent or Deceive? Can the Science of Deception Detection Help Detect Terrorists?", *Nature*, vol. 465, pp. 412–415.

——— (2011) 'Terrorist "Pre-crime" Detector Field Tested in United States', *Nature.com*, 27 May, [Online], Available: www.nature.com/news/2011/110527/full/ news.2011.323.html [Accessed 10 May 2015].

——— (2012) 'Microwave Weapons: Wasted Energy', *Nature.com*, 12 September, [Online], Available: www.nature.com/news/microwave-weapons-wasted-energy-1.11396 [Accessed 11 December 2015].

——— (2014a) 'Mind Control Moves Into Battle', *BBC.com*, 18 November, [Online], Available: www.bbc.com/future/story/20120704-mind-control-moves-into-battle [Accessed 16 January 2016].

——— (2014b) 'Building the Pentagon's "Like Me" Weapon', *BBC.com*, 18 November, [Online], Available: www.bbc.com/future/story/20120501-building-the-like-me-weapon [Accessed 10 August 2015].

Welsh, C. (2009) 'Outlaw Nonconsensual Human Experiments Now', *The Bulletin of Atomic Scientists*, 16 June, [Online], Available: thebulletin.org/outlaw-nonconsensual-human-experiments-now [Accessed 10 August 2015].

——— (2012) 'Cold War Non-Consensual Experiments: The Threat of Neuroweapons and the Danger That It Will Happen Again', *Essex Human Rights Review*, vol. 9, no. 1, pp. 1–32.

Wendt, A. (2003) 'Why a World State Is Inevitable', *European Journal of International Relations*, vol. 9, no. 4, pp. 491–542.

Wensierski, P. (1999) 'Stasi: In Kopfhöhe ausgerichtet', *Der Spiegel*, 17 May, [Online], Available: www.spiegel.de/spiegel/print/d-13395385.html [Accessed 10 August 2015].

West, L.J., Pierce, C.M. and Thomas, W.D. (1962), 'Lysergic Acid Diethylamide: Its Effects on a Male Asiatic Elephant', *Science*, vol. 138, no. 3545, p. 1102.

White, S.E. (2008) 'Brave New World: Neurowarfare and the Limits of International Humanitarian Law', *Cornell International Law Journal*, vol. 41, pp. 177–210.

White House (2013) 'Fact Sheet: BRAIN Initiative', *Press Release*, 2 April, [Online], Available: www.whitehouse.gov/the-press-office/2013/04/02/fact-sheet-brain-initiative [Accessed 4 May 2015].

Willis, G. (2010) *Bomb Power: The Modern Presidency and the National Security State*, New York: Penguin.

Willon, P. and Mason, M. (2015) 'California Gov. Jerry Brown Signs New Vaccination Law, One of Nation's Toughest', *Los Angeles Times*, 30 June, [Online], Available: www.latimes.com/local/political/la-me-ln-governor-signs-tough-new-vaccination-law-20150630-story.html [Accessed 15 January 2016].

Wilson, C. (2004) 'Network-Centric Warfare: Background and Oversight Issues for Congress', *Congressional Research Service*, 2 June, [Online], Available: fas.org/man/crs/RL32411.pdf [Accessed 16 January 2016].

Wojtysiak, M. (2001) 'Another View of the Myths of the Gulf War', *Aerospace Power Journal* (fall), [Online], Available: www.airpower.maxwell.af.mil/airchronicles/apj/apj01/fal01/wojtysiak.html [Accessed 18 December 2015].

Wolpaw, J.R., McFarland, D.J., Neat, G.W. and Forneris, C.A. (1990) 'An EEG Brain-Computer Interface for Cursor Control', *Electroencephalography and Clinical Neurophysiology*, vol. 78, pp. 252–259.

Wolpert, S. (2010) 'Neuroscientists Predict Your Behavior Better Than You Can', *UCLA Press Release*, 22 June, [Online], Available: newsroom.ucla.edu/releases/neuroscientists-can-predict-your-160549 [Accessed 16 January 2016].

Wood, P. (2015) *Technocracy Rising: The Trojan Horse of Global Transformation*, Mesa, AZ: Coherent Publishing.

Yakymenko, I., Tsybulin, O., Sidorik, E., Henshel, D., Kyrylenko, O. and Kyrylenko, S. (2015) 'Oxydative Mechanisms of Low-Intensity Radiofrequency Radiation', *Electromagnetic Biology and Medicine*, 19 August, pp. 1–16.

Yong, E. (2012) 'Mind-Controlling Virus Forces Parasitic Wasp to Put All Its Eggs Into One Basket', *Discover Magazine*, 10 April, [Online], Available: blogs.discovermagazine.com/notrocketscience/2012/04/10/mind-controlling-virus-forces-parasitic-wasp-to-put-all-its-eggs-in-one-basket/#.VaVd619VhBc [Accessed 16 January 2016].

Yousef, N.A. (2015) 'Pentagon Fears It's Not Ready for a War with Putin', *The Daily Beast*, 14 August, [Online], Available: www.thedailybeast.com/articles/2015/08/14/pentagon-fears-it-s-not-ready-for-a-war-with-putin.html [Accessed 16 January 2016].

Yuhas, A. (2016) '"Mini-Brains" Could Revolutionise Drug Research', *The Guardian*, 12 February, [Online], Available: www.theguardian.com/science/2016/feb/12/mini-brains-could-revolutionise-drug-research-and-reduce-animal-use [Accessed 12 February 2016].

Zeman, A. (2001) 'Consciousness', *Brain*, vol. 124, no. 7, pp. 1263–1289.

Zik, J.B. and Roberts, D.L. (2014) 'The Many Faces of Oxytocin: Implications for Psychiatry', *Psychiatry Research*, no. 226, pp. 31–37.

Index

265